Land Surface Remote Sensing

Remote Sensing Observations of Continental Surfaces Set

coordinated by
André Mariotti

Land Surface Remote Sensing

Environment and Risks

Edited by

Nicolas Baghdadi
Mehrez Zribi

ISTE PRESS

ELSEVIER

First published 2016 in Great Britain and the United States by ISTE Press Ltd and Elsevier Ltd

ISTE Press Ltd
27-37 St George's Road
London SW19 4EU
UK

www.iste.co.uk

Elsevier Ltd
The Boulevard, Langford Lane
Kidlington, Oxford, OX5 1GB
UK

www.elsevier.com

British Library Cataloguing-in-Publication Data
A CIP record for this book is available from the British Library
Library of Congress Cataloging in Publication Data
A catalog record for this book is available from the Library of Congress
ISBN 978-1-78548-105-5

Contents

Foreword

I have been entrusted by ISTE Science Publishing with the responsibility for a multidisciplinary editorial line: Earth System – Environment, and within this framework it gives me great pleasure today to present a set of books dedicated to the topic of remote sensing, compiled and edited by Nicolas Baghdadi and Mehrez Zribi.

Both the content and the organization of this collection have largely been inspired by reflections, analyses and prospective works conducted by almost 200 authors and researchers with a high level of international expertise in this discipline.

This community, which is recognized for its scientific merit, has sought to expand its research activities under the direction the two editing authors founded on a solid effort in the area of acquisition and wider dissemination of knowledge within this field.

This represents a community characterized by the firm commitment to adopting a holistic or even an ecosystem approach within the context of an interdisciplinary science of the Earth system. In this scientific context where the complexity of natural systems is compounded with the complexity of societies, the authors have given careful consideration to depicting a finalizable and "public" type of discipline, open to decision makers, managers and all those beyond the scientific community who are interested in the future of our planet.

Two main tools are necessary in order to satisfy the requirements in terms of understanding and characterizing our environment and its evolution: process modeling and observation.

Remote sensing observations in conjunction with measurements and modeling constitute a discipline that makes it possible to understand the functional properties of the observed system and their dependence on its structural properties. This is one of the key disciplines that allow the analysis and provide access to the understanding of the functioning of our environment: in general, this is dedicated to aspects such as the analysis of climate change, the effects of anthropogenic and demographic pressure, natural disasters, the increasing decline in resources (water, etc.), the degradation of biodiversity across all environments, desertification, the need to nourish the planet (for example, mapping of crops and yield prediction), etc.

Analyzing and understanding these different types of problems is rendered possible:

– by analyzing the detected, structural and functional objects (soils, hydrosystems, vegetation, etc.);

– by understanding the main basic processes, which incorporate these main elements: water flow covering all scales and compartments, erosion, meteorology, crop development, soil pollution, etc;

– by developing indicators in order to evaluate the short-, medium- and long-term evolution of all environmental compartments and variables.

The importance of these scientific questions has led to a general mobilization of the international organizations by means of various international conventions and agreements to protect the environment and meet the specific requirements in terms of observation. Various international networks have been developed over the past few years with the purpose of conducting continuous measurements. However, these punctual measurements could not provide sufficient spatiotemporal monitoring, in particular in difficult-to-access regions. Within this context, spatial observation could be implemented to its full potential, both by means of considerable progress in terms of instrumentation and by means of the development of effective data processing and analysis methods, data whose provision becomes increasingly free of rights.

Under the initiative of numerous space agencies (in particular European, North-American, Japanese, etc.), important space missions were launched for the purpose of conducting Earth observations, among which the following may be mentioned:

– Sentinel, within the framework of the Copernius program (formerly referred to as GMES for Global Monitoring for Environment and Security) implemented in numerous areas such as land and marine environment monitoring, emergency management (for example, natural disasters) and climate change monitoring (radar and optical imaging);

– Landsat;

– ALOS, launched by the Japan Aerospace Exploration Agency, in particular for deforestation monitoring;

– SMOS and SMAP, in particular for the global mapping of soil moisture, etc.

Although remote sensing represents a field in which specialist knowledge is required in order to conduct a better analysis and interpretation of data, this programmatic development is undoubtedly associated with a significant progress with respect to the implementation of space-based Earth observations at the level of an increasing number of laboratories across both developed and emerging nations. This development is likewise associated with new disciplines and thematic backgrounds, among which numerous areas of the humanities and social sciences which enrich and extend the primarily physical foundations of remote sensing may be mentioned.

It would be superfluous to list all remote sensing applications along with the disciplines and scientific questions which adopted this concept, as this would also inevitably result in regrettable omissions: nonetheless, the extensive implementation of spatial observation grants the latter a strong interdisciplinary status.

The launch of new large-scale space missions, the higher degree of convenience, including financial convenience, as well as the access to data will facilitate an intensification and generalization of the use of spatial observation data and products: new scientific subjects, new users (managers, decision makers, etc).

The high demand for educational material containing updated information on the various remote sensing concepts and methods and the main applications thereof, in particular at the level of continental surfaces, are derived therefrom.

It is within this framework that this collection of books is proposed, which aims to provide researchers, students in masters, engineer and PhD programs, as well as decision makers, engineers specialized in management services on a territorial, departmental, regional or national scale and players in the decision-making authorities with a tool which incorporates both the foundations of the physical principles underlying various spatial applications and the implementation methods and exemplification at the level of various applications based on spatial observation.

In these six volumes, Nicolas Baghdadi and Mehrez Zribi have mobilized almost 200 internationally recognized researchers to propose a comprehensive "toolkit", describing the latest scientific methods and actions in terms of the implementation of spatial observation.

The first two volumes describe the physical principles underlying various techniques which cover the frequency spectrum ranging from visible to microwaves. The third volume illustrates the agricultural and forestry applications of spatial observation. The fourth volume presents the applications of spatial observation in the field of continental hydrology. The fifth volume is dedicated to the observation of urban and coastal areas, whereas the final volume presents the implementation of spatial observation within the context of risk assessment and understanding.

Thanks are due to Nicolas Baghdadi and Mehrez Zribi for taking the time to draft, harmonize and partially edit these volumes and committing to this effort in terms of putting this modern and high-quality knowledge across and making it accessible to a diverse and vast scientific audience.

I wish to thank them both for their altruism, perseverance and devotion in service of the success of this endeavor.

André MARIOTTI

June 2016

Acronyms

2D	Two dimensions
3D	Three dimensions
4AOP	Automatized Atmospheric Absorption Atlas Operational
6S	Second Simulation of Satellite Signal in the Solar Spectrum
AATSR	Advanced Along-Track Scanning Radiometer
ACORN	Atmospheric Correction Now
ADC	Analog-to-digital converter
ADCP	Acoustic Doppler Current Profiler
ADEME	French Environment and Energy Management Agency
ADEOS	Advanced Earth Observing Satellite
AERONET	Aerosol Robotic Network
AET	Actual Evapotranspiration
AFRITRON	African Tropical Rainforest Observation Network
AGB	Above-Ground Biomass
AGNES	Agglomerative Nesting
AHS	Airborne Hyperspectral Scanner
AHT	Astronomical High Tide
AIEM	Advanced Integral Equation Model
AirSAR	Airborne Synthetic Aperture Radar
ALB	Airborne LiDAR Bathymeter
ALEXI	Atmosphere-Land Exchange Inverse
ALS	Airborne Laser Scanning

AltBOC	Alternate Binary Offset Carrier
AMARTIS	Advanced Modeling of the Atmospheric Radiative Transfer for Inhomogeneous Surfaces
AMMA	African Monsoon Multidisciplinary Analysis
AMSR	Advanced Microwave Scanning Radiometer
ANA	*Agência Nacionalde Aguas* (Brazilian National Water Agency)
AOL	Airborne Oceanographic LiDAR
APD	Avalanche Photodiode
API	Antecedent Precipitation Index
APOM	Aerosol Plume Optical Model
ARVI	Atmospherically Resistant Vegetation Index
ASAR	Advanced Synthetic Aperture Radar
ASCAT	Advanced Scatterometer
ASDF	Averaged Square Difference Function
ASI	*Agenzia Spaziale Italiana* (Italian Space Agency)
ASI	Agriculture Stress Index
ASTER	Advanced Spaceborne Thermal Emission and Reflection Radiometer
ATCOR	Atmospheric and Topographic Correction
ATLAS	Advanced Topographic Laser Altimeter System
ATM	Airborne Topographic Mapper
ATREM	Atmospheric Removal (atmospheric correction method)
ATSR	Along Track Scanning Radiometers
AUC	Area Under Curve
AVHRR	Advanced Very High Resolution Radiometer
AVIRIS	Airborne Visible/Infrared Imaging Spectrometer
AWiFS	Advanced Wide Field Sensor
BEIDOU/ COMPASS	Chinese Navigation Satellite System
BELA	BepiColombo Laser Altimeter
BGB	Below-Ground Biomass
BIL	Band Interleaved by Line
BIOMOD	Biomolecular Design (Computer platform implemented in R)
BIP	Band Interleaved by Pixel

BLUE	Best Linear Unbiased Estimator
BOC	Binary Offset Carrier
BPSK	Binary Phase Shift Keying
BPT	Binary Partition Tree
BRDF	Bidirectional Reflectance Distribution Function
BRGM	*Bureau de Recherches Géologiques et Minières* (French Geological Survey)
BSA	Back Scatter Alignment
BSQ	Band Sequential
BUI	Build-up Index
BWI	Bassist Wetness Index
C/A	Coarse Acquisition
CaCO3	Calcium Carbonate
CAIC	Consistent Akaike's Information Criterion
CALIOP	Cloud-Aerosol LiDAR with Orthogonal Polarization
CALIPSO	Cloud-Aerosol LiDAR Pathfinder Satellite Observation
CanEx-SM	Canadian Experiment for Soil Moisture
CART	Classification and Regression Trees
CASI	Compact Airborne Spectrographic Imager
CBERS	China–Brazil Earth Resources Satellite
CBOC	Composite Binary Offset Carrier
CCD	Charged Coupled Devices
CCDAS	Carbon Cycle Data Assimilation System
CCI	Climate Change Initiative
CCRS	Canada Center for Remote Sensing
CDF	Cumulative Density Function
CDMA	Code Division Multiple Access
CDOM	Colored Dissolved Organic Matter
CEC	Cation-Exchange Capacity
CEM	Constrained Energy Minimization
CEN	French Snow Studies Center
CEOS	Committee on Earth Observation Satellites
CERES	Crop Environment Resource Synthesis

CESBIO	*Centre d'Études Spatiales de la Biosphère* (Center for the Study of the Biosphere from Space)
CFC	Chlorofluorocarbons
CFD	Constant Fraction Discriminator
CFFDRS	Canadian Forest Fire Danger Rating System
cGNSS	Conventional Global Navigation Satellite Systems
CHAMP	Challenging Minisatellite Payload (German geosciences satellite)
CHL	Chlorophyll Content
CHM	Canopy Height Model
CHRIS	Compact High Resolution Imaging Spectrometer
CIRAD	Agricultural research for development (France)
CLASlite	Carnegie Landsat Analyse System Lite
CLM	Community Land Model
CLMGW	Community Land Model with a Ground Water Parameterization
CLPX	Cold Land Processes Field Experiment
CLS	Collecte Localisation Satellite
CLSM	Catchment Land Surface Model
CM	Code Moderate (for GNSS)
CMC	Canadian Meteorological Centre
CMEM	Community Microwave Emission Model
CNES	*Centre National d'Études Spatiales* (French Space Agency)
CNRM	*Centre National de Recherches Météorologiques* (National Centre for Meteorological Research, France)
CNRS	*Centre National de Recherche Scientifique* (National French Center for Scientific Research)
COD	Controlled Origin Designation
COG	Center of Gravity
CONUS	Contiguous United States
COST	Cosine Estimation of Atmospheric Transmittance
CR	Continuum Removed
CRI	Carotenoid Reflectance Index
Cryosat	Satellite Radar Altimeter
CSA	Canadian Space Agency
CSR	Centre for Space Research

CTA	Classification Tree Analysis
CTFC	Center for Tropical Forest Science (Cameroon)
CTFS	Center for Tropical Forest Science
CTMF	Cluster Tuned Matched Filter
CW	Continuous Waves
CWFIS	Canadian Wildland Fire Information System
CyGNSS	NASA's Cyclone Global Navigation Satellite System
CZCS	Coastal Zone Color Scanner
CZMIL	Coastal Zone Mapping and Imaging LiDAR
DAM	Dry Aerial Mass
DART	Discrete Anisotropic Radiative Transfer
DATAR	Interministerial Delegation for Territorial Planning and Regional Attractiveness
DBH	Diameter at Breast Height
DC	Drought Code
DCA	Dual Channel Approach
DCM	Digital Canopy Model
DDM	Delay Doppler Map
DEGRAD	Forest Degradation program
DEIMOS	Deep Imaging Multi-Object Spectrograph
DEM	Digital Elevation Model
DEOS	Delft Institute of Earth Observations and Space Systems
DERD	Double bounce Eigenvalue Relative Difference
DETER	*Detecção de Desmatamento em Tempo Real* (Near real-time deforestation detection system)
DGPS	Differential Global Positioning System
DHSVM	Distributed Hydrology Soil-Vegetation Model
DIACT	Inter-ministerial Agency for Spatial Planning and Competitiveness
DIANA	Divise Analysis
DIC	Digital Image Correlation
DIMAP	Digital Image MAP
DInSAR	Differential SAR Interferometry
DisALEXI	Disaggregated Atmospheric Land Exchange Inverse

DLIS	Desert Locust Information Service
DLR	*Deutsches Zentrum für Luft und Raumfahrt* (German Space Agency)
DM	Dry Matter
DMC	Disaster Monitoring Constellation or Duff Moisture Code, depending on the application
DMRT	Dense Media Radiative Transfer
DMSP	U.S. Air Force Defense Meteorological Satellite Program
DMU	De Monfort University
DOAS	Differential Optical Absorption Spectroscopy
DOS	Dark Object Substraction
DOY	Day-of-Year
DPSS	Diode-Pumped Solid-State
DS	Diffuse Scatterer
DSM	Digital Soil Mapping
DSM	Digital Surface Model
DTC	Dry Troposphere Correction
DTED	Digital Terrain Elevation Data
DTM	Digital Terrain Model
DVI	Difference Vegetation Index
DW	Dry Weight
DWBA	Distorted Wave Born Approximation
EAARL	Experimental Advanced Airborne Research LiDAR
ECDC	European Centre for Disease Prevention and Control
ECMWFMMT	European Centre for Medium-Range Weather Forecasts Mobile Mapping Technology
ECV	Essential Climate Variables
EDF	*Électricité de France* (French electric company)
EEA	European Environment Agency
EFA	Effective Field Approximation
EFFIS	European Forest Fire Information System
EGNOS	European Geostationary Navigation Overlay Service
EID-Méditerranée	Interdepartmental Agreement for Mosquito Control on the Mediterranean Coast

EKF	Extended Kalman Filter
ELBARA	ETH L-Band Radiometer
ELUE	Effective Light Use Efficiency
EM	Electromagnetic
ENEA	*Energia Nucleare ed Energie Alternative* (Italian National Agency for New Technologies, Energy and Sustainable Economic Development)
EnKF	Ensemble Kalman Filter
ENMAP	Environmental Monitoring and Analysis Program
ENSO	El Niño Southern Oscillation
ENVEO	Environmental Earth Observation Information
ENVISAT	Environmental Satellite
EOS	Earth Observing System
EPICA	European Project for Ice Coring in Antarctica
EPS	European Polar System
EQeau	Model developed by INRS-Ete with the objective to extract the soil water content from SAR images
ERM	Exact Repeat Missions
ERS	European Remote-sensing Satellite
ESA	European Space Agency
ESCAT	ERS Scatterometer
ESSA	Environmental Science Services Administration Satellite
ET	Evapotranspiration
ETM	Enhanced Thematic Mapper
EUFAR	European Facility for Airborne Research
EUMETSAT	European Organization for the Exploitation of Meteorological Satellites
EVASPA	Evapotranspiration Assessment from Space
EVI	Enhanced Vegetation Index
EWT	Equivalent Water Thickness
EZW	Embedded Zerotrees of Wavelet transforms
FAA	Federal Aviation Administration (USA)
FAI	Floating Algae Index
FAO	Food and Agriculture Organization of the United Nations

FAPAR	Fraction of Absorbed Photosynthetically Active Radiation
FBD	Fine-Beam Double Polarization
FBP	Fire Behavior Prediction
Fcover	Fraction of Vegetation Cover
FDTD	Finite Difference Time Domain
FFMC	Fine Fuel Moisture Code
FFT	Fast Fourier Transform
FI	Fine Particles Index
FIPAR	Fraction of Intercepted Photosynthetically Active Radiation
FLAASH	Fast Line-of-sight Atmospheric Analysis of Spectral Hypercubes (Atmospheric correction)
FLOE	Fish LiDAR Oceanographic Experimental
FMA	Fore-Mid-Aft
FMCW	Frequency Modulated Continuous Waves
FOTO	Fourier-based Textural Ordination
FOV	Field of View
FSA	Forward Scatter Alignment
FST	Spectro-Transfert functions
FT	Fourier Transform
FW	Full Waveform or Fresh Weight, depending on the application
FWHM	Full Width at Height Maximum
FWI	Fire Weather Index
FWS	Fraction of Water Surface
GAGAN	GPS Aided GEO Augmented Navigation system
GAI	Green Area Index
GALILEO	European Global navigation satellite system
GCOS	Global Climate Observing System
GCP	Ground Control Points
GDAL	Geospatial Data Abstraction Library
GDR	Geophysical Data Record
GEO	Group on Earth Observations
GEOCAPI	Geostationary Ocean Color Advanced Permanent Imager
GEOGLAM	Group on Earth Observations Global Agricultural Monitoring

GEOSS	Global Earth Observing System of Systems
GEOSUD	Geoinformation for Sustainable Development (France)
GEOTIFF	Geostationary Earth Orbit Tagged Image File Format
GEROS-ISS	GNSS Reflectometry, Radio Occultation and Scatterometry Onboard International Space Station
GF	Green Fraction
GFO	Geosat Follow-On (radar altimeter satellite)
GFZ	*GeoForschungs Zentrum* (German research centre for Geosciences)
GHG	Green House Gase
GIAM	Global Irrigated Area Map
GIEWS	Global Information and Early Warning System on Food and Agriculture
GIM	Global Ionospheric Model
GIMMS	Global Inventory Modeling and Mapping Studies
GIS	Geographic Information System
GLAI	Green Leaf Area Index
GLAS	Geoscience Laser Altimeter System
GLCM	Grey Level Co-occurrence Matrix
GLDAS	Global Land Data Assimilation System
GlobCover	Global Land-Cover Map
GLONASS	Global Navigation Satellite System (Russia)
GLP	Global Land Project
GLRT	Generalized Likelihood Ratio Test
GLS	Global Land Survey (Landsat)
GM	Geodetic Mission
GMES	Global Monitoring for Environment and Security
GMFS	Global Monitoring of Food Security
GMM	Gaussian Mixture Model
GNSS	Global Navigation Satellite System
GNSS-R	Global Navigation Satellite System – Reflectometry
Go	Giga-Octet (1,000,000,000 octets)
GOCE	Gravity field and Steady-state Ocean Circulation Explorer (ESA)
GOES	Geostationary Operational Environmental Satellite

GOFC Fire IT	Global Observation of Forest Cover Fire Implementation Team
GOM	Geometrical Optics Model
GORS	GNSS Occultation Reflectometry Scatterometry
GPCC	Global Precipitation Climatology Centre
GPP	Gross Primary Production
GPR	Gaussian Process Regression
GPS	Global Positioning System
GRACE	Gravity Recovery and Climate Experiment (satellite)
GRDC	Global Runoff Data Center
GRGS	Space Geodesy Research Group
GtC	Gigatonnes of Carbon (10^9 tons)
GUS	Ground Uplink Stations
GVI	Difference Vegetation Index
GWIS	Global Wildfire Information System
Ha	Hectare (= 10,000 m^2)
HCFC	Hydro-chlorofluorocarbons
HEC-RAS	Hydrologic Engineering Center River Analysis System (hydraulic model)
HgCdTe	Mercury Cadmium Telluride
HiRI	High Resolution Optical Imager
HITRAN	High-Resolution Transmission Database
HOG	Histogram of Oriented Gradient
HPC	High Performance Computing
HRG	High Resolution Geometry
HSC	Height-Scaled Crown
HSCOI	Height-Scaled Crown Openness Index
HSR	High Spatial Resolution
HYMAP	Hyperspectral Mapper (airborne hyperspectral sensor)
HYPXIM	Hyperspectral-X Imagery
HYSPEX	Hyperspectral Imaging System
HyspIRI	NASA's Hyperspectral Infrared Imager
IASI	Infrared Atmospheric Sounding Interferometer
ICA	Independent Component Analysis

ICARE	Inversion Code for urban Areas Reflectance Extraction
ICESat	Ice, Cloud and Land Elevation Satellite
ICF	Interferometric Complex Field
ICP	Iterative Closest Point technique
IDAN	Intensity Driven Adaptative Neighborhood
IEM	Integral Equation Model
IFN	French National Forest Inventory
IFOV	Instantaneous Field of View
IGN	French National Geographic Institute and Forest Information
iGNSS	Interferometric GNSS
IHS	Intensity, Hue, Saturation
IMU	Inertial Measurement Unit
INERIS	French National Institute for Environmental Protection and Industrial Risks
InGaAs	Indium Gallium Arsenide
INPE	*Instituto Nacional de Pesquisas Espaciais* (Brazilian Institute of Space Research)
INRA	French National Institute for Agricultural Research
INSAR	Interferometric Synthetic Aperture Radar
InSb	Indium antimonide
INSEE	French National Institute for Statistics and Economic Studies
IOD	Indian Ocean Dipole
IPCC	Intergovernmental Panel on Climate Change
IPT	Interference Pattern Technique
IRD	French Research Institute for Development
IRNSS	Indian Regional Navigational Satellite System
IRSTEA	French National Research Institute of Science and Technology for Environment and Agriculture
ISBA	Interactions Soil-Biosphere-Atmosphere (model)
ISDC	Integrated System Data Center
ISI	Initial Spread Index
ISODATA	Iterative Self-Organizing Data Analysis Technique
ISRO	Indian Space Research Organisation
ITC	Individual Tree Crown

ITCZ	Intertropical Convergence Zone
ITG	Institute of Theoretical Geodesy
IWPB	Institute of Water Problem of Bishkek, Kyrgyzstan
Jason	Radar altimeter
JAXA	Japan Aerospace Exploration Agency
JECAM	Joint Experiment for Crop Assessment and Monitoring
JPEG	Joint Photographic Experts Group (image format)
JPL	Jet Propulsion Laboratory
JRC	European Commission's Joint Research Centre
KBR	K-Band Microwave Ranging
KLT	Karhunen–Loeve transform
LaDAR	Laser Detection And Ranging
LADS	Laser Airborne Depth Sounder
LAGEOS	Laser Geodynamics Satellite
LAI	Leaf Area Index
LANDSAT	LAND + Satellite
LAUVA	Airborne Ultraviolet Aerosol LiDAR
LAX	Maximum LAI
LBAS	Local Based Augmentation System
LCCS	Land Cover Classification System
LCLU	Land Cover / Land Use
LDAS	Land Data Assimilation System
LEGOS	Laboratory for Studies in Geophysics and Spatial Oceanography (France)
LEnKS	Local EnKF Smoother
LEO	Low Earth Orbit
LEP	Leading Edge Position
LEWIS	L-band for Estimating Water In Soils
LFMC	Live Fuel Moisture Content (%)
LHCP	Left Hand Circular Polarization
LiDAR	Light Detection and Ranging
LISAH	Laboratory for Soil, Agrosystems and Water Systems (France)
LISFLOOD-FP	Two-dimensional Hydrodynamic Model

LMA	Leaf Mass per Area
LMM	Linear Mixed Model
LOADDT	Spatial planning and territorial development
LOLA	Lunar Orbiter Laser Altimeter
LOV	Villefranche Oceanography Laboratory (France)
LPCA	Laboratory for Physico-Chemistry of the Atmosphere
LPRM	Land Parameter Retrieval Model
LRM	Low Resolution Mode
LSCE	Climate and Environment Sciences Laboratory (France)
LSM	Land Surface Model
LSSM	Least Squares 3D Surface Matching
LST	Land Surface Temperature
LULCC	Land Use and Land Cover Change
LUT	Look-Up Table
LWIR	Long-Wave Infrared
LZW	*Lempel-Ziv-Welch* (compression algorithm)
MACCS	Multisensor Atmospheric Correction and Cloud Screening processor
MARS	Monitoring of Agriculture with Remote Sensing
MATISSE	Advenced Earth Modeling for Imaging and Scene Simulation
MaxEnt	Maximum Entropy Method
MBOC	Multiplexed Binary Offset Carrier
MCT	Mercury Cadmium Telluride
MEB	Microwave Emission of the Biosphere
MERIS	Medium Resolution Imaging Spectrometer
MESA	Monitoring of Environment and Security in Africa
METEOSAT	METEO + Satellite
METOP	Meteorological Operational Satellite Programme/Advanced Scatterometer
METRIC	Mapping Evapotranspiration at High Resolution with Internalized Calibration
MGVI	MERIS Global Vegetation Index
MHz	Mega-Hertz (= 1,000,000 Hz)
MIMR	Multichannel Microwave Imaging Radiometer

MIPERS	Multistatic Interferometric Polarimetric Electromagnetic model for Remote Sensing
MIR	Middle Infrared
MISDc	*Modello Idrologico Semi Distribuito in Continuo* (Continuous rainfall-runoff model)
MISR	Multi-angle Imaging Spectro Radiometer
MISTIGRI	Microsatellite for Thermal Infrared Ground Surface Imaging (CNES, France)
MLR	Multiple Linear Regression
MLS	Mobile LiDAR Scanner
MMD	Minimum–Maximum Difference
MMS	Mobile Mapping Systems
MMU	Minimum Mapping Units
MMV	Mobile Mapping Vehicle
MNDWI	Modified Normalized Difference Water index
MNF	Maximum Noise Fraction
Mo	Mega-octet (1,000,000 octets)
MODCOU	Hydrogeological model
MODIS	Moderate Resolution Imaging Spectroradiometer
MODTRAN	Moderate Resolution Atmospheric Transmission
MOLA	Mars Orbiter Laser Altimeter
MPE	Maximum Permissible Exposure
MSAS	Multi-functional Satellite-based Augmentation System
MSG	Meteosat Second Generation
MSI	Moisture Stress Index
MSI	Multispectral Instrument (Sentinel-2)
MTSAT	Multi-functional Transport Satellites
MVSA	Minimum Volume Simplex Analysis
MWIR	Mid-Wavelength Infrared
NAOMI	New AstroSat Optical Modular Instrument
NASA	National Aeronautics and Space Administration (USA)
NBR	Normalized Burn Ratio
NCC	Normalized Cross-correlation
NCEP	National Centers for Environmental Prediction

Nd:YAG	Neodymium-doped Yttrium Aluminium Garnet
NDSI	Normalized Difference Snow Index
NDVI	Normalized Difference Vegetation Index
NDVITM	Normalized Difference Vegetation Index Threshold Method
NDWI	Normalized Difference Water Index
NEBN	Noise Equivalent Beta Naught
NEDT	Noise Equivalent Delta Temperature
NEE	Net Ecosystem Exchange
NEF	Noise Equivalent Flux
NEM	Normalized Emissivity Method
NEP	Net Ecosystem Productivity
NEP	Noise Equivalent Power
NIR	Near Infrared
NLES	Navigation Land Earth Station
NLRI	Near Laser Ranging Investigation
NMC	National Meteorological Center (USA)
NMF	Non-negative Matrix Factorization
NOAA	National Oceanic and Atmospheric Administration
NOHD	Nominal Ocular Hazard Distance
NORUT	Norut Northern Research Institute (Norway)
NPV	Non-photosynthetic Vegetation
NPW	Numerical Weather Prediction models
NSC	Naryn Syr Darya Cascade
NSCAT	NASA Scatterometer
NSIDC	National Snow and Ice Data Center (USA)
NWP	Numerical Weather Prediction
OA	Overall Accuracy
OBIA	Object Based Image Analysis
OLCI	Ocean and Land Colour Instrument
OLI	Operational Land Imager
OLS	Operational Linescan System
OM	Organic Matter
OMI	Ozone Monitoring Instrument

ONERA	French Aerospace Research Agency
ONF	French National Forest Office
OS	Open Service
OSCAT	OceanSat-2 Scatterometer
OTB	OrfeoToolBox
PA	Producer's Accuracy
PACE	Pre-Aerosol, Clouds and Ocean Ecosystems
PAH	Polycyclic Aromatic Hydrocarbons
PAI	Plant Area Index
PARIS-IoD	Passive Reflectometry and Interferometry System In Orbit Demonstrator
PCA	Principal Components Analysis
PERSIANN	Precipitation Estimation from Remotely Sensed Information using Artificial Neural Networks
PET	Potential Evapotranspiration
PFT	Phytoplankton Functional Types
PGN	Permanent GNSS Network
PHR	Panchromatic High Resolution
PIF	Pseudo-Invariant Features
piGNSS-R	Partial Interferometric GNSS-R
PLF	Polarization Loss Factor
PLOF	Local Land Tenure Plan (France)
PLSR	Partial Least Squares Regression
PMT	Photomultiplier Tubesv
POA	Polarization Orientation Angle
PODAAC	Physical Oceanography Distributive Active Data Center
POLDER	Polarization and Directionality of the Earth's Reflectances
PolInSAR	Polarimetry-interferometry Synthetic Aperture Radar
PolSAR	Polarimetry SAR
POLYMER	Polynomial Based Algorithm Applied to Meris
POM	Physical Optics Model
PPCDAM	Plan for Preventing and Controlling Deforestation in Amazônia Legal
PRESS	Prediction Sum of Squares

PRF	Pulse Repeat Frequency
PRISMA	*Precursore Iperspettrale della Missione Applicativa* (Italian hyperspecral mission)
PRN	Pseudo Random Noise
PROBA	Project for On-board Autonomy
PRODES	*Programa de Cálculo do Desflorestamento da Amazônia* (Brazilian Amazon Deforestation Monitoring Program)
PROSPECT	Leaf Optical Properties Spectra (radiative transfer model)
PRS	Public Regulated Service
PS	Persistent Scatterers
PSI	Persistent Scatterers Interferometry
PSR	Penalized-Spline Regression
PV	Photosynthetic Vegetation
PVI	Perpendicular Vegetation Index
PWC	Plant Water Content
QBO	Quasi-Biennal Oscillation
QCA	Quasi-Crystalline Approximation
Qgis	GIS software (open source)
QPSK	Quadrature Phase Shift Keying
QUIKSCAT	Quick Scatterometer (NASA)
QZSS	Quasi-Zenith Satellite System
RADAR	Radiodetection and Ranging
RAF	French Altimetric System
RAINFOR	Amazon Forest Inventory Network
RAMSES	ONERA Airborne Multi-frequency SAR Imaging System (France)
RAN	Royal Australian Navy
RANSAC	Random Sample Consensus
RCA	Radio Corporation of America
RCM	RADARSAT Constellation Mission
RCS	Radiometric Control Sets
REDD	Reduction of Emissions from Deforestation and Forest Degradation

REDDAF	Reducing Emissions from Deforestation and Degradation in Africa (European Project)
RENAG	French National GNSS Permanent Networks
RESIF	French Seismologic and Geodetic Network
RF	Random Forests (classifier)
RFI	Radio Frequency Interference
RG	Relative Greenness
RGB	Red Green Blue
rGNSS-R	Reconstructed GNSS-R
RHCP	Right Hand Circular Polarization
RMSD	Root Mean Square Difference
RMSE	Root Mean Square Error
RPAS	Remotely Piloted Aircraft System
RPC	Rational Polynomial Coefficients
RPD	Ratio of Performance to Deviation
RPIQ	Ratio of Performance to Inter-Quartile range
RTK	Real Time Kinematic
RTM	Radiative Transfer Model
RUE	Rain Use Efficiency
RVI	Radar Vegetation Index
RVoG	Random Volume over Ground
RXD	Reed-Xiaoli Detector
SAFY	Simple Algorithm For Yield estimate
SAM	Spectral Angle Mapper
SAMIR	Satellite Monitoring of Irrigation (model)
SAR	Synthetic Aperture Radar
SAR in	SAR interferometric
Saral/Altika	Radar Altimeter (French–Indian altimetry mission)
SASS	SEASAT Advanced Scatterometer System
SAVI	Soil-Adjusted Vegetation Index
SBAS	Satellite Local-based Augmentation System
SBAS	Small Baselines
SCA	Snow Cover Area

SCARAB	Scanning Radiometer for Radiation Balance
SCF	Snow Cover Fraction
SDC	Snow Depletion Curve
SDS	Science Data System (USA)
SEAS	Survey of the Environment Assisted by Satellite
SeaWiFS	Sea-viewing Wide Field-of-view Sensor (satellite)
SEBAL	Surface Energy Balance Algorithm for Land
SEBS	Surface Energy Balance System
SEC	Standard Error of Calibration
SEKF	Self-extended Kalman Filter
SEP	Standards Error of Prediction
SER	Section Efficace Radar
SERD	Single bounce Eigenvalue Relative Difference
SEVIRI	Spinning Enhanced Visible and Infrared Imager
SFCW	Stepped Frequency Continuous Waves
SfM	Surface-from-Motion
SFT	Strong Fluctuation Theory
SGBM	Semi-Global Block Matching algorithm
SHALOM	Spaceborne Hyperspectral Applicative Land and Ocean Mission
SHI	State Hydrological Institute (St. Petersburg, Russia)
SHOALS	Scanning Hydrographic Operational Airborne LiDAR Survey
SHOM	French Navy's Hydrographic and Oceanographic Service
SID	Spectral Information Divergence
SIERRA	Spectral Reflectance Image Extraction from Radiance with Relief and Atmospheric Correction
SIFT	Scale Invariant Feature Transform
SIGMA	Simulation Innovation for Global Monitoring of Agriculture
SLA	*Scanner LiDAR aérien* (aerial LiDAR scanner)
SLC	Single Look Complex
SLR	Single-Lens Reflex
SM	Soil Moisture
SMA	Spectral Mixture Analysis
SMAC	Simplified Method for Atmospheric Correction

SMAP	Soil Moisture Active and Passive mission (Radiometer)
SMAPVEX	SMAP Validation Experiment
SMEX02	Soil Moisture Experiment 2002
SMF	Spectral Matched Filter
SMLR	Stepwise Multiple Linear Regression
SMMR	Scanning Multichannel Microwave Radiometer
SMOS	Soil Moisture and Ocean Salinity mission (satellite)
SNAS	Chinese Satellite Navigation Augmentation System
SNR	Signal to Noise Ratio
SNSB	Swedish National Space Board
SNV	Standard Normal Variate
SPAD	Single-Photon Avalanche Diode
SPM	Small Perturbation Model
SPM	Suspended Particulate Matter
SPOT	Satellites for Earth Observation
SRM	Snowmelt-Runoff Model
SRTM	Shuttle Radar Topography Mission
SSA	Surface Specific Area
SSC	Soil Surface Characteristics
S-SEBI	Simplified Surface Energy Balance Index
SSM	Soil Surface Moisture
SSMI	Special Sensor Microwave Imager (satellite)
SST	Sea Surface Temperatures
STF	Spectrotransfer Functions
STICA	Socio-technical Information and Communication Arrangements
STICS	Crop model
SUCROS	Simple and Universal Crop Growth Simulator
SUHI	Surface Urban Heat Islands
Suomi-NPP	Suomi National Polar-Orbiting Partnership
SURFEX	Surface model platform (Météo France)
SVAT	Soil–Vegetation–Atmospheric Transfer
SVM	Support Vector Machine
SVMR	Support Vector Machine Regression

SVR	Support Vector Regression
SWAT	Soil and Water Assessment Tool
SWE	Snow Water Equivalent
SWI	Soil Wetness Index
SWIR	Short-wave Infrared
SWOT	Surface Water Ocean Topography (satellite)
SYSIPHE	Airborne hyperspectral imaging system
T/P	Topex/Poséidon (Franco-American altimeter)
TDR	Time Domaine Reflectometry
TEC	Total Electron Content
TES	Emissivity Separation algorithm
THEIA	French Land Data Centre
THIRSTY	Thermal Infrared Spatial System (satellite project by CNES and NASA)
TIFF	Tag Image File Format
TIN	Triangular Irregular Network
TiO_2	Titanium dioxide
TIR	Thermal Infrared
TISI	Temperature Independant Spectral Indices
TLS	Terrestrial Laser Scanning
TM	Thematic Mapper
TMBOC	Time Multiplexed Binary Offset Carrier
TNT2	Topography based Nitrogen Transfer and Transformation
To	Tera-octet (1,000,000,000,000 octets)
TOA	Top of Atmosphere
TOC	Top of Canopy
TomoSAR	Tomography SAR
Topex/ Poseidon	Radar altimeter
TOPLATS	Topographic Land Atmosphere Transfer Scheme
TRIP	Total Runoff Integrating Pathways
TRMM	Tropical Rainfall Measuring Mission (satellite)
TSAVI	Transformed Soil Adjusted Vegetation Index

TSEB	Two-Source Energy Balance (model)
TTL	Transistor-Transistor Logic
TWAP	Transboundary Water Assessment Program
TWS	Terrestrial Water Storage
UA	User's Accuracy
UAA	Utilized Agricultural Area
UAV	Unmanned Aerial Vehicles
UHI	Urban Heat Islands
ULICE	Ultraviolet LiDAR for Canopy Experiment
UNEP	United Nations Environment Program
USDA	United States Department of Agriculture
USGS	United States Geological Survey
USO	Ultra-Stable Oscillator
UTC	Coordinated Universal Time
UTM	Universal Transverse Mercator
UV	Ultra-violet
VARI	Visible Atmospherically Resistant Index
VCA	Vertex Component Analysis
VD	Virtual dimensionality
VHF	Very high frequency
VHI	Vegetation Health Index
VHSR	Very High Spatial Resolution
VIC	Variable Infiltration Capacity
VIIRS	Visible Infrared Imager Radiometer Suite
VIS	Visible
VISAT	Video, Inertial, and Satellite GPS
Vis-NIR	Visible and Near Infrared
VITO	Flemish Institute for Technological Research (Belgium)
VLA	Very Large Array
VLBI	Very Large Baseline Interferometry
VOS	Volatile Organic Compound
VPD	Vapor Pressure Deficit
VSDI	Visible and Shortwave Infrared Drought Index

VWC	Vegetation Water Content
WALID	Water LiDAR Simulation Model
WASS	Wide Area Augmentation System
WDI	Water Deficit Index
WGHM	Water GAP Global Hydrology Model
WGS	World Geodetic System
WMA	Winter Metric Anomaly
WMO	World Meteorological Organization
WMS	Wide-area Master station
WRS	Wide-area Reference Stations
WSI	Water stress index
WTC	Wet Troposphere Correction
XML	Extensible Markup Language
ZSSD	Zero-mean Sum of Squared Difference

Introduction

Our civilization is becoming fragile and vulnerable to increasingly frequent environmental challenges. Our environment is suffering from a lack of resources, deforestation, drought, pollution, etc. The changes in our environment that have been observed in conjunction with climate change and anthropogenic pressure must be monitored and mapped. Moreover, in the face of the human and economic costs of frequent natural disasters, there is an urgent need to act. Today, the question of risks and our planet's environment is the subject of several international treaties and agreements.

Spatial observation is certainly an essential tool for monitoring these changes and responding to the need for alerts with regard to different types of events. In this context, over the past decade, European and international institutions have committed to significant funding for earth observation to encourage the transition from R&D (research and development) to operational services. The Copernicus Program (formerly called GMES – "Global Monitoring for Environment and Security"), which is a joint initiative of the European Space Agency (ESA) and the European Union, seeks to promote services in numerous fields including observing the earth and the marine environment, emergency management (of a natural disaster, for example) and the monitoring of climate change. The Sentinel missions (1–5) show Europe's desire to develop the earth observation satellites necessary for the Copernicus Program (multi-spectral sensors).

The aim of this volume is to present the major themes in applying remote sensing to the environment and to risk. This work, supported by scientists

---

Chapter written by Nicolas BAGHDADI and Mehrez ZRIBI.

who are internationally renowned in their fields, will keep up-to-date knowledge and describe the research and development issues in the years to come. It is intended for remote sensing research teams and for second- (engineering schools, masters, etc.) and third-cycle (doctorate) university students. The first part of this work is full of environmental applications, with the first chapter on desertification and the second on monitoring deforestation. Chapter 3 deals with potential of remote sensing in relation to wildfires.

The second part of this work is full of applications for risks, beginning with Chapter 4 on the characterization of industrial plumes by passive remote sensing. Chapter 5 presents optical image correlation for quantification of earth surface motion and the investigation of tectonic and geomorphologic processes. Chapter 6 is dedicated to optical remote sensing's applications in the mining environment, from exploitation to post-mining. Chapter 7 shows radar imaging's strong potential for volcanology and the study of urban and mining subsidences. Chapters 8 and 9 deal with the use of remote sensing in locust management and the contribution of remote sensing to the epidemiology of infectious diseases, respectively.

We would like to thank those who have contributed to the preparation of this volume. First and foremost, there are the authors of the chapters, but also the members of the scientific committee for their critical reading of the chapters and for the corrections they provided. This project could not have been completed without the support of IRSTEA (the French Institute for Research in Science and Technology for Environment and Agriculture), CNRS (the French National Center for Scientific Research) and CNES (the French National Center for Space Studies).

We also would like to thank our families for their support and Messrs André Mariotti (Professor Emeritus, Pierre and Marie Curie University) and Pierrick Givone (Scientific Director, IRSTEA) for their confidence.

1

Drylands and Desertification

1.1. Drylands

Regions with dry climates cover a large part of the world's landmass (more than 40% [ADE 05], Figure 1.1). Collectively known by the English term "drylands", they encompass a wide range of ecosystems, including the American "deserts" of California, Nevada and Mexico; the vast steppes of Central Asia and North Africa; the Sahelian savannah and real deserts such as the Sahara in Africa or the Gobi in China.

1.1.1. *Common features*

The primary defining feature of drylands is a climate with average annual precipitation levels (P) that are substantially less than the evaporative capacity of the air (ETP, evapotranspiration). When using the aridity index P/ETP, *drylands* are defined as regions where this ratio is less than two-thirds [ADE 15]. In addition, these are further subdivided into classes using this ratio (for example hyper-arid regions or "real deserts" correspond to an aridity index of less than 0.05). There can be over-variability both in space and in time, even if these climates also experience seasons. Therefore, to roughly summarize their differences, the drylands in North Africa experience rain, especially in the winter, and the natural vegetation is the steppe type, dominated by low woody shrubs, while south of the Sahara

---

Chapter written by Richard ESCADAFAL.

there is savannah, whose entire landscape quickly develops and grows vegetation in the summer monsoon season, which largely disappears during the long dry season, leaving a landscape that is almost bare.

Although each of the world's drylands has its own particular characteristics, the irregularity of precipitation and consequently the patchy aspect of vegetation cover over time and/or space is by far the most common feature. The world's large regions with dry climates also have a very recognizable physiognomy: large open spaces with local dune formations, places subject to the action of the wind and associated phenomena. These are also territories that have mostly been used as grazing lands by pastoral societies, some of whom adopted a nomadic lifestyle, since the Neolithic. Drylands are therefore mostly used for pastoral activity on at least two-thirds of their surface area [ADE 15], while less than 10% is used for rain-fed crops and irrigated areas represent barely 3% [LHO 07].

The characteristics that distinguish drylands have implications for the environmental problems that affect them as well as for the use of remote sensing to study them.

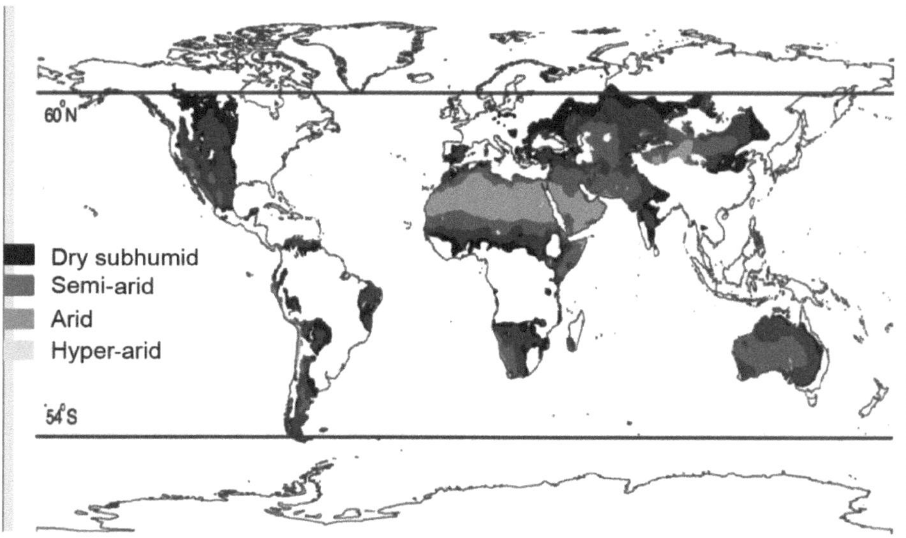

Figure 1.1. *World map of the distribution of drylands; source: Millennium Ecosystem Assessment (2005)*

1.1.2. *Desertification: a form of degradation specific to drylands*

In drylands, the soil is more directly exposed to changes in the climate, and particular forms of land degradation arise, collectively referred to as "desertification". Although this term does not adequately describe the phenomena involved and often leads to an erroneous interpretation in which desertification is equated with invasion by the desert, it is nevertheless now commonly used. The United Nations convention in charge of combating this phenomenon provides the following definition: "desertification is land degradation in arid, semi-arid and dry sub humid zones resulting from diverse factors including climatic variations and human activity"[1].

This definition emphasizes both the climatic context and the land management aspect. As with the majority of land, arid or not, degradation generally occurs following maladaptive and unsustainable practices that reduce the environment's capacity to provide ecosystem services, of which the most important is the production of food.

Numerous studies have been done to understand the mechanisms of desertification and to counteract it [REY 11, DOD 13, ESC 15][2]. For this, a diagnosis of the state of various environments and the methods for monitoring their changes are crucial to programs for action. In this chapter, we will explore cases where data and remote sensing methods have been implemented.

1.2. Dryland vegetation observed by satellite

With regard to remote sensing, drylands have a decisive advantage: minimal cloud cover. This makes the use of optical imaging much easier, especially for observing and monitoring vegetation cover. In addition, the use of satellite images to study these vast areas, which are often difficult to observe directly on the ground, has appeared to be promising since the beginning of civilian remote sensing. If optical data are dominant in the image providers' catalogs, the most recent development of other spectral ranges, especially infrared and microwaves, allows for the diversification of

1 www.unccd.int.

2 See also the summary published by the CSFD (www.csf-desertification.org).

the information obtained, and in particular the development of a multisensor approach.

1.2.1. *Vegetation cover observed by satellite*

Whether the sensor used to observe drylands is passive or active, the radiation that it receives is reflected or emitted by land surfaces primarily made up of soil and vegetation, with the latter generally being the less dominant element. The term "soil" is used here according to the broad definition used in French soil science, where rocky outcroppings and accumulations of sand (dunes, for example) are raw mineral soils (while in the North American classification system, these are "non-soils"). Here, the soil is a very important component, even if for remote sensing used for ecological monitoring the emphasis is on monitoring green vegetation and is done mostly with the NDVI (*normalized difference vegetation index*). In comparison with Figure 1.1, we can see the low NDVI index values for the drylands in Figure 1.2.

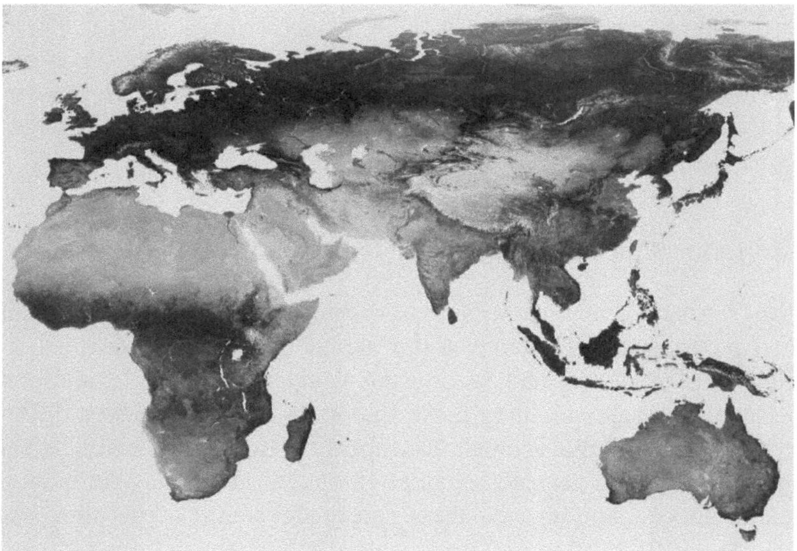

Figure 1.2. *NDVI vegetation index values, varying from gray to dark green (National Oceanic and Atmospheric Administration – Advanced Very High Resolution Radiometer (NOAA – AVHRR) satellite data, 2012 summary, copyright NOAA). For a color version of this figure, see www.iste.co.uk/baghdadi/6.zip*

To correctly characterize the vegetation of these regions, it should be noted that the vegetation is discontinuous over time and/or space. This is an important point, especially in choosing dates for observation. In cases where herbaceous vegetation covers the landscape after it rains, the NDVI is well-adapted to evaluating the cover or the biomass of that vegetative cover and characterizing its fluctuations, especially in the long term.

Therefore, NOAA-AVHRR satellite images have been used by numerous teams to measure the changes in vegetation cover at the global level and to make diagnoses of the state of the environment. One of the difficulties present is the ability to link satellite measurements with the reality on the ground: a measuring point of the satellite sensors used here covers 9 km × 9 km, and it is often difficult to obtain vegetation measurements in the field of the same area at the same time when the satellite is passing over. In addition, the atmosphere's transparency is affected by clouds and dust, which complicates the comparison of two types of measurement: that done on the ground and that done by satellite. The most recent large-field instruments, such as SPOT Vegetation (SPOT-VGT), MODIS and MERIS, have a finer spatial resolution, allowing for a more detailed monitoring, but over a shorter amount of time since they have been observing the planet for less time.

Among the recent work done, a French team was able to compare the level of vegetation coverage on the ground with the vegetation index measured by satellite on two reference sites: one in the Gourma region of Mali and one in the Fakara region of Niger [DAR 14a]. They were able to make this comparison for the years 1981 to 2011 using field measurements that have been carried out at these sites for almost that entire period (Figure 1.3).

a) b)

Figure 1.3. *Amount of vegetation cover on a site with sandy soil in Sahelian Africa: a) dry season; b) wet season (photos: P. Hiernaux)*

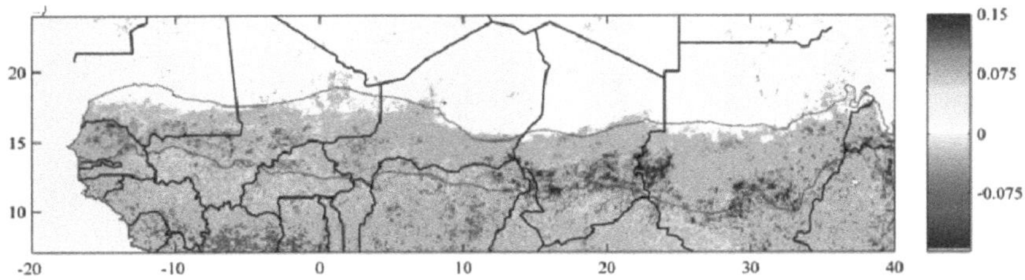

Figure 1.4. *Average change in the satellite vegetation index (NDVI-AVHRR) during the 1981–2011 period (in red: decrease in vegetation cover; in green: increase in vegetation; in gray: clear trend). For a color version of this figure, see www.iste.co.uk/baghdadi/6.zip*

Among the results of this study, Figure 1.4 shows the change in the NDVI during the same period for the entire Sahelian zone. It clearly shows the "re-vegetation" recorded in this region, since a large proportion of green is visible in the image, corresponding to an increase in the vegetation index during the 30 years being studied. Therefore, there is an overall re-vegetation of the Sahel, and we can say that it has not been desertified. Experts agree that this is primarily due to an increase in long-term average rainfall.

Nevertheless, disparities are visible, which show that the two sites have had different fates. The majority of the vegetation cover is made up of therophytes, and although those in Mali have increased on average, the opposite has happened at the site in Niger. In both cases, the satellite measurements are corroborated by observations in the field. An analysis of the different studies monitoring vegetation by satellite has shown that their findings are often inconsistent [HUT 05].

These time series of satellite images make it easy to monitor vegetation cover and detect trends, but they must be corroborated with appropriate field observations and care must be taken when interpreting them (the period chosen for establishing a trend is important). Figure 1.5 shows that, depending on the period being studied, the conclusions are completely different. There is no clear trend over the course of the entire series, but on the other hand, we can conclude that there has been re-vegetation in the 1984–1999 period, while during the 1999–2004 period, there has been a degradation of the vegetation cover.

To interpret variations in the vegetation cover observed from space in terms of performance of ecosystems, their abundance is compared to precipitation, especially using the *rain use efficiency* (RUE) value. In this concept, the primary production per millimeter of rain is used as an indicator of the state of the environment [LEH 84]. However, it is not universal; its application at the continental level does not seem practical [FEN 11], and it must be contextualized [DAR 14b]. The monitoring of RUE remains a usable approach at the level of a drainage basin for monitoring its environmental status, even if the main difficulty is having access to information about precipitations and their actual distribution that is sufficiently detailed.

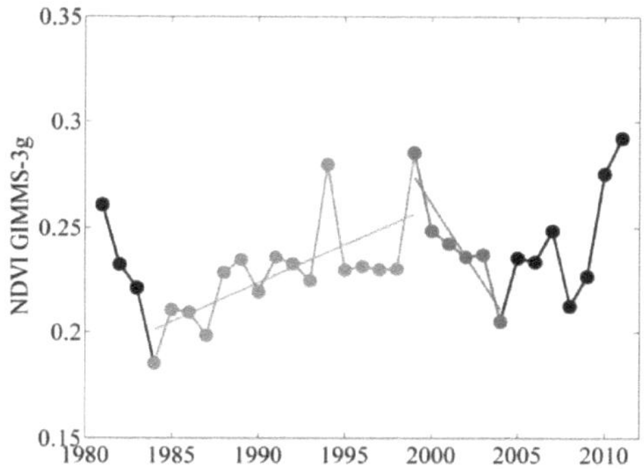

Figure 1.5. *Time series of the vegetation index (NDVI GIMMS-3g of the AVHRR sensor) between 1981 and 2011, in the Gourma region in Mali (two contrasting periods are highlighted: the decrease in vegetation cover is in red and the increase is in green). For a color version of this figure, see www.iste.co.uk/baghdadi/6.zip*

In the case of steppe-type vegetation, the growing period after it rains is also reflected by the development of green biomass. However, this is generally less abundant and sparser. Therefore, monitoring of these formations using the NDVI is more difficult, especially since a large portion of the plants are woody, not green [KEN 89].

The detection and monitoring of non-photosynthetic vegetation (NPV) is much more difficult. This "non-green" component primarily includes the woody parts of vegetation, standing dead biomass or in the form of litter. The use of spectral bands other than those used for the usual vegetation indexes (visible and near infrared) has undergone new developments, especially by using mid-infrared. They have nevertheless not yet been able to succeed in creating an index as universal as the NDVI, even if convincing results have been obtained in some cases [JAC 14]. Focus is now on new sensors such as LiDAR, for more easily distinguishing NPV [LI 15] whose importance in the drylands is considerable, not only at the global level due to its role in the carbon cycle, but also on the local level, for its role in habitat and soil protection.

1.2.2. *Discontinuity of vegetation cover (patchiness)*

Beyond the measure of the vegetation mass and its changes, which is essential for understanding the functioning of ecosystems and the performance level of agro-ecosystems, the spatial distribution of this vegetation is also useful information and can be obtained using remote sensing.

In the Sahel, the organization of the vegetation into bands creates formations known as "tiger bush", which quickly attracted the attention of naturalists. Mathematicians have also studied these particular patterns and have offered different models to explain their existence. Field studies of the ecologic and hydrologic functioning of these environments has revealed that they represent an excellent adaptation to aridity [RIE 04]. It should also be noted that this type of organization of bands of vegetation, separated by bands of bare ground, is found, for example in the slope agriculture traditionally established by farmers in southern Tunisia (Figure 1.6).

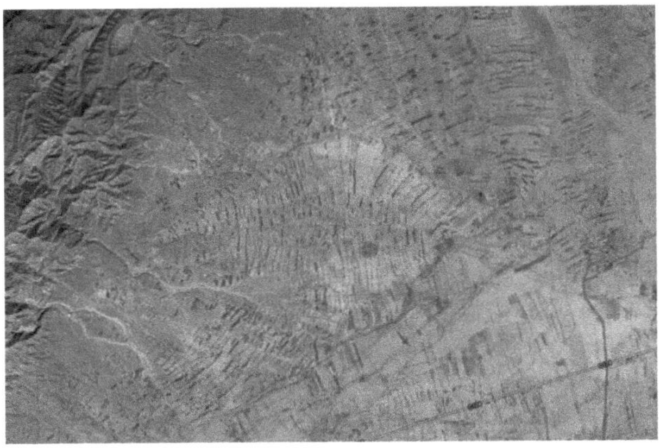

Figure 1.6. *Fields organized in perpendicular bands, foothills near Bou Hedma National Park, southern Tunisia (from Google Maps, 34°42'N, 9°36'E)*

More recently, studies analyzing the distribution of clumps of vegetation have shown that in some cases, the law of distribution reflects the state of the environment [KEF 07]. Therefore, at test sites in Spain, Greece and Morocco, the size of the clumps follows a power law distribution pattern over the three sites when these environments are in good shape, while the coverage level varies between sites. It has also been shown that the distribution no longer follows this law when the environments have been degraded. At the moment, these observations have not been able to be generalized and used as a routine diagnostic criterion for the environmental status of drylands. Recent results nevertheless seem to confirm that analyzing distributions could permit measuring this state using remote sensing based on the measurement of distance between a random distribution of plants observed on high-resolution images obtained via Google Earth [XU 15].

Whether it follows identified patterns or not, the distribution of vegetation on the soil surface has a strong impact on the water cycle, as much by the redistribution of surface water (interception) as by facilitating infiltration, and above all the partition between direct evaporation and ETP of the water stored in the soil. It is clear that maximization on the part of soil water that returns to the atmosphere via plants is the key to optimal primary production. Therefore, the structure of the surface is an interesting element to observe by satellite, since it determines this partition of the water and nutrients that it transports and allows for a diagnosis of the state of ecological functioning.

1.2.3. *Ligneous plants: a key element in discontinuous plant cover*

With regard to low woody vegetation, which makes up the majority of steppes and pseudo-steppes of the dry Mediterranean climates of the world (North Africa, the Middle-East, Chile, etc. [LEH 86]), their influence on the NDVI is low and limited to the growing season. On the other hand, they change the albedo significantly, and their distribution in spaced clumps (*clumped vegetation*, example in Figure 1.7) produces an anisotropy of the radiation reflection. Therefore, they present a specific distribution function of the specific bidirectional reflectance, a property that can be used to distinguish surfaces with clumped vegetation from others. Promising results have been obtained that way with experimental instruments that provide multiangular data, including MISR and CHRIS-PROBA [CHO 08], but these are no longer in use and there are currently few instruments available that can routinely provide this type of measurement [HE 12]. Among the operational data, the MODIS data taken from different viewing angles depending on the transit time have been used to create a global map of a clumping index (Figure 1.8). Note that the arid regions at the northern and southern borders of the Sahara (which appear in white) have elevated index values, while the NDVI is low (Figure 1.2), this combination therefore seems usable as a criterion for distinguishing the discontinuous vegetation cover of drylands.

Figure 1.7. *Steppe vegetation in southern Tunisia, overhead view on the right (photos: R. Escadafal)*

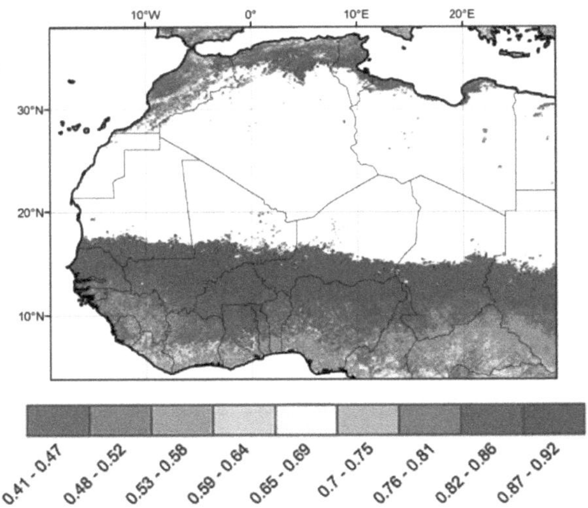

Figure 1.8. *Extract from the of the global clumping index map established by [HE 12] using MODIS data from the year 2006 (spatial resolution 500 m) showing Western Africa. For a color version of this figure, see www.iste.co.uk/baghdadi/6.zip*

Likewise, in Australia, processing MODIS multiangular data using a simple model of surfaces with discontinuous cover made it possible to distinguish types of arid vegetation undifferentiated in traditional NDVI [MCC 12]. There is confirmed interest in multiangular approaches for characterizing the vegetation in these areas, but operational applications that allow any real ecological monitoring still need to be developed.

The case of trees is particularly interesting, since there are no large forests in drylands, there are nevertheless numerous trees and shrubs that play a considerable role both from an ecological perspective and an economic and cultural one. These trees are generally spaced out; they do not form continuous cover or a canopy, but the extent of their stands can be rather large and maintained by humans. This is the case with the "wooded parks" of the Sahel, or more generally of agroforestry, which is being increasingly promoted as a sustainable land management technique [BAY 14].

Remote sensing of these scattered trees was at first hampered by the low spatial resolution of the images available. However, since the availability of very high resolution, it has seen a sharp increase. Access to old aerial photographs, especially declassified photos from military satellites (the Corona series, for example), makes it possible to analyze the changes between the years 1960 and 1970 when they were taken and the current images [AND 06]. As many authors like to emphasize [REI 08], the changes observed are often surprising: while as a whole the Sahel has been affected by desertification, some regions have seen a spectacular increase in the number of trees. In certain cases, this results from farmers developing a type of agriculture involving trees, especially by assisted natural regeneration that allows for better maintenance of land quality. However, this is not the case for the entire Sahel, and the promotion of agroforestry in these regions has the clear goal of intensifying agriculture, in a sustainable way, by multiplying and diversifying the establishment of trees "outside the forest".

Properly speaking, there are no wooded parks in the Northern Sahara, but self-sown open stands of trees adapted to the dryness can occupy vast areas (as is the case with the arganeraie in southwestern Morocco) and make up a key element of a traditional agro-sylvo-pastoral economy. Trees are also planted as soon as local conditions allow it, and the systems of small dams in talwegs (jessours) and banks on slopes (tabias) that retain runoff rainwater are a remarkable rainfed agriculture technique that has been used for centuries in Southern Tunisia and recently revived. The number of trees and their growth thanks to these thousands of small constructions is certainly an indication of changes in the state of the environment [JED 09].

Finally, there are irrigated tree plantations, in particular the palm groves predominantly used for cultivating dates in the Northern Sahara.

In general, surveillance of irrigated areas is somewhat similar to the monitoring of cultivated areas in other climates, with the advantage of being able to more easily follow the development of crops because of the lower cloud cover. New satellites with a higher overflight frequency allow for refined temporal monitoring, which is the case with the Sentinel-2 constellation (first satellite launched in June 2015); they are particularly adapted to environmental and agricultural monitoring. For example, preliminary pilot programs carried out in semi-arid zones in Morocco have shown that it greatly helps in calculating water consumption by crops; a crucial piece of information when it comes to managing limited water resources, as is the case in the majority of drylands [CHE 08].

1.3. Soils and surface materials: the dominant components of drylands

Due to the low level of precipitation and resulting minimal plant cover, soils are the major component of dryland surfaces. Weak primary production does not allow for a net enrichment of the soil in organic matter, so their content is generally low, around 1% with local exceptions, according to the topography and the forms of erosion present. They are most often exposed to precipitation that causes water erosion, but wind erosion is also one of the major characteristics of drylands. It changes soils by “winnowing” them in some places (removing of fine elements by the action of the wind) and by accumulating sand in other locations. Moreover, the very intense evaporation causes precipitation of solutes in the gaps in the soil, forming concretions, nodules and calcareous and/or gypseous crusts; sodium salts also precipitate in endorheic parts of the landscape. All of these elements lead to the specific characteristics of the soils in these regions, which are often sandy and poor in organic matter, which gives them an unstable structure, and sometimes rich in carbonates.

As a result, these soils generally appear light when viewed from space. The exception is the soil of depressions, which is often more clayey and moist. Eolian materials deposited in varying quantities (sand veils, coatings, nebkhas, dunes, etc.) change the composition of the soil surface and can also change as a result of sandstorms.

On a larger scale, dune formations, sometimes very imposing, form the ergs encountered in almost every desert on the planet, which are very emblematic of drylands; their distribution and evolution have been studied since the first satellite images became available. This has made it possible to dispel a certain number of myths, confirming the fact that desertification is not characterized by the general movement of dunes that would increase the size of the desert by spreading. Although the phenomenon of sand accumulation does occur locally, and warrants the sand fixation projects that are implemented here and there, it is not happening on a continental scale. Observation from space has made it easier to understand the origin of ergs and their fate: generally speaking, they are very stable on a human timescale.

1.3.1. *Sandstorms and the production of desert aerosols*

Surfaces covered with fine particles are an important source of the mineral aerosols that, on entering atmospheric circulation, have an impact on the climatic functions. Sandstorms, haze and, more generally, the transport of dust into the air and the atmosphere are phenomena typical of drylands. Nevertheless, their origin, evolution and size all remain difficult to monitor. Since their impact is important not only on the environment (for example on crops), but on human health, air navigation and the climate in general, Saharan dust and aerosols (especially) have been studied by scientists since the 1950s. Satellite images quickly proved capable of identifying sandstorms and measuring their extent, which is very difficult to do from the ground (Figure 1.9).

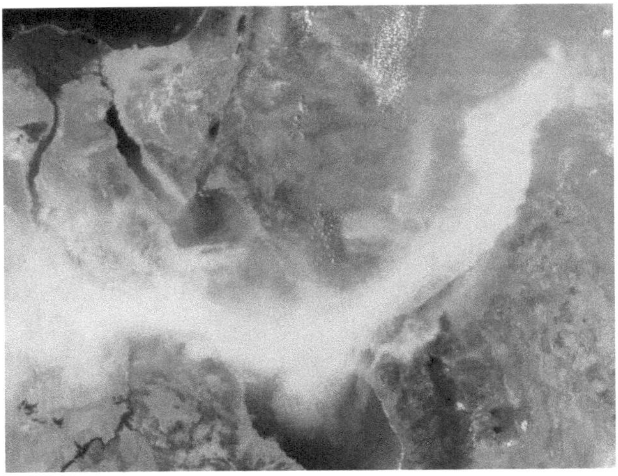

Figure 1.9. *Sandstorm crossing the Red Sea, observed on May 13, 2005, by the MODIS sensors of the Aqua satellite (NASA image courtesy of Jacques Descloitres, MODIS Rapid Response Team, Goddard Space Flight Center)*

By looking at successive images of the regions where these storms originate, it is also possible to identify areas that are sources of dust and to spot which ones are most often active. Figure 1.10 shows the main source areas in Saharan regions. It can be seen that those that most often emit dust are located in Chad. Field observations and recent studies indeed confirm that the Bodélé Depression is the zone that produced by far the most dust, this dried up part of the ancient Lake Mega-Chad even enriches the soil of the Amazon with nutrients since tons of its fine dust particles cross the Atlantic. This is an unexpected effect of dust storms, a positive one, which has been measured with the help of NASA-MODIS satellites [KOR 04].

Predicting these storms is much more complicated, since they form when strong winds encounter dry terrain covered by fine particles. Of course the wind speed on the ground, which causes the removal of these particles, depends on the meteorological conditions and on the roughness of soil surfaces. Modelers that develop prediction methods use satellites to estimate aerodynamic roughness, one of the key parameters of their models (Figure 1.11).

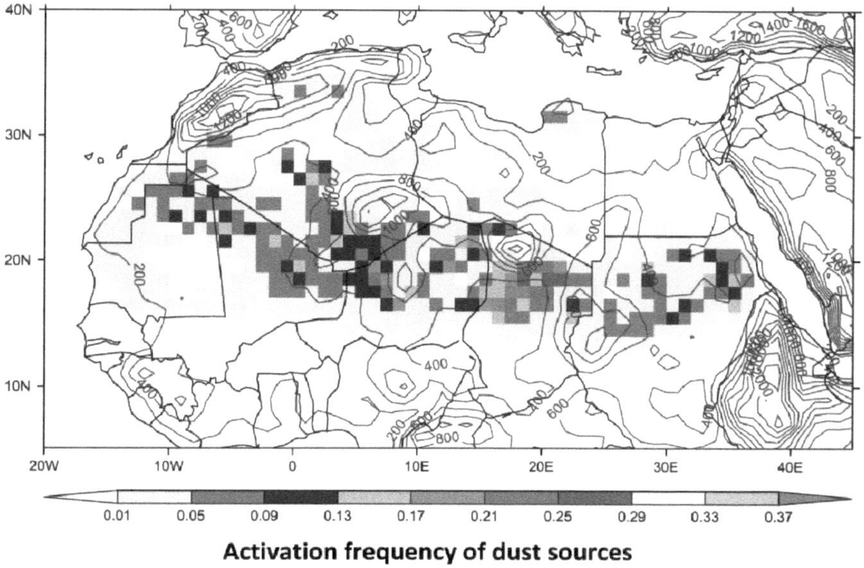

Figure 1.10. *Frequency of activation of dust sources detected with the infrared bands of the SEVIRI sensor on the European satellite METEOSAT [SCH 12]. For a color version of this figure, see www.iste.co.uk/baghdadi/6.zip*

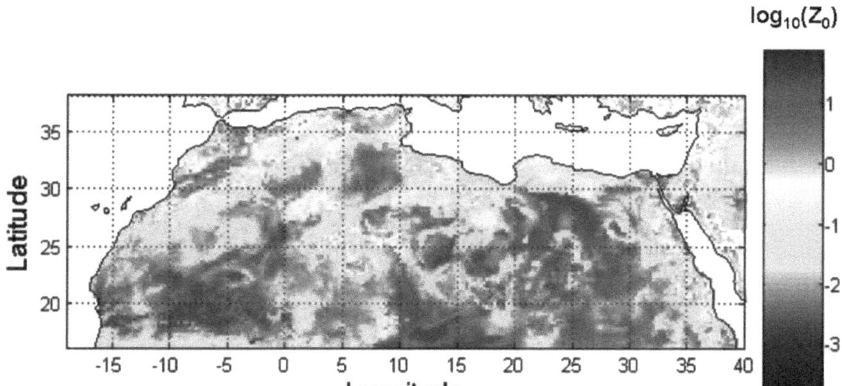

Figure 1.11. *Map of aerodynamic roughness length (Z0) at the resolution (0.25° × 0.25°) taken from POLDER-1 satellite data, and completed from a geomorphological analysis of surface conditions [MAR 04]. For a color version of this figure, see www.iste.co.uk/baghdadi/6.zip*

1.3.2. *The first millimeters of the surface*

As a result of the presence of materials with different characteristics, the soil surface of drylands shows varying colors corresponding to variations in the spectral characteristics detectable in the data from satellites with optical sensors. Color index for soils were then able to be developed based on the contrast between bands in the visible spectral domain [ESC 88], and used in certain cases to distinguish different types of surface materials. This variability in spectral characteristics is also expressed in the near-infrared region, which means that some soils show non-null values on the vegetation index in the absence of any photosynthetic component on their surface. This causes a "noise" that must be taken into account during the remote sensing of low plant cover, a frequent situation in these regions. It is linked to the coloration of the soil: the more a soil is colored (reddish), the stronger the noise signal. An index for soil coloration was created and used for a simple method for correcting this noise [ESC 91]. Figure 1.12 shows this noise in SPOT-VGT images.

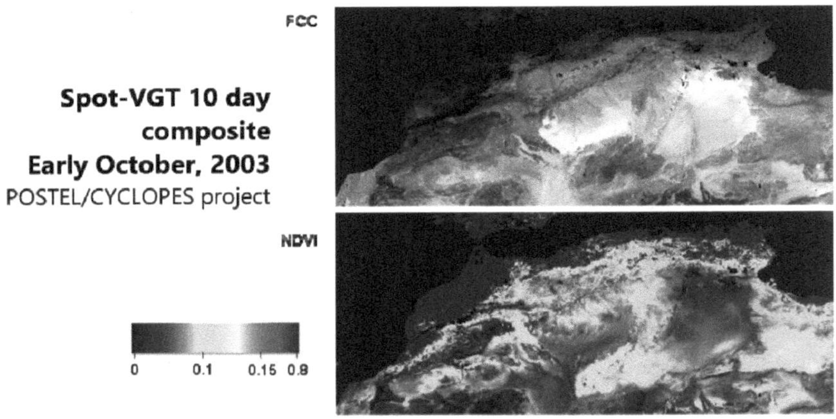

Figure 1.12. *A color composite in false colors of visible and near-infrared channels (high) and NDVI values (low) in North Africa. For a color version of this figure, see www.iste.co.uk/baghdadi/6.zip*

Moreover, other elements present on the soil surface can modify the spectral characteristics. *Biocrust* (Figure 1.13) is a widespread component of arid natural environments, the study of which has revealed interesting properties [BEL 01]. It can be made up of cyanobacteria, mosses, algae and other microorganisms. These are involved in the carbon cycle and change the permeability and consistency of the surface [ROD 14]. Well-developed algal crusts are considered a diagnostic element of good health of the ecosystem. Conversely, their alteration and/or degradation should be avoided. There have even been attempts to restore degraded environments by reintroducing the crust via inoculation.

Figure 1.13. *Photograph showing the biocrust on the soil of an arid region (photo: N. Baghdadi)*

The first characterization tests of the optical properties of biocrust have shown that when they are moist, for example after it rains, they have NDVI values reaching 0.30, which is a potential source of error in estimating the abundance of green vegetation by satellite [KAR 99]. Due to the presence of specific pigments in components of biocrust, Figure 1.14 shows the appearance of characteristic absorption bands, especially between 600 and 700 nm, potentially usable in hyperspectral remote sensing [UST 09].

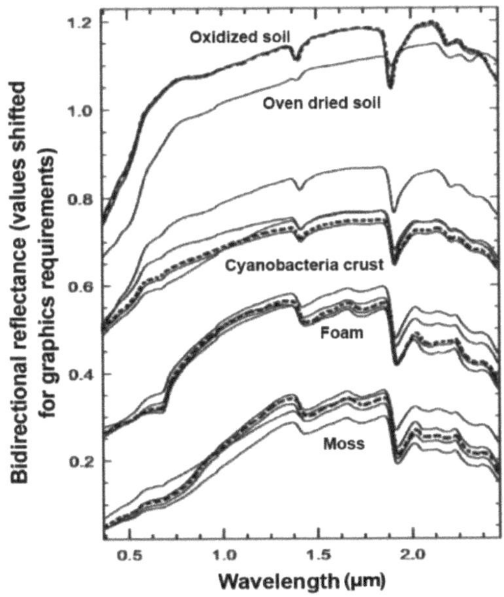

Figure 1.14. *Reflectance spectrums of soil samples without biocrust (above) and the components of biocrust that covers the soil [UST 09]*

1.3.3. *Monitoring of surface water: ponds, floods, ravines, etc.*

If these regions are referred to as dry, precipitation is not necessarily absent: it is very rare in hyper-arid areas, although they are very much present in the Sahel, for example during the rainy season (or monsoon, studied by the international program AMMA[3]). Characterized by periods of strong intensity, these rains fill temporary lakes and ponds where the monitoring by satellites has revealed a paradox: the ponds are filled more in dry years [GAR 10].

Violent rain causes floods and leads to erosion that can be catastrophic locally. The traces of this hydric erosion, in the form of incision, ravines and sediment deposits, can be measured with very high spatial resolution images, especially those obtained from drones [OLE 12], whose use is expanding rapidly. Again, the weakness of the vegetation cover makes it easy to observe changes in the soil surface.

3 http://www.amma-catch.org.

1.3.4. *Variable characteristics of surfaces: roughness, water content and temperature*

Beyond identifying the relatively permanent components of surfaces and monitoring their evolution, other more labile parameters, such as roughness and water content, can be usefully measured by remote sensing. In this respect, drylands are not that different from other regions, and the same microwave or thermal sensors can be used for agricultural and hydrological applications (Volumes 3 and 4 of this series).

1.3.5. *Other applications*

Beyond the methods reviewed, based on the surface reflectance it should be noted that the microwave sensors, in particular those with decametric wavelengths, have the ability to detect underlying geological materials through dry sandy surfaces [SCH 86]. This characteristic revealed by the first space shuttle missions has since been used to identify fossil hydrographic networks [PAI 09].

Many of the remote sensing methods, which are described in other chapters of this book, are used in drylands. We have focused here on those which are more specific to drylands and based on optical imagery, which is most widely available and usable in these regions. Nevertheless, the other sensors, especially those functioning in the microwave region, also provide interesting and complementary images. As a result, the data from ERS-2 radar have been used to successfully map typical vegetation cover of the arid zone in Tunisia [ZRI 03].

1.4. Conclusions and outlook

Thanks to satellite data and their processing, especially those mentioned in this chapter, the monitoring of the evolution of the state of arid regions is continually improving. These regions play a major role in the functioning of the biosphere and more generally in the life of our planet. The degradation of these environments, desertification, is a real and serious threat that led to the creation of the United Nations Convention to Combat Desertification in 1997. Currently, this convention has developed recommendations to its member countries for environmental monitoring of these environments[4], including the use of satellite imagery, with the goal of seeing its concerns taken into consideration in the international agenda [THO 12].

In terms of outlook, the use of remote sensing data will very certainly increase because of a combination of easier access (a large amount of data are now pretreated and free), an increase in types of data as well as wavelengths and spatial resolution of the sensors, and the accumulation of historical data that provide a view of the changes over three to four decades. This last item has been made easier by the emergence of processing methods for large data series (Big Data), some of whose results are already accessible to the general public (*Google Earth Engine*, for example).

Among the methods already tested that should be developed for operational use in drylands, regular production of anisotropy maps of vegetation or "density of clumps" appears promising. Estimation of RUE will become more reliable when we have access to high-quality rainfall data at a scale as detailed as that of the green vegetation cover data (obtained from the NDVI of the MODIS instrument, for example, with a spatial resolution of 250 m). They will be effectively complemented by monitoring and evaluation methods of non-chlorophyll vegetation masses. This will allow much more precise diagnostics on changes in the state of these dry environments, usable for managing territories and providing more accurate statistics at the country level, and at higher levels through aggregation (global monitoring by large international organizations).

1.5. Key points

Drylands are particularly well observed by sensors onboard satellites, which "see" them easily. The images obtained are very useful since they provide access to large areas, many of which are difficult to access by land. The specific characteristics of these regions – sparse vegetation that is discontinuous over time, soils exposed to atmospheric factors and the associated eolian phenomena – require approaches adapted to the study of these environments.

---

4 www.unccd.int.

Satellites whose sensors function in the optical region (such as NOAA-AVHRR, NASA-LANDSAT, MODIS, SPOT HRV and SPOT-VGT, Sentinel-2, etc.) can measure the photosynthetic activity of the soil surface and the amount of green vegetation with vegetation indexes (NDVI, for example) according to a revisit time and spatial resolution that are more or less fine, and covering shorter or longer periods of time. Taking into account the interannual variability of the climate, it is necessary to make long-term observations to diagnose trends, for which the data series from the oldest instruments are the most useful (NOAA series, Landsat, since 1971 and SPOT since 1986), even if the quality of the information is inferior to that of more recent instruments.

It should nevertheless be remembered that since vegetation is not abundant in drylands, the surface is mostly made up of bare soils locally covered by sand, stones and/or biocrust. Monitoring using the NDVI is not appropriate in zones where green vegetation is not very dense, and other approaches must be used, such as the monitoring of non-photosynthetic vegetation (dry, woody parts, etc.) using a custom index (combination of bands including mid-infrared). We can also use the directional properties of surfaces, namely the impact of the distribution of clumps on the angular radiation reflection, whether it is solar (*clumping index*) or radar (backscatter), for example. New instruments that have recently appeared (SMOS) and are in development (BIOMASS) will fill out the existing range, especially for monitoring the water content of the soil. Hyperspectral sensors, which have proven effective on airplanes, will be more and more numerous on space platforms in the years to come.

Of course, remote sensing can be used in drylands for traditional applications not specific to these regions (monitoring irrigation, infrastructure, urban planning, health, etc.), but in ecological terms it has an irreplaceable role in monitoring the state of dry environments, following their evolution, including desertification, and to ensure that their management practices are sustainable.

1.6. Bibliography

[ADE 05] ADEEL Z., SAFRIEL U., NIEMEIJER D. *et al.*, *Ecosystems and Human Well-being: Synthesis. Millennium Ecosystem Assessment*, Island Press, Washington, DC., p. 155, 2005.

[AND 06] ANDERSEN G.L., "How to detect desert trees using CORONA images: discovering historical ecological data", *Journal of Arid Environments*, vol. 65, no. 3, pp. 491–511, 2006.

[BAY 14] BAYALA J., SANOU J., TEKLEHAIMANOT Z. *et al.*, "Parklands for buffering climate risk and sustaining agricultural production in the Sahel of West Africa", *Current Opinion in Environmental Sustainability*, vol. 6, pp. 28–34, 2014.

[BEL 01] BELNAP J., LANGE O.L., *Biological Soil Crusts: Structure, Function, and Management*, Springer, Berlin, 2001.

[CHE 08] CHEHBOUNI A., ESCADAFAL R., DUCHEMIN B. *et al.*, "An integrated modelling and remote sensing approach for hydrological study in arid and semi-arid regions: the SUDMED Programme", *International Journal of Remote Sensing*, vol. 29, nos. 17–18, pp. 5161–5181, 2008.

[CHO 08] CHOPPING M., "Terrestrial applications of multiangle remote sensing", in LIANG S. (ed), *Advances in Land Remote Sensing*, Springer, Berlin, 2008.

[DAR 14a] DARDEL C., KERGOAT L., HIERNAUX P. *et al.*, "Re-greening Sahe: 30 years of remote sensing data and field observations (Mali, Niger)", *Remote Sensing of Environment*, vol. 140, pp. 350–364, 2014.

[DAR 14b] DARDEL C., KERGOAT L., HIERNAUX P., *et al.*, "Rain-use-efficiency: what it tells us about the conflicting Sahel greening and Sahelian paradox", *Remote Sensing*, vol. 6, no. 4, pp. 3446–3474, 2014.

[DOD 13] D'ODORICO P., BHATTACHAN A., DAVIS K.F. *et al.*, "Global desertification: drivers and feedbacks", *Advances in Water Resources*, vol. 51, pp. 326–344, 2013.

[ESC 88] ESCADAFAL R., GIRARD M.C., COURAULT D., "Modeling the relationships between Munsell soil color and soil spectral properties", *International Agrophysics*, vol. 4, no. 3, pp. 249–261,1988.

[ESC 91] ESCADAFAL R., HUETE A., "Improvement in remote-sensing of low vegetation cover in arid regions by correcting vegetation indexes for soil noise", *Comptes Rendus de l'Académie des Sciences de Paris*, vol. 312, no. 11, pp. 1385–1391, 1991.

[ESC 15] ESCADAFAL R., BARBERO-SIERRA C., EXBRAYAT W. *et al.*, "First appraisal of the current structure of research on land and soil degradation as evidenced by bibliometric analysis of publications on desertification", *Land Degradation & Development*, vol. 26, no. 5, p. 413–422, 2015.

[FEN 11] FENSHOLT R., RASMUSSEN K., "Analysis of trends in the Sahelian 'rain-use efficiency' using GIMMS NDVI, RFE and GPCP rainfall data", *Remote Sensing of Environment*, vol. 115, no. 2, pp. 438–451, 2011.

[GAR 10] GARDELLE J., HIERNAUX P., KERGOAT L. *et al.*, "Less rain, more water in ponds: a remote sensing study of the dynamics of surface waters from 1950 to present in pastoral Sahel (Gourma region, Mali)", *Hydrology and Earth System Sciences*, vol. 14, no. 2, pp. 309–324, 2010.

[HE 12] HE L., CHEN J.M., PISEK J. *et al.*, "Global clumping index map derived from the MODIS BRDF product", *Remote Sensing of Environment*, vol. 119, pp. 118–130, 2012.

[HUT 05] HUTCHINSON C.F., HERRMANN S.M., MAUKONEN T. *et al.*, "Introduction: the 'greening' of the Sahel," *Journal of Arid Environments*, vol. 63, no. 3, pp. 535–537, 2005.

[JAC 14] JACQUES D.C., KERGOAT L., HIERNAUX P. *et al.*, "Monitoring dry vegetation masses in semi-arid areas with MODIS SWIR bands", *Remote Sensing of Environment*, vol. 153, pp. 40–49, 2014.

[JED 09] JEDDI K., CHAIEB M., "Using trees as a restoration tool in Tunisian arid areas: effects on understory vegetation and soil nutrients", *Rangeland Journal*, vol. 31, no. 4, pp. 377–384, 2009.

[KAR 99] KARNIELI A., KIDRON G. J., GLAESSER C. *et al.*, "Spectral characteristics of Cyanobacteria soil crust in semiarid environments", *Remote Sensing of Environment*, vol. 69, no. 1, pp. 67–75, 1999.

[KEF 07] KEFI S., RIETKERK M., ALADOS C.L. *et al.*, "Spatial vegetation patterns and imminent desertification in mediterranean arid ecosystems", *Nature*, vol. 449, no. 7159, pp. 213–217, 2007.

[KEN 89] KENNEDY P.J., "Monitoring the phenology of Tunisian grazing lands", *International Journal of Remote Sensing*, vol. 10, nos. 4–5, pp. 835–845, 1989.

[KOR 04] KOREN I., KAUFMAN Y.J., "Direct wind measurements of Saharan dust events from Terra and Aqua satellites", *Geophysical Research Letters*, vol. 31, no. 6, pp. L06122, 2004.

[LEH 84] LE HOUÉROU H.N., "Rain use efficiency – a unifying concept in arid-land ecology", *Journal of Arid Environments*, vol. 7, pp. 213–247, 1984.

[LEH 86] LE HOUÉROU H.N., "The desert and arid zones of northern Africa", *Hot Deserts and Arid Shrublands*, Elsevier, Amsterdam, pp. 101–147, 1986.

[LHO 07] LHOSTE P., "Pastoral societies and desertification in the Sahel", *Bois et forêts des tropiques*, vol. 239, no. 3, pp. 49–59, 2007.

[LI 15] LI Z., GUO X., "Remote sensing of terrestrial non-photosynthetic vegetation using hyperspectral, multispectral, SAR, and LiDAR data", *Progress in Physical Geography*, vol. 40, no. 2, pp. 276–304, 2015.

[MAR 04] MARTICORENA B., CHAZETTE P., BERGAMETTI G. *et al.*, "Mapping the aerodynamic roughness length of desert surfaces from the POLDER/ADEOS bi-directional reflectance product", *International Journal of Remote Sensing*, vol. 25, no. 3, pp. 603–626, 2004.

[MCC 12] MCCAMLEY G., GRANT I., JONES S. *et al.*, "Characterising vegetated surfaces using MODIS multiangular satellite data", *The ISPRS Annals of the Photogrammetry, Remote Sensing and Spatial Information Sciences*, vol. I–7, pp. 13–18, 2012.

[OLE 12] OLEIRE-OLTMANNS S., MARZOLFF I., PETER K., *et al.*, "Unmanned aerial vehicle (UAV) for monitoring soil erosion in Morocco", *Remote Sensing*, vol. 4, no. 11, pp. 3390–3416, 2012.

[PAI 09] PAILLOU P., SCHUSTER M., TOOTH S. *et al.*, "Mapping of a major paleodrainage system in eastern Libya using orbital imaging radar: the Kufrah River", *Earth and Planetary Science Letters*, vol. 277, nos. 3–4, pp. 327–333, 2009.

[REI 08] REIJ C.P., SMALING E.M.A., "Analyzing successes in agriculture and land management in sub-Saharan Africa: is macro-level gloom obscuring positive micro-level change?", *Land Use Policy*, vol. 25, pp. 410–420, 2008.

[REY 11] REYNOLDS J.F., GRAINGER A., BASTIN G. *et al.*, "Scientific concepts for an integrated analysis of desertification", *Land Degradation & Development*, vol. 22, pp. 166–183, 2011.

[RIE 04] RIETKERK M., DEKKER S.C., DE RUITER P.C. *et al.*, "Self-organized patchiness and catastrophic shifts in ecosystems", *Science*, vol. 305, no. 5692, pp. 1926–1929, 2004.

[ROD 14] RODRIGUEZ-CABALLERO E., CANTON Y., LAZARO R. *et al.*, "Cross-scale interactions between surface components and rainfall properties. Non-linearities in the hydrological and erosive behavior of semiarid catchments", *Journal of Hydrology*, vol. 517, pp. 815–825, 2014.

[SCH 86] SCHABER G.G., MCCAULEY J.F., BREED C.S. *et al.*, "Shuttle imaging radar: Physical controls on signal penetration and subsurface scattenng in the eastern Sahara", *IEEE Transactions on Geoscience and Remote Sensing*, vol. GE-24, no. 4, pp. 603–623, 1986.

[SCH 12] SCHEPANSKI K., TEGEN I., MACKE A., "Comparison of satellite based observations of Saharan dust source areas", *Remote Sensing of Environment*, vol. 123, pp. 90–97, 2012.

[THO 12] THOMAS R.J., AKHTAR-SCHUSTER M., STRINGER L.C. *et al.*, "Fertile ground? Options for a science–policy platform for land", *Environmental Science & Policy*, vol. 16, pp. 122–135, 2012.

[UST 09] USTIN S.L., VALKO P.G., KEFAUVER S.C. *et al.*, "Remote sensing of biological soil crust under simulated climate change manipulations in the Mojave Desert", *Remote Sensing of Environment*, vol. 113, pp. 317–328, 2009.

[XU 15] XU C., HOLMGREN M., VAN NES E.H. *et al.*, "Can we infer plant facilitation from remote sensing? A test across global drylands", *Ecological Applications*, vol. 25, no. 6, pp. 1456–1462, 2015.

[ZRI 03] ZRIBI M., LE HEGARAT-MASCLE S., TACONET O. *et al.*, "Derivation of wild vegetation cover density in semi-arid regions: ERS2/SAR evaluation", *International Journal of Remote Sensing*, vol. 24, no. 6, pp.1335–1352, 2003.

2

Remote Sensing and Measuring Deforestation

2.1. Introduction: forests, deforestation and forest degradation

Forests occupy one-third of the planet's landmass, covering nearly 4 billion hectares. The largest forested areas are located in the boreal and equatorial zones. This current distribution is not fixed, but has continuously changed over time due to the influence of environmental change and anthropogenic impact. One-third of current forested areas are made up of forest that is considered primary or intact [FAO 10]. The remaining two-thirds are subject to anthropogenic activity and have, as a result, an uncertain future.

Each year, several million hectares of forest disappear and vast areas are subject to degradation. Deforestation is the complete destruction of forest cover. Conversion into agricultural or pastoral land is the principal cause of deforestation. Once certain areas are abandoned, forest cover can once again develop, leading to what are known as secondary forests. Unlike deforestation, degradation is partial destruction of the forest cover. It is defined as a loss of the forest's capacity to provide ecosystemic services (carbon storage, forest products, etc.) following anthropogenic activity [THO 13]. This degradation is characterized by the breaking up of forested areas following illegal land clearing, overexploitation of wood (Figure 2.1), fires, etc. Degraded forests include a large diversity of forest types depending on the nature, intensity and frequency of the degradation.

Chapter written by Lilian BLANC, Valery GOND and Dinh HO TONG MINH.

Forests provide a diversity of products and ecosystemic services. The living biomass of forests stores around 300 pg of carbon, which means it plays an important role in the potential mitigation of climate change. Forests also represent a reservoir of biodiversity, especially tropical forests. Knowing the extent of forested areas, the state of forests and how they change over time is therefore particularly important with regard to current environmental concerns linked to climate change.

Figure 2.1. *An example of information collected by remote sensing: a) moderate forest degradation with cutting sites in the process of regeneration in a Landsat image in a color composite with mid-infrared (IR), near-IR and red (in the red, green and blue channels, respectively); b) intense degradation with large portions of the ground visible following forest exploitation; c) average degradation with traces of exploitation and old roads (source: used with the permission of Brazil's National Institute for Space Research (INPE, http://www.obt.inpe.br/prodes/index.php]). For a color version of this figure, see www.iste.co.uk/baghdadi/6.zip*

Remote sensing has emerged as the only tool capable of continuously providing reliable data on the regional or continental scale. Unlike other types of land use, forests are a type of vegetation that is easily recognizable on satellite images. Despite the advent of remote sensing in the 1970s, estimates of forest area and its change over time on a planetary scale have only been available since the beginning of the 2000s.

In this chapter, we present an overview of the different sources of images and techniques for processing and analysis used to document deforestation and measure its extent. Monitoring programs on a planetary scale are presented, as well as a regional program (Brazilian Amazon). Finally, the problem of forest degradation, a new and significant challenge, is demonstrated using a tool developed in French Guiana.

2.2. Which images should be used for measuring deforestation? Limits and advantages

The satellite data used to provide estimates of deforestation can be of two types. Optical data, acquired in visible and infrared (IR) wavelengths, are sensitive to operating variables of the vegetation (chlorophyll activity, structure and internal water content of the leaves). Radar data, acquired in the field of multifrequencies with longer wavelengths, are sensitive to the structure of the vegetation (basal area, stand structure and presence of water).

2.2.1. *The contribution of optical data*

2.2.1.1. *Optical data at low and medium spatial resolution*

The observation of deforestation began with data from NOAA's (National Oceanic and Atmospheric Administration) series of satellites with an advanced very high resolution radiometer (AVHRR). These images were frequently used during the 1980s and 1990s [TUC 84, MAL 89, MAY 98]. The AVHRR scanning radiometer detects radiation in the visible, near-IR, mid-IR and thermal IR parts of the spectrum, with a spatial resolution of 1.1 km and a daily temporal frequency. These low spatial resolution data have been used to produce regional evaluations of deforestation, especially in the Brazilian Amazon [NEL 87, SHI 94] and in Central Africa [MAL 95]. At the end of the 1990s, SPOT-Vegetation images, with a spatial resolution identical to the NOAA-AVHRR images, were used to produce estimates of deforestation, notably in Southeast Asia [STI 04].

However, the low spatial resolution of these images is a major limitation when monitoring deforestation, since they cannot measure impact on areas smaller than 1 km^2 which are nevertheless very common. Images with a higher resolution have therefore been used more frequently, such as those from MODIS (moderate resolution imaging spectroradiometer) sensors of the TERRA and AQUA platforms. With a resolution of 250 m, 500 m or 1 km in optical wavelengths, these images are available for free and summarized every 16 days. The processing makes it possible to minimize the influence of clouds. However, these measurements remain subject to problems with the angular configuration of the images [SAL 07, SAM 10].

2.2.1.2. *Optical data with high and very high spatial resolution*

Since the mid-1980s, Landsat data with two MSS (multispectral scanners) and a TM (thematic mapper) has become the principle source of data used to provide estimates of deforestation [SKO 93, DEF 02, ACH 04, HAN 08]. The high spatial resolution (30–60 m), historical depth (data collected since 1972) and the open access to 2 million images since 2009 (http://glovis.usgs.gov/) make it a unique dataset. Nevertheless, these images have a number of limitations when monitoring deforestation: limited availability for the first decades (before 1990), clouding over of images common in tropical zones and a low frequency of images (a satellite takes 18 days to return to the same place). To correct this problem, these images can be combined with MODIS images [MOR 05]. For example, because of the temporal acquisition of the sensors at low and medium resolutions, it is possible to identify sectors where degradation can be studied in more detail with data at high and very high spatial resolutions [KAT 15]).

On the local scale, images with very high spatial resolution are occasionally used for estimating the deforestation and degradation of forests. This is the case with SPOT-6/7 (1.5 m), RapideEye (5 m), Ikonos (0.81 m), Pleiades (0.5 m) and GeoEye (0.34 m; http://www.satimagingcorp.com/satellite-sensors/) images. Surfaces with a small area, including those with relatively limited impact, are then analyzed in detail.

This type of image can, for example, be used to estimate the impacts on vegetation of a rapid increase in population as a result of migration (one recent example is the events in the Central Africa Republic in 2014).

All of these data are not free, and in general the finer the resolution, the higher the price. However, all of the medium-resolution images from Landsat satellites are free and have open access. Each source of information therefore has a cost that needs to be calculated according to goals, needs and resources.

2.2.2. *Contribution of radar data*

Measuring deforestation with optical satellite data is often very difficult due to the persistence of clouds in wet tropical zones. Because of its characteristic "all-weather sensor", SAR (synthetic aperture radar) radar data have a strong potential for observation in tropical zones.

For mapping and monitoring deforestation with radar images, two approaches are often used:

– an approach based on the intensity of the radar backscatter [SHI 14];

– an approach taking into account the coherence of the radar signatures and the in-phase component [TOU 99].

In the latter approach, the basic principle consists of comparing coherence before and after the impacts on the forest, coherence being sensitive to the stability of the forest [ZEB 92]. This approach is, however, strongly dependent on the availability of suitable images, which are quite limited. As a result, the majority of studies are based on intensity.

As a rule of thumb, the performance of radar sensors in characterizing a forest increases with the wavelength [LET 11]. This is linked to the capacity for signatures with long wavelengths to penetrate the canopy cover, and as a result a better identification of the state of the forest (dense/open, intact/degraded). C-band ($\lambda \sim 5.6$ cm – RADARSAT-2, Sentinel) and X-band ($\lambda \sim 3.1$ cm – TerraSAR-X, COSMO-SkyMed, etc.) waves cannot penetrate very far into the vegetation cover, since they interact with a large number of forest components such as leaves, twigs and small branches. L-band waves ($\lambda \sim 23.6$ cm – Advanced Land Observing Satellite (ALOS)/PALSAR) are not affected by the fine structure of the surface and can easily make it possible to distinguish forests from deforested areas [SHI 14]. The P-band ($\lambda \sim 69$ cm) is also adapted to forestry applications. The future BIOMASS space mission, with a P-band radar sensor, will not only make it possible to distinguish and estimate forest cover, but also to estimate the biomass of tropical forests [HOT 14].

2.3. Methods for monitoring deforestation

2.3.1. *Using optical data*

The use of optical data to identify deforestation or forest degradation takes into account spatial and temporal variations in reflectance. The combination of channels for obtaining indicators of photosynthetic activity such as the NDVI (*normalized difference vegetation index*[1]) or the EVI (*enhanced vegetation index*[2]) makes it possible to locate the pixels corresponding to forest and the pixels corresponding to bare soil. The sharp decline in the values clearly show that the forest has been degraded or deforested. This can occur from one pixel to the other in the same image or the same pixel with images acquired on different dates. This section presents a technique for monitoring deforestation, developed by the Carnegie Institution for Science. A semi-automatic system for analyzing the environment is offered by the team led by Gregory Asner (Carnegie Landsat Analysis System Lite – CLASlite [ASN 09]). Using this analysis system, the user can process nine different satellites to produce maps of deforestation and degradation in tropical forests. CLASlite performs several functions:

– radiometric and atmospheric calibrations to obtain surface reflectance (top of canopy (TOC));

– unmixing spectral data (spectral mixture analysis (SMA)). This approach considers pixels made up of more than one type of soil cover to be "mixed". For example, a pixel located on the edge of a forest contains a portion of bare soil and a portion of vegetation. In contrast, the pixels referred to as "pure" are made up of a single type of ground cover. The SMA method aims to determine[3] the internal composition of mixed pixels on the

1 NDVI = (NIR – R)/(NIR + R), where NIR is the near-IR reflectance and R is red reflectance.

2 EVI = G [(NIR – R)/((NIR) + (C1.RED) – (C2.B) + L)] where G = 2.5; C1 and C2 are the coefficients for mitigating the contamination of aerosols (from a value of 6 and 7.5, respectively), B is the reflectance of the blue band and L is an adaptation coefficient (equal to 1).

3 For each spectral band of the pixel, reflectance is seen as a linear combination of the reflectance of each endmember.

basis of the spectral characteristics of pure pixels known as endmembers. These pure pixel endmembers must show all of the classes of soil cover in one image. CLASLite distinguishes between three pure classes of soil cover: photosynthetic vegetation (PV), non-photosynthetic vegetation (NPV), which especially represents senescent vegetation and wood, and bare soil (S). The spectral signatures of these three pure elements are stored in a vast library owned by CLASLite, constructed from field measurements (spectrometer) and hyperspectral satellite observations (Hyperion sensor). By using an unmixing model (automated Monte Carlo unmixing model), CLASlite determines the proportion of PV, NPV and S for each pixel in the processed image, with the sum of the three equaling 100%;

– calculating the level of forest cover. Measuring tropical deforestation is crucial for understanding the global functioning of the biosphere and its interactions with the atmosphere. Establishing baseline measurements in order to calculate changes in tropical forests over time is therefore a challenge of the utmost importance. To establish these references, the FAO (Food and Agriculture Organization of the United Nations) provides a forest resources assessment (FRA). It was found that for the reference year (1990), the surface area of tropical forests in 90 countries was 1.756 billion hectares [FAO 95], which was then corrected to 1.926 billion [FAO 01] and finally to 1.949 billion [FAO 06]. All of these estimates, made by the same international organization (which specializes in the subject), for the same reference year, varied by 10%, which suggests imprecision and errors [GRA 08]. Using a model (automated Monte Carlo unmixing model) supplied by a database of spectral signatures is the best solution to the problem:

– classifying the images into two classes (forest and non-forest), depending on the proportion of PV, NPV and bare soil contained within each pixel;

– detecting changes in forest cover between the dates of the satellite images to determine the evolution of deforestation and degradation over time.

CLASlite's spectral analysis algorithm is capable of detecting changes in the forest cover in the order of 1%, which corresponds to a Landsat pixel with an area of 10 m^2. CLASlite offers a potential for very detailed detection of tropical forest degradation, in particular allowing detection of the damage done to the crowns of trees during forest exploitation (Figure 2.2). CLASlite software has been free and usable everywhere since December 2013. All that is necessary is to download it and complete a short online training course.

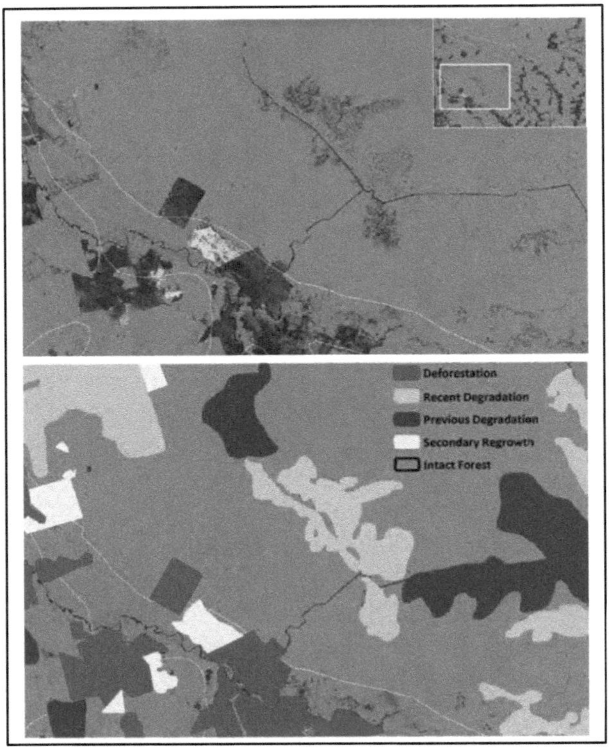

Figure 2.2. *Illustration of a local map of deforestation and forest degradation using CLASlite software: a) the raw output of processing from a Landsat image; b) sketch map with CLASlite showing the different states of the forest (deforestation in red, recent degradation in orange, previous degradation in blue and secondary regrowth in yellow) (source: used with the permission of Gregory P. Asner, Carnegie Institution for Science). For a color version of this figure, see www.iste.co.uk/baghdadi/6.zip*

2.3.2. *Using radar data*

When using radar data, preprocessing is necessary. The sensitivity of the backscatter of radar microwaves to the parameters of the forest can be disrupted by several sources, including the speckle effect, effects of the forest structure (density, biomass, leaf surface, coverage level and crown cover) and environmental conditions (topography and soil moisture content). In the presence of these various sources of noise, mapping the parameters of the forest is a difficult task that cannot be done without comprehensive methodology. Image preprocessing consists of reducing geometric and radiometric distortion while taking topography into account. First there is the preprocessing of images that consists of reducing the geometric and radiometric distortion caused by the geometry of radar viewing angles and topography. Detecting deforestation is based on methods for detecting changes in backscatter radar, which require radiometric normalization, topographic correction and the reduction of *speckle* [QUE 00].

2.3.2.1. *Radiometric normalization*

Radiometric normalization of SAR images is considered necessary for forestry applications since the images used are generally acquired at different times and with different methods (incidence angles in particular). Radiometric normalization makes it possible to work on images where the thresholds used for separating forest/non-forest are relevant and robust.

2.3.2.2. *Topographic correction*

In order to minimize the effect of topography on radar backscatter, it is necessary to use a digital elevation model (DEM) to minimize dependency on backscatter at local incidence angles [VIL 15].

2.3.2.3. *Reducing speckle*

It is often necessary to apply a filter to reduce the effect of speckle in radar images. These are adaptive filters that are applied to reduce speckle while retaining the fine structures present in the images as much as possible. Among the filtering methods used, there is multi-image filtering (use of a time series of images) developed by Quegan *et al.* [QUE 00], which makes it possible to get corrected images with a better smoothing out of speckle. This filter is based on the following relation:

$$P_n(x,y) = \frac{\langle P_n(x,y)\rangle}{N}\sum_{i=1}^{N}\frac{P_i(x,y)}{\langle P_i(x,y)\rangle} \quad \text{with} \quad n = 1, 2, \ldots N \qquad [2.1]$$

– $P_n(x,y)$ is the radar backscatter of the output image *n* for the pixel (*x*,*y*);

– $P_i(x,y)$ is the radar backscatter of the input image for the pixel (*x*,*y*);

– $<P_i(x, y)>$ is the average local intensity of the radar backscatter (within a given window) of the input image i for pixel (x,y);

– N is the number of images.

With this filter, the radiometric resolution of the images is improved without damaging the spatial resolution. A map can be desirable to make it possible to visualize the deforestation/degradation evident in a radar image.

2.4. Programs for estimating deforestation on a global scale

Until very recently, the only figures available on the evolution of forest cover on a global scale were those from the FAO. Since 1946, the FAO has compiled and analyzed statistics and national inventories provided by each country. A report is published every 5 years [FAO 12]. The evolution of forested areas is a report on the loss of forested areas through deforestation and the gains in surface area by tree plantations or by expansion of the forest. These data are nevertheless very heterogeneous between countries because of the diversity of methods, terms and definitions employed. It is generally accepted that estimates of the evolution of forested areas made from these national inventories are marred by numerous errors and imprecision. In 2007, to compensate for these problems, the FAO and the European Commission's Joint Research Centre (JRC) launched a work program based on remote sensing to evaluate the evolution of forested areas. The worldwide results were published in 2012 [FAO 12], then the results specific to tropical areas [ACH 14]. Another source of statistical information on the evolution of forest resources on a global scale using remote sensing has been provided by Hansen *et al.* [HAN 13]. Finally, the Japan Aerospace Exploration Agency (JAXA) also offers estimates of forested areas worldwide at a 25 m resolution (http://www.eorc.jaxa.jp/ALOS/en /palsar_fnf/data/index.htm). The data are available for 2007–2010 with ALOS-1. These estimates should resume following the launch into orbit of ALOS-2 in May 2014. These three programs, presented in the following, provide estimates of deforestation on a global scale using remote sensing.

2.4.1. *FAO/JRC program*

Assessing forest cover on the global scale is based on a systematic sampling based on each intersection of a degree of longitude and latitude. This sampling includes around 13,000 measuring points each one covering a surface of 10 km × 10 km, for a total 1% of the Earth's land area. For each sample, Landsat images from the years 1990, 2000 and 2005 (optical bands 1–5 and 7) have been interpreted and classified. They come from the *Landsat Global Land Survey* (GLS) of the *United States Geological Survey*. The processed images (correction of radiometric differences) have been segmented and classified with the help of eCognition® software. The segmentation consists of identifying a polygon, or set of pixels similar from a spectral perspective. Around 7 million polygons were analyzed for each period and classified according to the land cover (by automated supervised classification). These classifications have been interpreted, corrected and validated by forest experts who have a solid understanding of the terrain and by experts in remote sensing, who used images with very high resolution (http://www.fao.org/forestry/fra/fra2010/fr/). A map unit is considered to be classified as "forest" when the tree cover is more than 30%.

2.4.2. *Global Forest Change program*

Unlike the previous program that relies on a sampling of points, analysis by Hansen *et al.* [HAN 13] is exhaustive and worldwide. The database is open access (http://earthenginepartners.appspot.com/science-2013-global-forest). This analysis constitutes the main source of information of the *Global Forest Watch* (http://www.globalforestwatch.org/) site. The observation period is more limited than that of the JRC since it extends over 12 years (2000–2012). Since February 2014, this program provides high-resolution data (30 m) detailing the annual expansion, and loss and gains of forests worldwide. This database is the result of analyzing 654,178 Landsat 7 ETM+ scenes. The processing was carried out using the *Google Earth Engine* environmental analysis platform. Each Landsat footprint was temporally made into a mosaic to provide data (forest, loss of forest and increase in forests) with 30 m spatial resolution. Various spectral metrics have been calculated from the optical reflectance values. Four classes of forest are distinguished according to the percentage of tree cover (0–25,

25–50, 50–75 and 75%). The loss of forest area is the sum of these four classes. On the other hand, the increase in surface area is the sum of the two classes with tree cover of more than 50% (Figure 2.3).

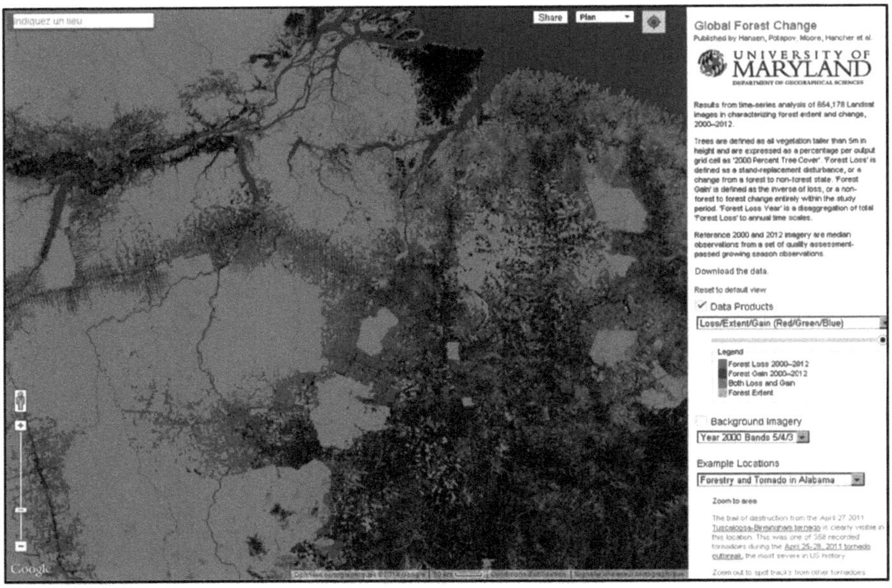

Figure 2.3. *Illustration homepage (Amazon region) for downloading data produced by the Global Forest Change (GFC, http://earthenginepartners.appspot.com/science-2013-global-forest). The extent of mature forest is represented by green, forest loss is in red and forest gain is in blue. The unclassified pixels are shown in black. The information is available per year between 2000 and 2012 at 30 m of spatial resolution (source: used with the kind permission of Hansen/UMD/Google/USGS/NASA). For a color version of this figure, see www.iste.co.uk/baghdadi/6.zip*

The estimates of forest areas for the two periods in consideration (before and after 2000) vary by a factor of 1–3 depending on the data (national inventories vs. images) and methods (FAO/JRC and [HAN 13] see Table 2.1). A comparison of the estimates of net loss of forest area produced by the two remote sensing programs (Table 2.1) shows a sizeable difference between the two programs: 6.2 million hectares for FAO/JRC [FAO 12] and 12.4 million hectares for Hansen *et al.* [HAN 13] for the years after 2000. These results can be explained principally by the differences in defining "forest". Indeed, considering only Hansen's work where forest is considered to have a canopy cover more than 50%, we obtain a net loss of 5.4 million hectares (data not presented here), closer to the 6.2 of FAO/JRC [FAO 12]. The main problem is in measuring tree cover of partially wooded land such as of tree savannah or dry open woodland (such as in the Sudano-Sahelian zone or in Southern Africa, for example) and whether they can be categorized (or not) as forests. The different definitions of these two approaches make comparison of the methods difficult. If the availability of images and the means employed to analyze them have long been an obstacle to measuring deforestation, it now seems that the priority is to find a common and shared definition of what can be classified as a forest.

In addition to the problem of definition, differences in estimation can also be explained by methods of analysis and validation that are still not consolidated (Table 2.1). Using the same methodological framework as that of the FAO and JRC [FAO 12], Achard *et al.* [ACH 14] provide higher estimates of deforestation in tropical zones (5.9 million hectares) estimated on a global scale (5.2 M·ha) for the 2000–2010 period.

	Zone studied	Period studied	Loss	Gain	Net loss (in millions of hectares)
FAO [FAO 10]	Globe	1990–2000			**8.3**
		2000–2010			**5.2**
FAO/JRC [FAO 12]	Globe	1990–2000	9.5	6.8	**2.7**
		2000–2005	13.5	7.3	**6.2**
Hansen *et al.* [HAN 13]	Globe	2000–2012	19.1	6.7	**12.4**
	Regions (climatic zones)				
Hansen *et al.* [HAN 13]	Tropical	2000–2012	9.2	2.1	**7.2**
	Boreal		5.1	1.7	**3.3**
	Subtropical		2.5	1.6	**0.9**
	Temperate		2.3	1.3	**1.0**
Achard *et al.* [ACH 14]	Tropical	1990–2000	8.0	1.9	**6.1**
	Tropical	2000–2010	7.6	1.5	**5.9**

Table 2.1. *Annual estimates of gains, losses and net loss (in bold) in forest area in millions of hectares*

2.4.3. *Japanese Space Agency program for mapping forest cover*

Mapping deforestation on the global scale has always been a priority of Japan's space agency (JAXA) since the launch of the first JERS-1 L-band radar in 1992, followed by the ALOS-1/PALSAR radar in 2006 and recently, in 2014, the launch of ALOS-2/PALSAR [SHI 14]. JAXA has provided global coverage forested areas for the years 2007–2010 with a spatial resolution of 25 m (http://www.eorc.jaxa.jp/ALOS/en/palsar_fnf/data/index.htm, Figure 2.4). These maps have been validated with a precision of 90% in comparison with ground data.

In the near future, maps of forest cover will be derived from PALSAR sensors in full band dual mode, with 10 m of spatial resolution.

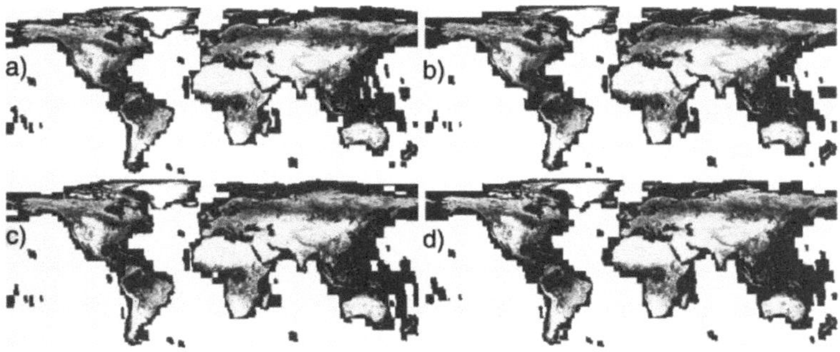

Figure 2.4. *Worldwide mosaics of forest cover (in dark green) at 25 m resolution from ALOS-1/PALSAR images: a) 2007, b) 2008, c) 2009, d) 2010 (http://www.eorc.jaxa.jp/ALOS/en/palsar_fnf/data/index.htm). For a color version of this figure, see www.iste.co.uk/baghdadi/6.zip*

2.5. Programs evaluating deforestation and degradation on a regional scale: some case studies

2.5.1. *Estimating deforestation on a regional scale: the case of Brazil*

2.5.1.1. *Application based on optical data in the Brazilian Amazon*

Brazil is a unique case among countries in tropical zones, having developed a satellite checking tool for monitoring deforestation and made the data available to the public.

A program monitoring deforestation in Amazônia Legal[4], PRODES, was launched in 1988. This program was integrated, starting in 2005, into an arsenal of measures forming the action plan for preventing and controlling deforestation in Amazônia Legal (PPCDAM). The plan is a set of restrictions, control measures and application of new public policies. It was implemented by the federal government and mobilized 14 ministries as well as all of the Amazônian states.

Program	Year of implementation	Product	Images	Minimum surface detected
PRODES: monitoring deforestation in Amazônia Legal	1988	Map and yearly deforestation rate	Landsat/CBRES (resolution 30 m)	> 6.25 ha
DETER: monitoring deforestation in real time	2004	Deforestation and degradation alert (bimonthly)	MODIS	25 ha
DEGRAD: mapping forest degradation	2008	Map and yearly degradation rate	Landsat/CBRES (resolution 30 m)	> 6.25 ha

Table 2.2. *Technical specifications of programs monitoring deforestation in the Brazilian Amazon*

The National Institute for Space Research (INPE, http://www.obt.inpe.br) has developed different programs for monitoring deforestation (Table 2.2). They have two main objectives:

– providing spatial and temporal estimates of deforestation and forest degradation (the PRODES and DEGRAD programs, respectively, Figure 2.5); these estimates are widely disseminated and made available to every level of the country's government (federal, federated and district) as key components of environmental policies. Since 2005, Brazil has recorded an 82% reduction of its annual rate of deforestation rate (Figure 2.6);

4 Amazônia Legal extends over nine Brazilian states. The administrative area includes the seven states in the northern region of the country (Acre, Amazonas, Amapá, Pará, Rondônia, Roraima and Tocantins) as well as the two states of Mato Grosso and a part of Maranhão. All of the monitoring indicators for vegetation provided by the National Institute for Space Research (INPE) always refers to this territory.

– establishing a monitoring tool for the area with information about deforestation in real time that allows rapid control and intervention on the ground. The DETER program provides bimonthly alerts on a minimum area of 25 ha using analysis of MODIS images.

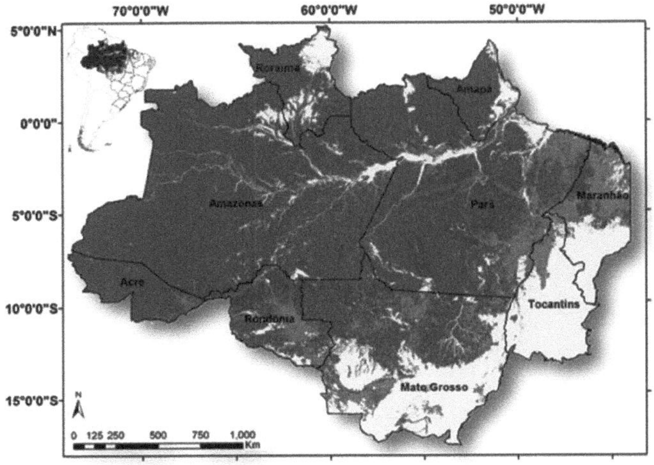

Figure 2.5. *This map shows the forest cover in the Brazilian Amazon in green, deforestation between 2000 and 2010 in red and the previously deforested forests (before 2000) in gray. The non-forest zones are in white (source: used with the permission of Brazil's National Institute for Space Research (INPE, http://www.obt.inpe.br/prodes/index.php)). For a color version of this figure, see www.iste.co.uk/baghdadi/6.zip*

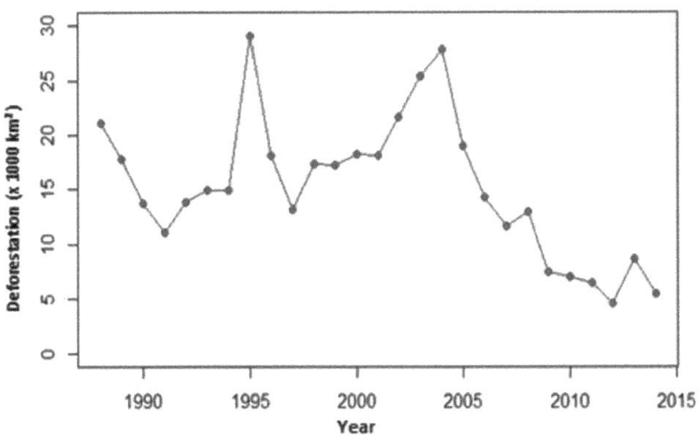

Figure 2.6. *Yearly deforestation rate in the Brazilian Amazon (source: http://www.obt.inpe.br)*

Using data provided by the PRODES program, the Brazilian Agricultural Research Institute (Embrapa) and the INPE have analyzed changes in land use (http://www.inpe.br/cra/projetos_pesquisas/terraclass2010.php). The assessment made in 2010 shows that 18.2%, an area of 739,672 km^2, of the Amazon had been deforested. Close to two-thirds of this surface (61.9%) has been turned into grazing land in various states of degradation. A total of 22.2% of the deforested surface is once again colonized by the forest (secondary forests). Aguiar [AGU 12] estimates this deforestation in the Amazon emits 0.15–0.28 pg of carbon annually, which is almost one-third of the planet's carbon emissions, due to changes in land use. The phases of this typology are not fixed in time, but pass from one state to the other depending on management practices. For example, degradation of grazing lands causes forest regrowth, which can once again be burned to create new grazing land. This mosaic of landscapes is not fixed but is constantly evolving over time and space.

The non-governmental organization Imazon has developed a program for monitoring deforestation and forest degradation, which produces monthly bulletins and annual maps of deforestation and forest degradation in Brazil (http://imazon.org.br/). Its system is based on the spectral analysis of mixed pixels Landsat. The goal is to find the proportion of active vegetation, inactive vegetation, bare soil and drop shadow in each pixel to determine its class. The analysis benefits from a large database of spectral signatures precisely indicating the different qualities of the desired elements. The principle is therefore similar to the one developed in CLASlite, but the tool is used continuously and operationally to provide information about the state of the forests of the Brazilian Amazon.

2.5.1.2. *Application based on radar data (State of Pará)*

Figure 2.7 shows an example in Brazil (State of Pará) using the L-band to distinguish the forest from non-forest zones. For forested areas, a large amount of microwave backscatter is recorded by the sensor (light color). On the other hand, deforested surfaces act like mirrors that reflect the radar signal in the opposite direction of the sensor, which produces low values (in black). By comparing the image acquired by JERS-1 in 1993 with the one acquired by ALOS-1/PALSAR in 2010, we can identify deforestation over a period of 19 years. The images show that deforestation was not significant in the 1990s and that the phenomenon has increased since the 2000s.

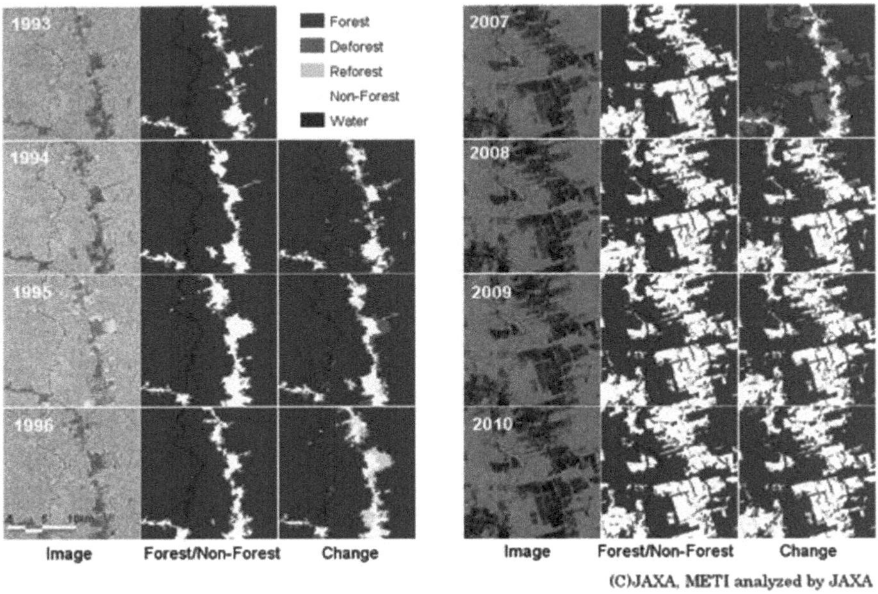

Figure 2.7. *Deforestation in the Amazon between 1993 and 2010 (State of Pará, Brazil). From left to right, L-band backscatter, forest (in dark green)/non-forest (yellow) and map of changes in reference to 1993. The red indicates deforestation, while the green shows reforesting (source: http://www.eorc.jaxa.jp/ALOS/en/new/ amazonFNF_1993-2010.jpg). For a color version of this figure, see www.iste.co.uk/ baghdadi/6.zip*

2.5.2. *Estimating forest degradation in French Guiana*

Anthropogenic impact on forests is not limited to deforestation. Forest degradation also represents a serious threat. In the Amazon, deforestation has decreased significantly since 2005, but we can see a concomitant increase in forest degradation. According to Asner *et al.* [ASN 05], the surface area of forests degraded by forest exploitation represents between 12,000 and 20,000 km^2/year between 1999 and 2002. Forest degradation is therefore a significant source of carbon emissions. Identifying this degradation remains problematic, complicated and costly [ASN 09, ASN 05, SOU 13]. It represents a serious challenge for remote sensing, both for estimating the surface area of degraded forests and for quantifying its impact on ecosystem services.

New methods are being developed to meet this challenge. In French Guiana, the ONF (National Forestry Commission) and CIRAD (French Agricultural Research Centre for International Development) have developed a tool for monitoring the impacts of forest exploitation. The impact of timber cutting on tropical rainforests appears on optical images as small isolated objects in an ocean of verdure. These objects can be linear (roads), in the form of points (clearings) or both (mining operations are linear alternating with areas of bare soil and sedimentation basins full of water that is more or less clean). Detecting clearings is possible with satellite acquisition from 6 to 12 months after exploitation [GON 09]. The simplest technique is based on the local contrast between photosynthetically active surfaces identified using the NDVI (the canopy) and impacted surfaces that no longer have this capacity. By using wavelengths dedicated to observing vegetation (red (0.61–0.68 μm), near IR (0.79–0.89 μm) and mid-IR (1.58–1.75 μm)), the contrast is enough to identify the impacted pixels by using reflectance thresholds. The use of indexes also provides access to this information. Applying thresholds to the NDVI or normalized difference water index (NDWI[5]) values and mid-IR (1.58–1.75 μm) makes it possible to identify pixels impacted by forest exploitation. The advantage of these robust methods is that they are able to use data from SPOT and Landsat satellites [PIT 13], as well as future data from Sentinel-2, in a semi-automatic way.

In French Guiana, selective logging of the forest causes gaps in the canopy that are detected operationally by SPOT images at 10 m resolution (Figure 2.8). The figure shows a plot of land of 400 ha with a main logging track running through it. Following a mini-tornado after logging, part of the stand has been significantly damaged by large openings (colored pink surrounded by a blue border). In the logged parts that were not disturbed by the tornado (below), the gaps have been identified on a color composite because of value thresholds applied to the NDVI, the NDWI and mid-IR (NDVI in red, NDWI in green and mid-IR in blue). These thresholds make it possible to isolate the impacts (gaps/roads, outlined in red).

---

5 NDWI = (NIR – MIR)/(NIR + MIR), where NIR is the near-IR reflectance and MIR is the mid-IR reflectance (according to [GAO 96]).

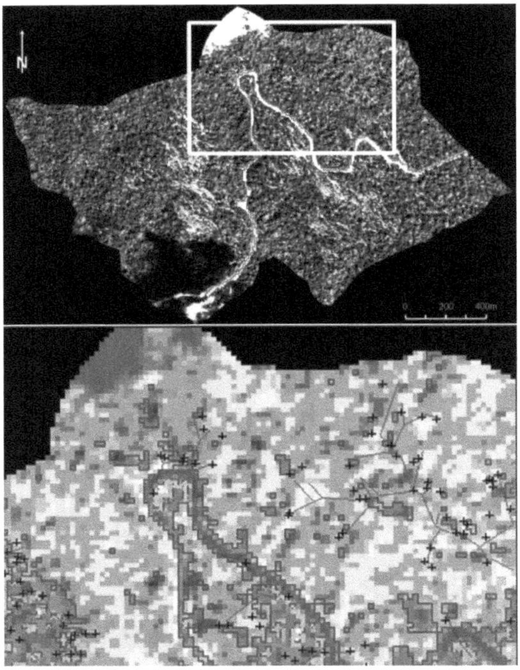

Figure 2.8. *Example of logging operation observed by satellite in French Guiana (latitude 5.33N longitude 53.22W): a) general view of the operation from a color composite at 10 m resolution (SPOT-5); b) part of a color composite (NDVI in red, NDWI in green and mid-IR in blue) after processing, identifying the clearings (pixels with a red margin). Validation data from the ground have been added: the black crosses show the stumps and the skidding tracks are in green [GON 09]. For a color version of this figure, see www.iste.co.uk/baghdadi/6.zip*

Similarly, detecting roads and tracks is also necessary for forest management. The opening, use and abandonment of road networks are geographic indicators of the opening and degradation of the forest. Documenting this dynamic is possible using SPOT archives and Landsat. Laporte *et al.* [LAP 07] used photointerpretation to map the networks of roads in Central Africa in order to show scale of logging activities within the forest. When you show the red, near-IR and mid-IR spectral bands in color composites as red, green and blue, respectively, the active roads in brown and the intact forest in dark green (Figure 2.9) are clearly visible. In order to automatically process the archive over large areas, Bourbier *et al.* [BOU 13] have proposed a method using Landsat data to allow managers to visualize the local (in a concession) or national dynamic of logging road networks

(Figure 2.9). This example shows the assessment of the opening of the canopy using of Landsat data, where the networks of roads are visible through photointerpretation (1). Using a method based on thresholding (2), the "non-forest" pixels are identified. After cleaning, using a morphological filter that removes isolated pixels with a small range (3), a summary of several dates completes the detection of networks of roads and tracks (4). Finally, an aggregation of pixels in a cell of a chosen size (250, 500 or 1,000 m) allows for the evaluation of the percentage of "non-forest" pixels on a given surface or cell (5). This indicator can then be compared over time using the archive (6) and it is possible to analyze the impact of opening the road network in 2001 (going from 100 to 70% of forest for the red curve that represents a cell) and its gradual closure (up to 90% for the red curve and 100% for the green curve).

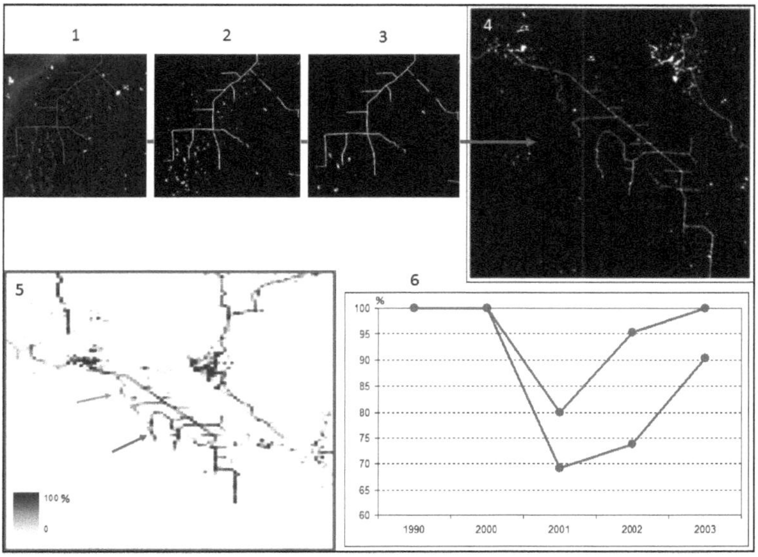

Figure 2.9. *Extraction of information concerning networks of logging roads in Central Africa: (1) color composite from a Landsat image with 30 × 30 m resolution (TM3 in red, TM2 in green and TM1 in blue); (2) following the application of thresholds on the channels, the non-forest pixels are identified (in white); (3) use of a spatial morphological filter to eliminate isolated pixels and connect the segments of the road network (in white); (4) general view of the study area; (5) application of a 500 × 500 m cell in order to estimate the percentage of pixels considered non-forest; (6) temporal evolution between 1990 and 2003 (x-axis) of the percentage of forest pixels (y-axis) of two 500 × 500 m cells (the curves corresponding to the cells are indicated by the arrows in 5). For a color version of this figure, see www.iste.co.uk/baghdadi/6.zip*

2.6. Outlook

Techniques for detecting deforestation by optical remote sensing and/or radar already offer usable and operational products (such as Brazil's PRODES). There are still a certain number of limits such as the need to consolidate the definition of the concept "forest" and analysis techniques for producing reliable validated data, especially on the planetary scale. Nevertheless, the production of statistical data on deforestation, in both spatial and temporal dimensions, should experience significant development in the next two decades because of several factors:

– The availability of images at high and very high spatial resolution continues to diversify following the launch of new satellites (especially the Sentinel-1 and Sentinel-2 satellite constellations), but also the continuation of the AVHRR, MODIS and Landsat programs in the United States. The platforms of the Sentinel-1 and Sentinel-2 constellations are equipped with sensors that allow observation of tropical forested areas every 5 days under stable angular conditions, notwithstanding the presence of clouds (http://www.esa.int/Our_Activities/Observing_the_Earth/Copernicus/Overview4). Under these conditions, it will be possible to observe logging areas in almost real time (Figure 2.10). The SPOT-4 (Take-5) experiment made it possible to simulate the viewing angles of the orbital conditions of the future Sentinel-2 satellite constellation with SPOT-4 data (http://www.cesbio.ups-tlse.fr/multitemp/?page_id=1685). The images were acquired every 5 days. Between February and June 2013 images were selected from sites around the world (http://www.cesbio.ups-tlse.fr/multitemp/?page_id=1685). The Northern Congo site was observed four times under good conditions (without clouds). This frequency is enough to observe the progress of a logging area with the opening of the road (March 4), the extension toward the south (March 19), the first side track (April 3), the second and the beginning of the exploitation (June 2), all using a simple color composite. The availability of images also involves radar since the existing programs such as the ALOS/PALSAR radar satellite series of Japan will continue. These observations will be completed in 2020 by the European Space Agency (ESA) Biomass P-band satellite, which will be dedicated to measuring vegetation biomass.

– The combination of ALOS/PALSAR radar data and Sentinel-1 with each other or with optical data of the Landsat and Sentinel-2 sensors is also an opportunity to improve the monitoring of deforestation. This combination will guarantee regular monitoring, reliable and continuous over time and space.

– Access to these images by scientists is facilitated by the dissemination policies of different space agencies: the ESA (http://www.esa.int/Our_Activities/Observing_the_Earth/Copernicus), the Japanese Space Agency (JAXA: http://www.eorc.jaxa.jp/ALOS/en/palsar_fnf/data/index.htm) or even the National Aeronautics and Space Administration (NASA: https://earthdata.nasa.gov/) in the United States.

– The large-scale development of airborne LiDAR measuring programs should allow more precision in characterization of the canopy structure and in the estimation of biomass. The repetition of such measurement programs will make very detailed estimates of degradation possible. However, the cost of these measurement programs would necessarily limit their repetition.

– Today the challenges linked to deforestation, and the necessity of developing reliable monitoring tools are recognized worldwide.

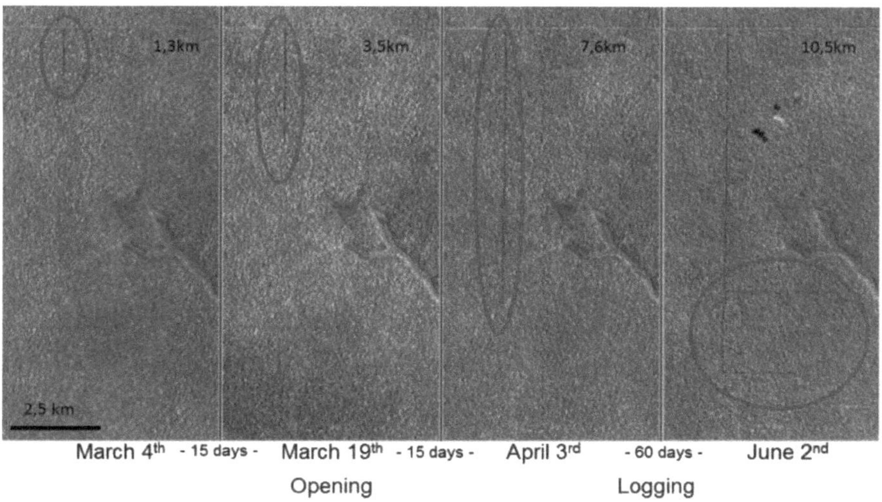

Figure 2.10. *Monitoring forest practices at north of Congo-Brazzaville (2.95° N – 17.1° E)*

2.7. Key points

On the planetary scale, forests cover one-third of the Earth's land area and play a crucial role in climate regulation and the production of goods. Knowing the extent of forested areas, the state of forests and how they change over time is therefore particularly important with regard to current environmental concerns linked to climate change. Remote sensing has emerged as the only tool capable of providing reliable data at multiple spatial and temporal scales. The optical data (in visible and IR wavelengths) and more recently radar data (multifrequencies with longer wavelengths) are the two main sources of data. Estimates of net deforestation resulting from the processing and analysis of satellite images on a planetary scale have been available for several years but remain very dependent on the definitions used and the analysis techniques employed. In the future, a greater diversity of optical images and radar should greatly contribute to the development of reliable tools for monitoring deforestation and forest degradation, something essential for responding to one of the greatest environmental challenges of the 21st Century.

2.8. Bibliography

[ACH 04] ACHARD F., EVA H.D., MAYAUX P. *et al.*, "Improved estimates of net carbon emissions from land cover change in the tropics for the 1990s", *Global Biogeochemical Cycle*s, vol. 18, no. 2, pp. 1–11, 2004.

[ACH 14] ACHARD F., BEUCHLE R., MAYAUX P. *et al.*, "Determination and tropical deforestation rates and related carbon losses from 1990 to 2010", *Global Change Biology*, vol. 20, pp. 2540–2554, 2014.

[AGU 12] AGUIAR A.P.D., OMETTO J.P., NOBRE C. *et al.*, "Modeling the spatial and temporal heterogeneity of deforestation-driven carbon emissions: the INPE-EM framework applied to the Brazilian Amazon", *Global Change Biology*, vol. 18, no. 11, pp. 3346–3366, 2012.

[ASN 05] ASNER G.P., KNAPP D.E., BROADBENT E.N. *et al.*, "Selective logging in the Brazilian Amazon", *Science*, vol. 310, pp. 480–482, 2005.

[ASN 09] ASNER G.P., KNAPP D.E., BALAJI A. *et al.*, "Automated mapping of tropical deforestation and forest degradation: CLASlite", *Journal of Applied Remote Sensing*, vol. 3, no. 1, p. 033543, 2009.

[BOU 13] BOURBIER L., CORNU G., PENNEC A. *et al.*, "Large scale forest opening estimation using remote sensing in Central Africa", *Bois et Forêts des Tropiques*, vol. 315, pp. 3–9, 2013.

[DEF 02] DEFRIES R.S., HOUGHTON R.A., HANSEN M.C., *et al.*, "Carbon emissions from tropical deforestation and regrowth based on satellite observations for the 1980s and 1990s", *Proceedings of the National Academy of Sciences of the United States of America*, vol. 99, no. 22, pp. 14256–14261, 2002.

[FAO 95] FAO, Forest Resources Assessment 1990, Global synthesis, FAO Forestry Paper, no. 40, Rome, 1995.

[FAO 01] FAO, Global Forest Resources Assessment 2000, Main report, FAO Forestry Paper, no. 124, Rome, 2001.

[FAO 06] FAO, Global Forest Resources Assessment 2005, Main report, FAO Forestry Paper, no. 147, Rome, 2006.

[FAO 10] FAO, Global Forest Resources Assessment 2010, Main report, FAO Forestry Paper, no. 163, Rome, 2010.

[FAO 12] FAO and JRC, Global Forest Land-use Change 1990–2005, FAO Forestry Paper, no. 169, Rome, 2012.

[GAO 96] GAO B.C., "NDWI – a normalized difference water index for remote sensing of vegetation liquid water from space", *Remote Sensing of Environment*, vol. 58, pp. 257–266, 1996.

[GON 09] GOND V., GUITET S., "Élaboration d'un diagnostic post-exploitation par télédétection spatiale pour la gestion des forêts de Guyane", *Bois et Forêts des Tropiques*, vol. 299, pp. 5–13, 2009.

[GRA 08] GRAINGER A., "Difficulties in tracking the long-term global trend in tropical forest area", *Proceedings of the National Academy of Sciences of the United States of America*, vol. 105, no. 2, pp. 818–823, 2008.

[HAN 08] HANSEN M.C., ROY D.P., LINDQUIST E. *et al.*, "A method for integrating MODIS and Landsat data for systematic monitoring of forest cover and change in the Congo Basin", *Remote Sensing of Environment*, vol. 112, pp. 2495–2513, 2008.

[HAN 13] HANSEN M.C., POTAPOV P.V., MOORE R. *et al.*, "High-resolution global maps of 21st-century forest cover change", *Science*, vol. 342, no. 6160, pp. 850–853, 2013.

[HOT 14] HO TONG MINH D., LE TOAN T., ROCCA F. *et al.*, "Relating P-band synthetic aperture radar tomography to tropical forest biomass", *IEEE Transactions on Geoscience and Remote Sensing*, vol. 52, no. 2, pp. 967–979, 2014.

[KAT 15] KATEMBERA-CIZA S., MIKWA J.F., CIRHUZA-MALEKEZI A. *et al.*, "Identification des moteurs de déforestation dans la région d'Isangi, République Démocratique du Congo", *Bois et Forêts des Tropiques*, vol. 324, pp. 28–37, 2015.

[LAP 07] LAPORTE N., STABACH J., GROSCH R. *et al.*, "Expansion of industrial logging in Central Africa", *Science*, vol. 316, no. 5830, pp. 1451–1451, 2007.

[LET 11] LE TOAN T., QUEGAN S., DAVIDSON M.W.J. *et al.*, "The BIOMASS mission: mapping global forest biomass to better understand the terrestrial carbon cycle", *Remote Sensing of Environment*, vol. 115, no. 11, pp. 2850–2860, 2011.

[MAL 89] MALINGREAU J.P., TUCKER C.J., LAPORTE N., "AVHRR for monitoring global tropical deforestation", *International Journal of Remote Sensing*, vol. 10, pp. 855–867, 1989.

[MAL 95] MALINGREAU J.P., ACHARD F., D'SOUZA G. *et al.*, "The AVHRR for global tropical forest monitoring: the lessons of the TREES Project", *Remote Sensing Reviews*, vol. 12, pp. 29–40, 1995.

[MAY 98] MAYAUX P., ACHARD F., MALINGREAU J.P., "Global tropical forest area measurements derived from coarse resolution satellite imagery: a comparison with other approaches", *Environmental Conservation*, vol. 25, no. 1, pp. 37–52, 1998.

[MOR 05] MORTON D.C., DEFRIES R.S., SHIMABUKURO Y.E. *et al.*, "Rapid assessment of annual deforestation in the Brazilian Amazon using MODIS data", *Earth Interactions*, vol. 9, no. 8, pp. 1–22, 2005.

[NEL 87] NELSON R., HORNING R., STONE. T.A., "Determining the rate of forest conversion Mato-Grosso, Brazil, using Landsat MSS and AVHRR data", *International Journal of Remote Sensing*, vol. 8, no. 12, pp. 1767–1784, 1987.

[PIT 13] PITHON S., JUBELIN G., GUITET S. *et al.*, "Statistical based method for logging-related canopy gap detection using high resolution optical remote sensing", *International Journal of Remote Sensing*, vol. 34, pp. 700–711, 2013.

[QUE 00] QUEGAN S., LE TOAN T., JIONG JIONG Y. *et al.*, "Multitemporal ERS SAR analysis applied to forest mapping", *IEEE Transactions on Geoscience and Remote Sensing*, vol. 38, no. 2, pp. 741–775, 2000.

[SAL 07] SALESKA S., DIDAN K., HUETE A. *et al.*, "Amazon forests green-up during 2005 drought", *Science*, vol. 318, no. 5850, pp. 612-612, 2007.

[SAM 10] SAMANTA A., GANGULY S., MYNENI R., "MODIS enhanced vegetation index data do not show greening of Amazon forests during the 2005 drought", *New Phytologist*, vol. 189, pp. 11–15, 2010.

[SHI 94] SHIMABUKURO Y.E., HOLBEN B.N., TUCKER C.J., "Fraction images derived from NOAA AVHRR data for studying the deforestation in the Brazilian Amazon", *International Journal of Remote Sensing*, vol. 15, no. 3, pp. 517–520, 1994.

[SHI 14] SHIMADA M., ITOH T., MOTOOKA T. *et al.*, "New global forest/non-forest maps from ALOS PALSAR data (2007-2010)", *Remote Sensing of Environment*, vol. 155, pp. 13–31, 2014.

[SKO 93] SKOLE D.L., TUCKER C.J., "Tropical deforestation and habitat fragmentation in the Amazon: satellite data from 1978 to 1988", *Science*, vol. 260, pp. 1905–1910, 1993.

[SOU 13] SOUZA C.M. JR, SIQUEIRA J.V., SALES M.H. *et al.*, "Ten-year Landsat classification of deforestation and forest degradation in the Brazilian Amazon", *Remote Sensing*, vol. 5, no. 11, pp. 5493–5513, 2013.

[STI 04] STIBIG H.J., ACHARD F., FRITZ S., "A new forest cover map of continental Southeast Asia derived from SPOT-VEGETATION satellite imagery", *Applied Vegetation Science*, vol. 7, no. 2, pp. 153–162, 2004.

[THO 13] THOMPSON I.D., GUARIGUATA M.R., OKABE K. *et al.*, "An operational framework for defining and monitoring forest degradation", *Ecology and Society*, vol. 18, no. 2, p. 20, 2013.

[TOU 99] TOUZI R., LOPES A., BRUNIQUEL J. *et al.*, "Coherence estimation for SAR imagery", *IEEE Transactions on Geoscience and Remote Sensing*, vol. 37, no. 1, pp. 135–149, 1999.

[TUC 84] TUCKER C.J., HOLBEN B.N., GOFF T.E., "Intensive forest clearing in Rondonia, Brazil, as detected by satellite remote sensing", *Remote Sensing of Environment*, vol. 15, pp. 255–261, 1984.

[VIL 15] VILLARD L., LE TOAN T., "Relating P-band SAR intensity to biomass for tropical dense forests in hilly terrain: gamma0 or t0", *IEEE Journal of Selected Topics in Applied Earth Observations and Remote Sensing*, vol. 8, no. 1, pp. 214–223, 2015.

[ZEB 92] ZEBKER H.A., VILLASENOR J., "Decorrelation in interferometric radar echoes", *IEEE Transactions on Geoscience and Remote Sensing*, vol. 3, no. 5, pp. 950–959, 1992.

3

Remote Sensing of Wildfires

3.1. Introduction

Wildfire is one of the most prominent disturbances in forest and grassland ecosystems and considered as a natural risk [LEB 13, SCO 14]. Although wildfires maintain ecosystem health and diversity by regulating plant succession and fuel accumulation, controlling age, structure and species composition, affecting insect and disease populations, influencing nutrient cycles and energy flows, and determining wildlife habitats [LEB 13], they can be also harmful and become a threat to property and human life, and thus the economy, particularly in densely populated regions [SCO 14, CHU 14]. Between 350 and 450 million ha burn annually in the world [GIG 10], consuming billion tons of dry matter and releasing 1.5 to 3 PgC/year (petagram of carbon per year) as well as other compounds that affect the composition and functioning of the global atmosphere and human health [VAN 10].

Given that fires across the world are expected to increase due to climate change, better fire danger prediction and detection methods will have significant benefits both from economical and safety points of views. Accurate assessment of burnt areas are also needed for managing postfire environmental impacts, such as regeneration and erosion as well as for accounting for greenhouse gas emissions by fires. Satellite imagery can play

Chapter written by Brigitte LEBLON, Jesús SAN-MIGUEL-AYANZ, Laura BOURGEAU-CHAVEZ and Martin KONG.

an important role in monitoring wildfires worldwide. It offers the advantages of extensive regional coverage, zero disturbances of the area to be viewed, as well as a method for acquiring data in less accessible areas, on a regular and cost-effective basis.

In the first part of this chapter, we will present how remote sensing has been used for monitoring conditions (mainly fuel moisture) favorable for fires. The second part of the chapter is dedicated to the use of remote sensing for active fire detection. The third part concerns burn scar mapping. This chapter will primarily focus on space-borne sensors, although the theory may also be applied to airborne sensors, which are more often used for fire suppression and detection activities rather than as a prefire management or burn scar mapping tool. In the chapter, optical remote sensing refers to wavelengths between 400 and 2,500 nm, in contrast to the thermal infrared wavelengths (3,000–15,000 nm), although both types of wavelengths are recorded by optical sensors.

3.2. Fuel moisture estimation

3.2.1. *Fire danger predicting systems*

Ignition and spread of wildfires depends on four main factors [LEB 05]:

– the state and nature of the fuel (proportion of live and dead vegetation, compactness, morphology, species, density, stratification, fuel arrangement, continuity of fuel and moisture content);

– the physical environment (weather conditions and topography);

– causing factors (man-made or natural, for example, due to lightning);

– fire prevention and suppression means.

The first two types of factors are input parameters of fire danger prediction systems that have been developed for fire management in fire suppression activities. These systems are, among others, the Canadian Forest Fire Danger Rating System (CFFDRS) [STO 89] (Figure 3.1). The CFFDRS is used not only in Canada but also in other parts of the world, including the United States (Alaska), Europe and Asia.

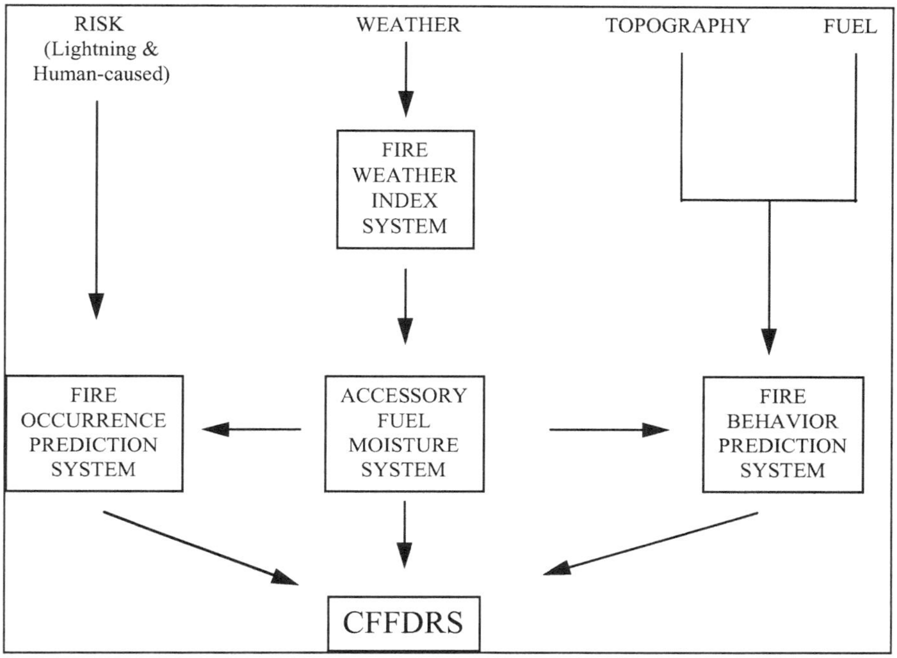

Figure 3.1. *Structure of the Fire Weather Index (FWI) system of the CFFDRS (modified from [STO 89])*

One of the CFFDRS subsystems, the Canadian Fire Weather Index (FWI) system, computes potential mid-afternoon fire danger from noon-time weather station records (Figure 3.2) [STO 89]. The FWI system is composed of three moisture codes: the fine fuel moisture code (FFMC) represents quick-drying surface litter fine fuel moisture contents, the duff moisture code (DMC)[1] represents slow-drying moderate depth loosely compact duff moisture contents and the drought code (DC) represents slow-drying deep layer compact organic matter moisture contents. These three moisture codes have time lags of 16 h, 15 days and 53 days, respectively [LAW 08]. Time lags represent the time necessary for a fuel type to lose about two-thirds of

1 In the FWI system, duff corresponds to loosely compact organic layers at moderate depth and mid-size woody materials.

its free moisture at equilibrium in standard drying conditions in July (air temperature of 21.1°C, relative humidity of 45% and wind speed of 13 km/h) [LAW 08]. Derived from these three moisture codes and wind speed, two indices provide a measure of the initial fire spread rate, the initial spread index (ISI), and the total fuel available for combustion, the build-up index (BUI). The FWI is then calculated from the combination of ISI and BUI [STO 89].

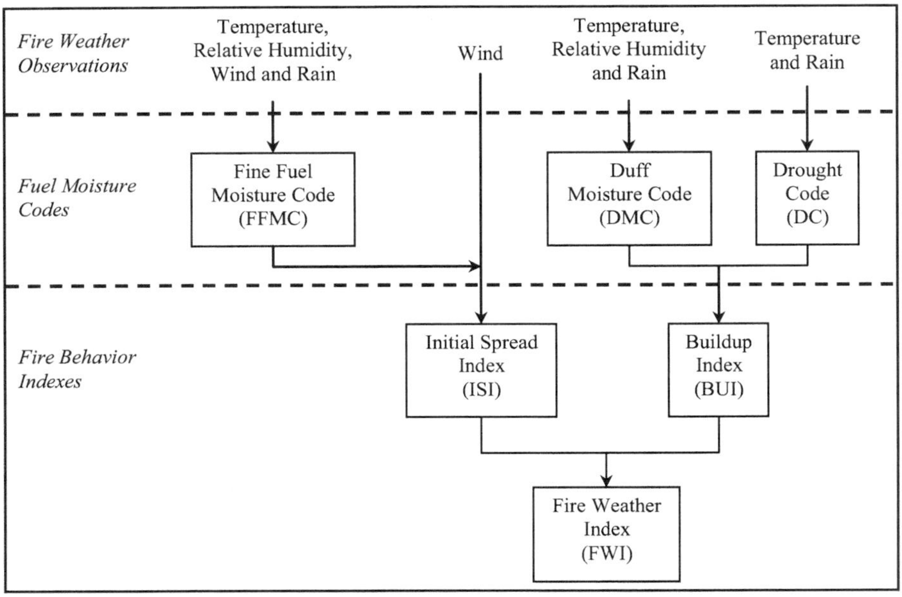

Figure 3.2. *Structure of the Fire Weather Index (FWI) system of the CFFDRS (from [STO 89])*

The FWI system is based primarily on point source weather parameters that are acquired in localized weather stations and the FWI maps are obtained by interpolation, which is a limitation when the weather data are acquired in a sparse network of weather stations. Because of their spatial nature, satellite images coupled with the development of geostatistics and spatial analyses using geographic information systems (GIS) allow us to obtain fire danger rating maps more easily.

Several prefire condition parameters can be mapped using remote sensing. The first ones are the fuel type and the fuel load. According to Merrill and Alexander [MER 87], fuel types are defined as: "identifiable associations of fuel elements of distinctive species, form, size, arrangement and continuity that will exhibit characteristic fire behavior under defined burning conditions".

Fuel type can be mapped from high spatial resolution optical (e.g. [BUR 98]) or radar images (e.g. [SAA 07]), such as classical vegetation mapping. Fuel load is the dry weight of combustible materials per unit area [MER 87] and can be mapped from high spatial resolution optical and radar images, such as vegetation biomass.

Maps of fuel types and loads can then be combined into a wildfire risk analysis system to other prefire condition parameters, such as topography, road proximity, in habited areas [BUR 98, CHU 10], etc. Another input variable of fire danger predicting systems related to prefire conditions and which can be estimated by remote sensing is the fuel moisture. There are two types of fuel moisture: live fuel and dead fuel moistures. Live fuel moisture can be quantified using two indices. The first one is live fuel moisture content (LFMC), which is defined as follows [LEB 02, YEB 13]:

$$LFMC = 100 \times \frac{FW - DW}{DW} \qquad [3.1]$$

where:

– $LFMC$ = live fuel moisture content (%);

– FW = fresh weight measured in the field (g or kg);

– DW = oven dry weight of the same sample, measured after drying in the laboratory (g or kg). The sample is dried at a temperature of 103 ± 2°C until there is no significant weight variation (Method A of [AST 09]).

The other index is the equivalent water thickness (EWT). At the leaf level, it is defined as the volume of water per unit leaf area [YEB 13]:

$$EWT_{leaf} = \frac{FW - DW}{A \times \rho_w} \qquad [3.2]$$

where:

– EWT_{leaf} = leaf equivalent water thickness (cm or m);

– FW = fresh weight measured in the field (g or kg);

– DW = oven dry weight of the same sample, measured in the laboratory (g or kg);

– A = leaf area (cm^2 or m^2);

– ρ_w = density of pure water = 1 g·cm^{-3} = 1 kg·m^{-3}.

At the canopy level, EWT_{canopy} is expressed by [YEB 13]:

$$EWT_{canopy} = \mathrm{LAI} \times \mathrm{EWT_{leaf}} \quad [3.3]$$

where:

– EWT_{canopy} = canopy equivalent water thickness (m);

– LAI = leaf area index (m^2·m^{-2}).

The intrinsic drying ability of species for the same drought stress can also be variable. For example, some Mediterranean species can have a LFMC varying from 50 to 200%, while other species that are better moisture savers have their moisture varying between 80 and 150% [CHA 15]. Live fuel moisture can also be quantified indirectly through the degree of water stress expressed in terms of evapotranspiration rates, as described in section 3.2.3. Another type of fuel moisture is the dead fuel moisture that can be more easily computed from weather data and dead fuel characteristics, because it is in balance with the surrounding atmosphere [BUR 98, NIE 10]. The FWI system allows for this computation (see the equation details in [VAN 87]).

3.2.2. *Optical remote sensing*

The first optical remote sensing studies on fuel moisture monitoring used relative greenness (RG) or normalized difference vegetation index (NDVI) images that are computed using the red and near-infrared images acquired by NOAA-AVHRR, LANDSAT-TM or SPOT-HRV [LEB 05]. NDVI or RG data were well correlated to FWI codes and indices [OLD 06, LEB 07], degree of curing in open ecosystems e.g. grasslands [BUR 98] and fire occurrences [BUR 98]. NDVI-based operational systems have been proposed to map fire potentials [BUR 98]. More recently, another visible band, the LANDSAT-TM blue band, has been combined with red and near-infrared data into the visible atmospherically resistant index, which was correlated with LFMC over Southern California and Australian ecosystems [YEB 13].

Studies using red, near-infrared, or blue images for fuel moisture mapping listed several problems, namely the saturation of relationships, the influence of site wetness on relationships, the influence of disease, senescence, shadowing or penumbra and the difficulty of using these images over forests, due to the spectral mixture between the overstory and the understory, both being different in nature and in moisture content [LEB 05]. In fact, these images are only indirectly related to fuel moisture conditions, because they rather measure the greenness and the chlorophyllous activity of the vegetation [LEB 05].

In a study on prefire conditions using NOAA–AVHRR over Northwest Territories boreal forests, Oldford *et al.* [OLD 03] showed that high FWI areas correspond to high values on the surface temperature image, indicating water stress, but also to high values over the NDVI image, which could indicate no drought conditions. Similarly, greenness and the chlorophyll activity of the vegetation explained the positive correlations between FWI codes and indices and cumulative NDVI found by Leblon *et al.* [LEB 07] over Canadian northern boreal forests, since both types of variables increase in parallel throughout the fire season, but for two different reasons: FWI codes and indices increase because of drought, and cumulative NDVI because of vegetation growth. For these reasons, a better use of NDVI images in forest fire management would be to map the time of deciduous leaf flushing, which is critical in fire management, because it is related to fire occurrence in mixed-deciduous forests. For example, one of the CFFDRS sub-systems, the fire behavior prediction system, predicts the behavior of wildfires for 11 standard fuel types. Five of these 11 types are defined as a function of the presence or absence of leaves, namely the D1 (leafless aspen stands without leaves), M1 (mixed boreal forests without leaves), M2 (mixed boreal forests with leaves), M3 (dead mixed balsam fir stands without leaves) and M4 (dead mixed balsam fir stands with leaves) types. Indeed, the leaf presence strongly reduces the fire propagation. For example, in the case of mixed boreal forests, when the overstory and understory have leaves (fuel M2), the fire propagation is less than one-fifth of the one observed in spring and autumn when these stories have no leaves (fuel M1), all other environmental conditions being similar [FOR 92]. Leaf presence is also important for the wildfire danger in other regions of the world. For example, in the Mediterranean region, deciduous species such as the spiny broom (*Calicotome villosa* (Poir.) Link) and semideciduous species such as the cistus of Montpellier (*Cistus monspeliensis* L.) lose their leaves during the summer drought period and are therefore more susceptible to fire.

Fuel moisture is theoretically better related to the shortwave infrared (SWIR) (1,300–2,500 nm) because these SWIR wavelengths are related to water absorption bands [PIE 90]. Several SWIR-related indices were related to LFMC for various ecosystems (see the review of Yebra *et al.* [YEB 13]) and to fire occurrences in boreal and Mediterranean forests [AKT 11, BIS 14]. There are however several limitations in using SWIR bands for fuel moisture monitoring. Relationships can be established, but only when the water stress is already important [PIE 90]. In addition, reflectance changes associated with moisture content changes can be smaller than those associated with the leaf area index (e.g. [PIE 90, YEB 13]). Finally, shortwave bands are highly disturbed by atmospheric effects.

Hyperspectral data have been also tested for fuel moisture retrieval. They have the advantage of allowing for the computation of derivative variables that are useful to remove the spectral effect of leaf structure, background and atmosphere as well as to resolve overlapping spectra to better separate components of the global spectrum [LEB 05]. Hyperspectral data can be related to plant water content through empirical relationships or radiative transfer models [YEB 13]. Radiative transfer models were also tested to estimate LFMC from multispectral MODIS raw data and related indices over various ecosystems [YEB 13].

However, from the operational point of view, both SWIR and hyperspectral data are only acquired by a limited number of space-borne sensors. SWIR data are only provided by sensors such as LANDSAT-TM, SPOT-VEGETATION, NOAA-16 and MODIS. Hyperspectral data are only provided by airborne sensors or the EO-1 HYPERION space-borne sensor. Moreover, both SWIR sensors and hyperspectral sensors have a long revisit period, which may be longer during cloudy periods, although most recent optical sensors, such as SPOT-VEGETATION, MODIS and NOAA-16 AVHRR, have the advantage of daily image acquisition. The performance of these new sensors is still under evaluation. By contrast, thermal infrared data are collected more often and mostly at the same time as the visible and near-infrared data, by several existing polar-orbiting space-borne sensors (e.g. NOAA-AVHRR, LANDSAT-TM, ATSR-2, RESURS-01 or MODIS) and geostationary space-borne sensors (e.g. METEOSAT and Geostationary Operational Environmental Satellite (GOES)). Geostationary sensors acquire images with high frequency (every 15 or 30 min), but at a spatially coarser scale than geosynchronous sensors and it is optimal to use both types of data.

3.2.3. *Thermal infrared remote sensing*

Surface temperature (T_S) measured by thermal infrared sensors was better correlated than NDVI to codes and indices of the FWI system [OLD 03, OLD 06, LEB 07], LFMC [YEB 13] and shrub water potential [GOU 91]. Detecting water-stressed coniferous stands is also useful when there are extreme differences in canopy water content [PIE 90]. In fact, the difference between surface and air temperatures ($T_S - T_a$) is a better image-based indicator of plant water status than T_S, because T_S is too sensitive to weather conditions [DUC 99].

Plants respond to water stress by stomata closure, thereby decreasing latent heat transfer from leaf surface to the atmosphere and causing an increase in the leaf surface temperature [PIE 90]. In the energy budget equation, if we express the sensible heat flux (H) as a function of the difference between surface and air temperatures ($T_S - T_a$), we obtain the following analytical relationship that allows the computation of the actual evapotranspiration rate (E or AET) as a function of $T_S - T_a$):

$$LE = R_n - G - \frac{\rho C_p (T_s - T_a)}{r_a + r_c} \qquad [3.4]$$

where:

– LE = latent heat flux ($W \cdot m^{-2}$), with L is the energy used for the evapotranspiration and E is the actual evapotranspiration rate (AET in mm water per time unit);

– R_n = net radiation flux ($W \cdot m^{-2}$);

– G = soil heat flux ($W \cdot m^{-2}$);

– ρC_p = volumetric heat capacity of air = ±1,200 $J \cdot m^{-3} \cdot K^{-1}$;

– r_a = aerodynamic resistance that is the air resistance to the sensible heat flux propagation ($s \cdot m^{-1}$);

– r_c = canopy thermal resistance, including the leaf boundary layer resistance and an excess resistance to heat transfer from leaves to the boundary layer ($s \cdot m^{-1}$);

– T_S = surface temperature (K);

– T_a = air temperature (K).

Cumulative $T_S - T_a$ data were correlated to monthly fire start numbers over Mediterranean forests [PRO 95]. For the same ecosystem, Vidal *et al.* [VID 94] used equation [3.4] to compute the daily ratio between actual and potential evapotranspirations (AET/PET) from daily NOAA-AVHRR surface temperatures and air temperatures measured at synoptic stations, PET being computed for example by the Penman–Monteith method, which is the most widely used and is considered as a reference method. As detailed in Allen *et al.* [ALL 98], it requires the following input variables: air temperature and incoming solar radiation, relative humidity, vapor pressure and wind speed. The ratio was related to fire occurrences [VID 94] and to two shrub flammability variables (ignition delay and drought index) [DES 96] over Mediterranean forests, as well as to the FWI codes and indices over Canadian northern boreal forests [STR 01] (Figure 3.3). The ratio was used to operationally monitor fire danger in 1994 [VID 94].

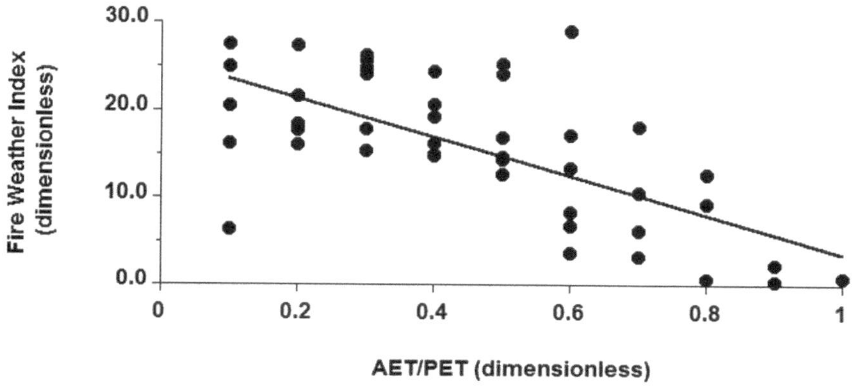

Figure 3.3. *Relationship between FWI and AET/PET in the case of boreal stands in northern Canada [STR 01]*

However, these studies also showed that estimating AET from $T_s - T_a$ using equation [3.4] is more problematic over forests than over crops [LEB 05] for the following reasons:

– canopy height is not comparable to a thin leaf surface, as supposed by equation [3.4], because of an additional level of radiation absorption and convective heat exchange between the ground and the superior stratum;

– the measured radiative surface temperature (T_{rad}) is different from the aerodynamic surface temperature (T_s) required by equation [3.4], because of an additional excess resistance (known as the kB^{-1} factor) to heat transfer from leaves, which increases with the canopy height;

– the aerodynamic resistance (r_a) is lower than the canopy resistance (r_c) and $T_s - T_a$ is thus less sensitive to moisture fluctuations. However, this lower sensitivity is compensated by the sensitivity of satellite signals to ground vegetation patches, which are generally shorter. For a short vegetation, the aerodynamic resistance (r_a) is higher than the canopy resistance (r_c). This understory vegetation often has an important role in the fire danger of several ecosystems.

– the clumped nature of canopy elements in tree crowns reduces wind speed near leaves and allows sunlit leaves to have temperatures well above T_a;

– wind can affect temporal fluctuations of $T_s - T_a$, but these fluctuations on the 1 km pixel basis of NOAA-AVHRR may be very small, because of the spatial integration over the pixel and because the eddies near the surface are on a scale of about 10 m.

Equation [3.4] requires an estimate for T_a. If synoptic T_a measurements are used, they should be corrected for shelter and tree height effects [PRO 95]. T_a can also be estimated as being equal to T_S of nearby well-watered canopies [DUC 99], since only water-stressed canopies have a difference between surface and air temperatures. T_a was also estimated as the T_S value corresponding to the extrapolation of the NDVI/T_S relationship to an NDVI of an infinitely thick vegetation canopy [GOW 94]. However, such an estimate requires that:

– the range of variation in NDVI and T_S is large enough to accurately define the slope;

– the images shall not be contaminated by clouds, snow, or standing water, because the slope can then be positive [GOW 94];

– the canopy should be well watered because the NDVI/T_S slope can changed as a function of the canopy moisture as NDVI/T_S was already shown to be related to moisture-related variables (see below).

Equation [3.4] also requires knowledge of the aerodynamic and canopy resistances (r_a and r_c), which are difficult to estimate [VID 94, STR 01] and whose estimates are valid only for small areas. Thereby, analytical models other than equation [3.4] were proposed for computing AET from T_S. The first is the Surface Energy Balance Algorithm for Land (SEBAL) [BAS 98]. As in equation [3.4], it computes LE as a residual quantity of the energy budget equation, but by contrast to equation [3.4], H is not computed from the difference between T_s and T_a, but it is derived from the vertical difference in air temperature (δT_a) between two heights: the surface roughness length to heat transport (z_{oh}) and the reference height (z_h), δT_a being directly inferred from T_S. SEBAL has been validated on both short and tall vegetation [BAS 98].

The second does not compute LE with the same method as equation [3.4]. It uses the Penman–Monteith approach, in which the vapor pressure deficit of the air (VPD) is estimated from the saturation vapor pressure at the mean daily surface temperature (VP^{*}_{Ts}) [GRA 97]. Indeed, feedback links between the surface and the overlying air are theoretically such that the observed surface temperature is a good indicator of the air humidity over the surface [GRA 97]. It is applicable to both short and tall canopies, because the VPD–VP^{*}_{Ts} relationship does not depend on the cover type. However, such an approach does not distinguish between vegetated and non-vegetated surfaces with the same surface roughness, temperature and air humidity, unless they have a different albedo leading to a different R_n. Its operational use thereby requires a careful land use mapping. There are other more sophisticated approaches to estimate AET from T_S, like soil-vegetation atmosphere transfer (SVAT) models [OLI 99]. SVAT models usually require a high number of input variables and thereby have little operational potentials in fire management.

3.2.4. *Synergisms between optical and thermal infrared remote sensing*

Several empirical studies have showed that inclusion of thermal infrared data improved correlations between optical indices and drought-related variables [CHU 03, OLD 03, OLD 06, LEB 07]. Oldford *et al.* [OLD 06] showed that, compared to weather station data interpolation, improved spatial resolution mapping of slow-drying fuel moisture, such as the DC, can be achieved at the pixel level, when DC is computed from T_s and NDVI images acquired by the NOAA-AVHRR sensor (Figure 3.4). The fire shown in the center of the 15 × 15 pixel area in Figure 3.4 was a surface fire caused by lightning, according to the Alberta Sustainable Resource Development Department. The fire burned in a closed coniferous forest cover type, which has a high DC value on the NOAA-AVHRR image, but as having a moderate DC value on the weather station based map.

Combining optical vegetation indices with T_s data helps account for the influence of the ground cover rate over the composite surface temperature measured by the thermal sensor. The inverse relationship between NDVI and T_S was related to fire occurrences in Mediterranean forests [PRO 95] and to moisture-related variables, such as canopy resistance [LEB 05, OLI 99] leaf water potential [GOW 94], accumulated rainfall [DUC 99], FWI codes and indices [OLD 03, OLD 06, LEB 07] and LFMC [CHU 03]. NDVI and T_S images were used to define several drought indices [CHU 03, AKT 11], some of them being used for inferring forest fire occurrences in Mediterranean areas [NIE 10] or in Alberta [AKT 11]. Other synergisms between optical and thermal infrared data can also be considered when estimating AET using equation [3.4]. Indeed, the required net radiation flux (R_n) can be computed from the solar irradiance at the surface or from the surface albedo, both variables being inferred from optical data [GRA 97, BAS 98]. Also, the G/R_n ratio can be analytically derived from optical vegetation indices [BAS 98, LEB 05].

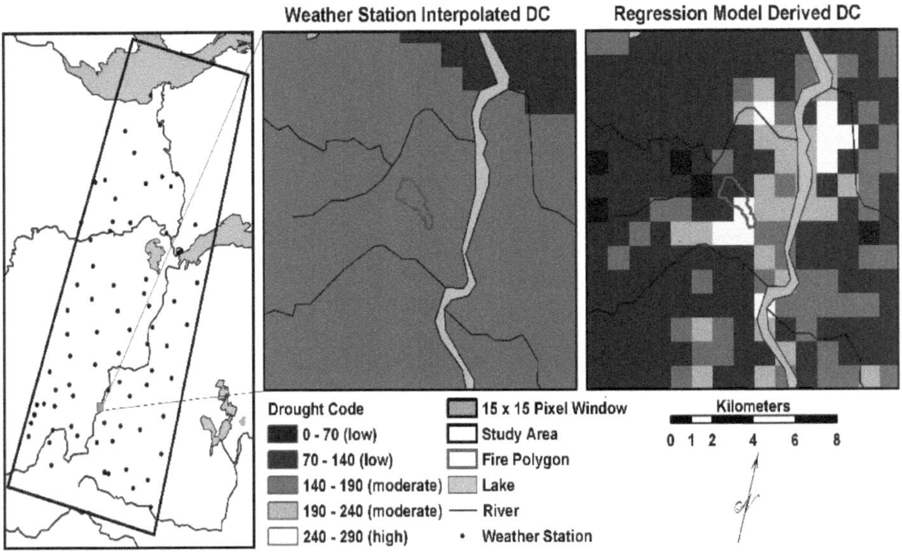

Figure 3.4. *Comparison between DC maps computed from weather station interpolation and from NOAA-AVHRR NDVI and Ts images for an area, where a surface fire starts. The image was acquired 3 days before fire ignition (according to [OLD 06]). For a color version of this figure, see www.iste.co.uk/baghdadi/6.zip*

Using both thermal infrared and optical images improves the correlation with fuel moisture variables, but these images have several operational limitations:

– cloud cover and other sun illumination conditions that limit image acquisition [LEB 05];

– there is an effect of the dry mater content on the EWT or on reflectance values [YEB 13];

– the optical sensors have a low spatial resolution [LEB 05].

As reviewed in Leblon [LEB 05], many strategies can be applied to overcome the problem of cloudy days, like the interpolation of evaporation fractions for the cloudy days, or the use of images acquired by passive or active microwave (radar) sensors, which are able to penetrate cloud cover. There are currently passive microwave sensors such as SSM/I, AMSRE and SMOS, but their images have a coarser spatial resolution than NOAA-AVHRR images. These sensors are still under evaluation, such as the active microwave sensors (synthetic aperture radar (SAR)). The use of SAR images for fuel moisture estimation will be now presented.

3.2.5. *Radar remote sensing*

SAR sensors acquire images of a greater spatial resolution than most other optical and thermal sensors. These sensors are currently on board multiple recent satellites, including, ERS-1/2, ENVISAT, RADARSAT-1/2, ALOS-PALSAR, Sentinel-1, ALOS-PALSAR-2, TerraSAR-X and COSMO-SkyMed.

Studies reviewed in Leblon *et al.* [LEB 02] and Abbott *et al.* [ABB 07] have shown that radar backscatter ($\sigma°$) measurements over forested areas depend on the following factors:

– vegetation type, species and structure;

– vegetation biomass [VIL 16];

– topography, surface roughness and canopy height;

– presence/absence of standing water and flooding;

– vegetation and soil moisture.

Three sources of moisture variation may contribute to forest radar backscatter:

– from dead vegetation;

– from the canopy (including its woody elements);

– from rain events.

In boreal forests, good relationships were observed between the FWI DC and radar backscatters retrieved from ERS-1 C-VV SAR images [LEB 02, BOU 07] and RADARSAT-1 C-HH SAR images [ABB 07] (Figure 3.5). While these studies produced encouraging results, they also showed that single channel C-band SAR images are restricted in their applicability, because of the effect of surface roughness and biomass.

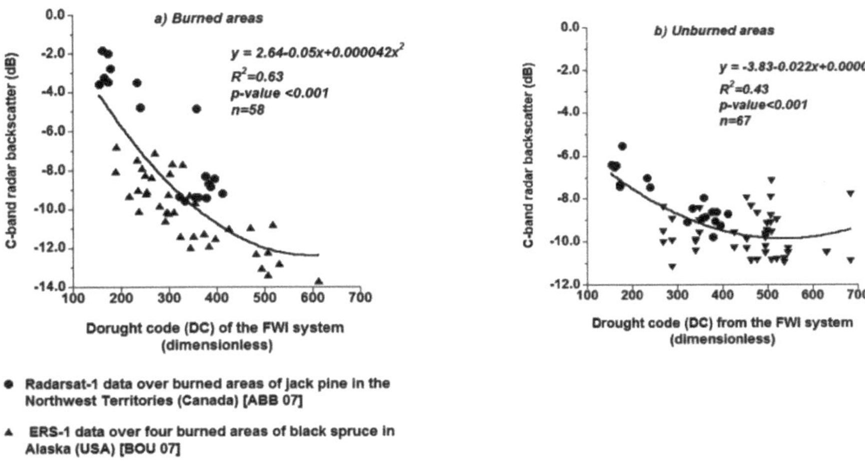

Figure 3.5. *Relationship between the Fire Weather Index drought code (DC) and ERS-1 C-VV and RADARSAT-1 C-HH SAR images acquired over burned and unburned boreal forests located in Alaska and in the Northwest Territories, Canada (data from [LEB 02, ABB 07, BOU 07]). Adapted from [ABB 07]*

More recently, fully polarimetric X-, C- and L-band SAR sensors were launched, such as ALOS-1/PALSAR in 2006, RADARSAT-2 in 2007, Cosmo-SkyMed in 2010 and ALOS-2/PALSAR in 2014. Polarimetric SAR images allow advanced SAR analysis that includes computing polarimetric variables such as those proposed in Touzi *et al.* [TOU 92] and various decomposition techniques [CLO 97, FRE 98, VAN 11]. These variables describe the backscattered energy into dominant scattering mechanisms (surface, double-bounce and volume) and thus can be useful for reducing the influence of surface roughness and for improving fuel moisture estimation in the absence of ancillary information.

Table 3.1 compares RADARSAT-2 polarimetric SAR variables calculated from images acquired under the same incident angle, but during an extreme dry date and a wet date over two types of savannahs (shrub/tree and herbaceous) [KON 14] and a chronosequence of Alaskan boreal black spruce ecosystems (recent burns, regenerating forests dominated by shrubs, open canopied and moderately dense forest cover) [BOU 13a]. In both cases, there was a significant difference between the wet and the dry dates for each single-polarized backscatter. Among the linear singular polarized backscatter, HH polarization was among the most sensitive to moisture changes (Table 3.1). For the circular-polarized backscatter, the most sensitive polarization depends on the vegetation type. Among all the polarimetric decomposition parameters, the parameters from the Freeman–Durden (FD_{vol}, FD_{surf}, FD_{dbl}) [FRE 98] and van Zyl decompositions (vZ_{vol}, vZ_{surf}, vZ_{dbl}) [VAN 11] were more different between the wet and dry dates than the parameters from the Cloude–Pottier decomposition (CP-H, CP-A, CP-α) [CLO 97], particularly for the double-bounce scattering in the case of the savannahs (FD_{dbl} and vZ_{dbl}) and to the surface scattering in the case of the Alaskan boreal forests (FD_{surf} and vZ_{surf}) (Table 3.1). The low sensitivity of the Cloude–Pottier decomposition parameters can be explained by the fact that these parameters work with the polarimetric state only and do not consider the span information (i.e. radar intensity). In particular, the Cloude–Pottier entropy (CP-H) and anisotropy (CP-A) describe the scattering randomness and the importance of one scattering mechanism to the others and do not represent intensity information. In contrast, the Freeman–Durden (FD) and Van Zyl (vZ) decompositions provide scattering intensity estimates from known physical mechanisms (surface, double-bounce and volume scattering). Among all the polarimetric discriminators, the minimum and maximum of scattered intensity (S_{min} and S_{max}) and the minimum and maximum of received power intensity (Pr_{min} and Pr_{max}) produced significant positive differences on both savannahs and boreal forest sites (except for the dense mature boreal forest). The minimum of the degree of polarization (d_{min}) produced stronger differences than the maximum of the degree of polarization (d_{max}). In both cases, the differences are negative for both the savannah sites and the spare or dense spruce forests. Both parameters were found to be related to biomass structure and surface roughness variations, as they are a measure of the polarization state of the received signal, which is highly related to the vegetation structure. Except for the dense spruce forest, the fractional polarization (FP) produced a smaller difference than its individual components (Pr_{min} and Pr_{max}).

Several polarimetric variables from multidate RADARSAT-2 C-band images that were acquired across a range of soil moisture conditions were used to develop empirical algorithms to estimate volumetric soil moisture maps over the Alaskan boreal test area [BOU 13b]. A mean error of 6.7% between observed and estimated values for the soil volumetric moisture content was achieved through a regression model that uses, as independent variables, one single channel SAR C-VH backscatter intensity and two polarimetric variables: the maximum of degree of polarization (d_{max}) and the maximum of the completely unpolarized component ($Unpol_{max}$). The model also showed improvement from 27 to 33% in the accuracy of the soil volumetric moisture content retrieval by comparison with the model using only single polarized C-HH data. By providing information on surface roughness and/or biomass, d_{max} appears to be helpful for extracting surface soil moisture from SAR data.

Although radar images are theoretically available independently of the weather conditions, the availability of images acquired with the old radar sensors was limited because of the long revisit cycle of the satellites, such as ERS-1/2 (35 days). Fortunately, the revisit period is shorter for the new radar satellites, such as RADARSAT-1/2 (3–6 days). The pointing capability of the RADARSAT-1/2 sensor allows for the acquisition of two images per day as shown in Abbott *et al.* [ABB 07]. Such revisiting time will be improved in the near future with the planned RADARSAT-3 constellation and the new Sentinel missions. In addition, radar images often have a finer spatial resolution (of the order of a magnitude of 10 m) than optical or thermal infrared images. Thus, we can conclude that for monitoring wildfire danger, radar remote sensing represents a data source that is complementary to optical or thermal infrared data.

Variable	Savannah		Boreal forest					
	Shrub/Tree	Herbaceous	Burn area by a fire of			Regeneration	Sparse black spruce forest	Dense black spruce forest
			Low intensity	Moderate intensity	High intensity			
C-HH	40.53	64.12	54.93	62.71	62.27	41.11	43.72	17.34
C-HV	45.45	72.55	45.16	51.09	60.24	30.42	31.42	21.02
C-VV	17.12	41.46	53.64	58.92	61.92	34.68	38.71	15.83
C-RR	48.04	75.15	55.78	53.01	60.15	21.70	39.33	15.64
C-LR	34.42	58.39	55.33	62.49	61.20	43.04	39.90	11.55
C-LL	54.54	75.13	50.07	53.59	62.29	29.12	34.64	15.21
CP-H	8.08	8.83	–1.86	–11.40	–0.67	–9.68	–1.74	1.42
CP-A	0.49	1.91	11.44	3.89	–1.41	–6.75	5.42	7.95
CP-α	16.09	21.73	0.09	–13.03	1.65	–15.74	–1.55	3.40
vZ_{vol}	44.33	71.39	49.95	53.53	62.10	29.01	34.61	15.24
vZ_{surf}	36.07	62.38	56.28	63.71	61.12	46.96	41.68	10.19
vZ_{dbl}	62.48	79.12	58.03	51.41	58.53	13.62	41.73	17.65
FD_{vol}	46.24	73.02	50.07	53.59	62.43	28.82	34.62	15.19
FD_{surf}	36.91	56.07	57.35	65.26	60.71	54.08	45.14	7.59
FD_{dbl}	77.28	79.00	54.06	29.21	43.71	–18.35	30.47	–9.42
d_{max}	–2.35	–2.59	5.34	0.76	1.78	8.76	0.26	–2.98
d_{min}	–44.01	–86.02	12.45	6.32	4.02	40.66	–8.97	–15.38
Δd	3.11	16.42	–6.25	–10.87	–3.38	–27.13	6.39	8.09
Pr_{max}	42.08	65.29	54.41	59.90	55.86	40.23	23.97	8.26
Pr_{min}	48.00	72.77	33.33	57.69	51.67	16.67	21.05	–2.30
FP	–2.39	–2.25	4.84	0.81	1.35	8.19	0.13	20.93
S_{max}	42.14	66.49	53.13	59.61	55.27	38.19	24.56	11.46
S_{min}	36.75	62.42	55.68	60.28	54.90	40.74	25.56	–3.03
ND_S	7.84	19.96	66.27	81.87	75.37	58.38	63.30	76.07

Table 3.1. *Normalized difference (%) between median values of polarimetric variables computed from RADARSAT-2 images acquired during a wet date and a dry date over South African savannahs [KON 14] and Alaska boreal forests [BOU 13a]*

3.2.6. *Synergisms between radar, optical and thermal infrared remote sensing*

Synergisms between optical or thermal infrared bands and single-polarized SAR data were also investigated. On partially covered cotton fields, Moran *et al.* [MOR 97] found a good relationship between NDVI and Ku-VV ground-based radar basckscatters. They theoretically justified this relationship using two reasons. First, Ku-VV radar backscatter is not sensitive to soil moisture. Second, both spectral variables can be related to the LAI through semiempirical models, based on the Lambert-Beer law for NDVI and on the *water cloud* model for Ku-VV radar backscatters. This semiempirical relationship works well for partially covered canopies, but not for fully covered canopies, because Ku-band and NDVI sensitivities exponentially decrease with increasing LAI. Troufleau *et al.* [TRO 97] estimated a semiempirical relationship between $T_S - T_a$ and m_S, where T_S is computed from airborne and LANDSAT-TM thermal infrared data and m_S is computed from ERS-1 SAR C-VV radar backscatter by inversion of the *water cloud* model. The results of Moran *et al.* [MOR 97] and Troufleau *et al.* [TRO 97] were obtained over low canopies, i.e. crops or semi-arid rangeland, where the sensitivity of NDVI and $T_S - T_a$ to water stress is more important than on forest canopies. There is very little studies on such synergisms over forest canopies.

3.3. Fire detection

Fire detection is critical for both firefighting or fire monitoring. For firefighting, early detection is essential. The revisit time provided by current satellite sensors is not considered sufficient for firefighting operations by forest fire mangers. Therefore, fire detection for firefighting is still based on human observation, fixed optical cameras and aerial surveys. However, the detection of wildfires and monitoring of wildfire effects for the vast territories of North America, Australia or West Eurasia is mainly based on satellite remote sensing. Wildfire detection is based on the spectral anomalies produced by fires that are detectable in several parts of the electromagnetic spectrum. Although fire detection is possible in the microwave wavelengths, these sensors are not used operationally, because of their high cost and their low nominal spatial resolution [KEM 06]. Therefore, we will focus here on the optical sensors.

Visible wavelengths can detect the light that active fires emit, but this is only possible at night. Elvidge *et al.* [ELV 01] developed an algorithm that uses visible images acquired by the operational linescan system (OLS) of the DMSP satellite. With a temporal image series, the algorithm defines a reference set of "stable" lights and fires are then identified as any light detected outside this set because fires are temporary by nature. The factors affecting fire detection with DMSP-OLS night images are moonlight and clouds. This method is applied mainly to night images such as the one shown in Figure 3.6. This is a strong operational limitation because most of the fires occur and have their highest intensity during the day. Such as the infrared wavelengths, visible wavelengths can also detect fire smoke plumes. This detection method is widely used as an alternative to human visual detection [BUG 14]. Image processing algorithms can be used to discriminate a smoke plume from its background. Although these systems eliminate false alarms produced by overheating of ground areas, they also present some limitations, because of two reasons. First, the smoke plume can only be detectable sometime after the fire has started; and second, smoke is often conducted along the surface and emerges in an area different from that where the fire started.

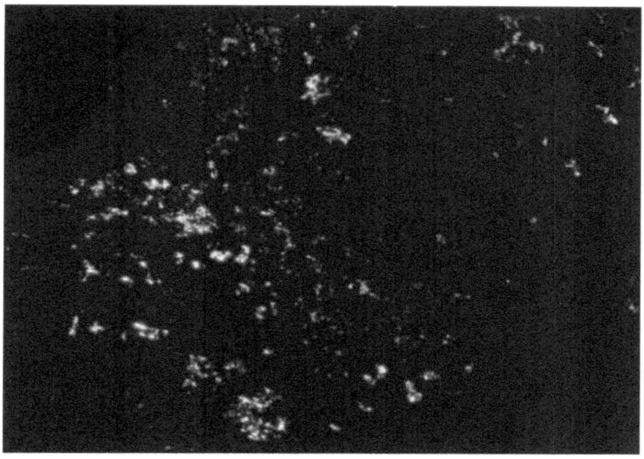

Figure 3.6. *Image acquired at night by the Visible Infrared Imaging Radiometer Suite (VIIRS) on board the Suomi National Polar-Orbiting Partnership (Suomi-NPP) satellite. The image was extracted from the "Black Marble" imagery series released by NASA and the National Oceanic and Atmospheric Administration in December 2012 showing active wildfires in Australia. The image is a composite of images acquired during nine nights in April 2012 and 13 nights in October 2012 (http://www.nasa.gov/mission_pages/NPP/news/aus-fires.html#.Vkt6Db-M4SR). For a color version of this figure, see www.iste.co.uk/baghdadi/6.zip*

Ground automatic detection systems consist of infrared and visible cameras mounted in towers, buildings or masts. The cameras can be fixed (attached to the structure) or mounted on a positioning system to vary the azimuth and elevation angles. A positioning system can be used to survey the entire environment by varying the scanning angles automatically. The detection delay depends on the scan velocity given by the motors and the optical system in the camera. Mobile sensors require high image processing capacities for automatic detection. Often the devices are coupled with intelligent algorithms that speed up the processing and recognition of signals from either smoke or distinctly hot temperatures produced by wildfires [ROS 14].

Most commonly, mid-infrared and thermal infrared wavelengths are used because, in these wavelengths, fires produce a high signal due to their high temperature. The radiant energy emitted per time and space unit by a black body (or total exitance) can be computed as function of its absolute surface radiative temperature T using the Stefan–Boltzmann's law as follows:

$$M_{\text{black}} = s\, T^4 \qquad [3.5]$$

where:

– M_{black} = total exitance ($W \cdot m^{-2}$) = energy that is emitted by a black body per time and space unit;

– T = absolute surface radiative temperature of the body (K) = t (°C) +273;

– s = Stefan–Boltzmann's constant = $5.6693 \times 10^{-8}\ w \cdot m^{-2} \cdot K^{-4}$.

The radiation emitted by black bodies is not uniform in all frequencies or wavelengths (λ). The spectral distribution of the emitted radiation, i.e. the variation of M as a function of λ, is given by the Planck's law as follows (Figure 3.7):

$$M_\lambda = \frac{2\pi h c^2 \lambda^{-5}}{\exp\left(\frac{hc}{\lambda k T}\right) - 1} \qquad [3.6]$$

where:

– M_λ = spectral exitance ($W \cdot m^{-2} \cdot m^{-1}$);

– c = light speed = 3×10^8 $m \cdot s^{-1}$;

– k = Boltzmann's constant, which is related to s, h and c. Its value is 1.38×10^{-23} $J \cdot K^{-1}$;

– h = Planck's constant = 6.625×10^{-34} $J \cdot s^{-1}$;

– λ = wavelength (m);

– T = absolute surface radiative temperature of the body (K).

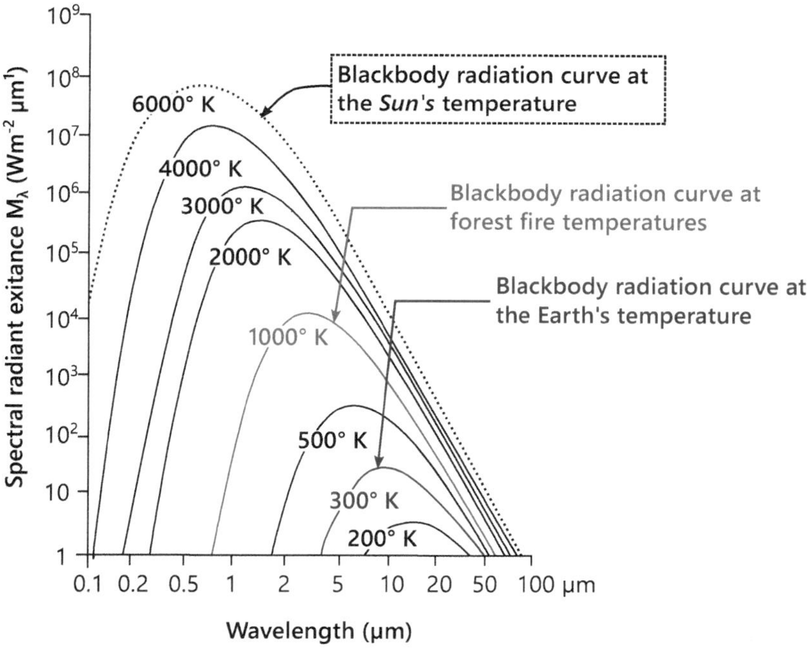

Figure 3.7. *Spectral radiant exitance as a function of the temperature of the black body. The figure shows that forest fires being hotter than the Earth's surface exhibit a peak in their spectral exitance at a shorter wavelength than the Earth's surface (adapted from [KEN 94]). For a color version of this figure, see www.iste.co.uk/baghdadi/6.zip*

Both axes of Figure 3.7 are logarithmic and the position of the maximal exitance is aligned along a straight line. The wavelength for which this maximum occurs only depends on the body's temperature, as described by the Wien's displacement law:

$$\lambda_{\max} = \frac{B}{T} \qquad [3.7]$$

where:

– $\lambda_{\max}$ = wavelength at which the spectral exitance is maximal (μm);

– B = constant = 2,898 μm·K;

– T = absolute surface radiative temperature of the black body (K).

Active fires produce temperatures ranging between 800 and 1,200 K, although they can reach up to 1,800 K. Such temperatures are significantly different from the Earth's temperature (17°C or 300 K). As a result, according to equation [3.7], the maximal emission of fires occurs in the mid-infrared spectral domain, around 2–3 μm, which is far from the peak of the Earth (9.7 μm) and solar (0.5 μm) radiations and thus suitable for fire detection (Figure 3.7)

The use of middle-infrared cameras for fire detection can be considered fully operational and is applied in many areas of the world. These devices are mainly used at the ground level, as they were shown to be highly efficient in monitoring large areas of fairly flat terrain or if installed in high look-out points, to provide a wide view of the surrounding areas. Mid-infrared bands for detecting fires can also be used by aerial or space borne sensors. Indeed, this spectra domain corresponds to an atmospheric window, where the atmospheric absorption is negligible. It is available on board several satellites, such as channel 3 of NOAA-AVHRR. Another atmospheric window is located in the thermal part of the spectrum, i.e. between 8 and 12 μm. For these wavelengths, fires also have some strong emissions, but the emission is closer to the peak of Earth radiation at a normal environmental temperature (300 K) (Figure 3.7).

Fires are detected on mid-infrared and thermal images using the so-called thresholding algorithms, where any area is considered a fire if its temperature is above a given threshold temperature, which is either a regional or an absolute maximum value. However, this method produces false alarms and/or undetected fires because of the differences in fire characteristics among regions in the world. Although fixed thresholding algorithms were used in the past, most current techniques for fire detection make use of the so-called contextual algorithms [GIG 03]. They use local maxima and other multispectral criteria to help differentiate active fire pixel temperatures from the environmental temperature in the proximity of the fire pixel [GIG 03, GIG 06].

Fire detection algorithms are applied to images acquired either by geostationary or geosynchronous satellite sensors. Geostationary sensors include the GOES Imager [HAO 14], the SEVIRI onboard the Meteosat Second-Generation (MSG) satellite [FRE 14] and the moderate-resolution optical imager onboard the Himawari Multifunciton Transport Satellite (MTSAT-2) [HYE 13]. GOES and SEVIRI are suitable for monitoring most wildfire processes, as they provide high frequency coverage, i.e. 30 and 15 min.

The most commonly used geosynchronous satellite sensors are the AVHRR onboard the NOAA meteorological satellite, the Along Track Scanning Radiometer (ATSR) onboard the ERS-1/2 and Envisat, as well as the Moderate Resolution Imaging Spectroradiometer (MODIS) onboard the Terra and Aqua satellites. Compared to ATSR, this is a relative advantage for monitoring fire activities with AVHRR that provides a maximum of one daily pass, or MODIS, which provides two daily passes per satellite. However, MODIS has become the standard sensor for active fire monitoring at regional to global scales because of its high spatial resolution (250–500 m), good fire detection capabilities and global coverage (Figure 3.8) [GIG 06, HAO 14]. The contextual algorithm for active fire detection is such that fires that are burning at the time of satellite overpass under relatively cloud-free conditions are detected by thresholds that are first applied to the observed mid-infrared and thermal infrared brightness temperature. False detections are then rejected by examining the brightness temperature relative to neighboring pixels [GIG 03].

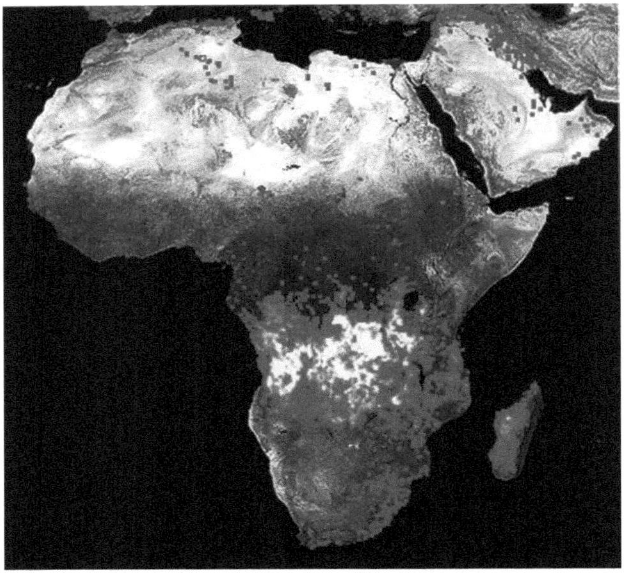

Figure 3.8. *Accumulated locations of active fires detected over Africa by MODIS onboard the Terra and Aqua satellites for the period between October 28, 2015 and November 6, 2015. Each colored dot indicates a location where MODIS detected at least one fire during the compositing period. Color ranges from red where the fire count is low to yellow where number of fires is large (from http://lance-modis.eosdis.nasa.gov/cgi-bin/imagery/firemaps.cgi). For a color version of this figure, see www.iste.co.uk/baghdadi/6.zip*

Active fire data from MODIS are distributed through NASA and are widely used in regional and global fire information systems, such as the Canadian Wildland Fire Information System[2] or the European Forest Fire Information System (EFFIS)[3]. In the case of EFFIS, postprocessing filters based on land cover ancillary data are applied to the MODIS product to reduce the number of false alarms produced by non-fire hot surfaces, such as industrial areas or warm ground soils. They allow, therefore, to increase the reliability of the active fire detection [SAN 12]. Images provided by the recently launched Visible Infrared Imaging Radiometer onboard the Suomi National Polar-Orbiting Partnership satellite [JUS 13] and by Sentinel-3 SLSTR [WOO 12] will provide continuity to the MODIS active fire products.

2 http://cwfis.cfs.nrcan.gc.ca.

3 http://effis.jrc.ec.europa.eu.

3.4. Burnt area mapping

Mapping of burnt areas and assessment of wildfire effects is one of the most successful applications of satellite remote sensing, as it provides comprehensive and harmonized information on the effects of wildfires over large territories at low costs. A wide variety of optical and radar sensors have been used for burnt area mapping from local to global scales.

Fires produce a significant change in the structure and the reflectance of vegetation and the soil properties within the burnt area that are noticeable in the visible, the infra-red and even in the microwave part of the electromagnetic spectrum. Since a wide variety of satellite sensors provide data on the above-mentioned spectral ranges, the choice of the sensor is driven by the required spatial coverage of the area of interests, i.e. global, regional or national/local, and the persistence of cloud coverage over this area. Generally, the lower the spatial resolution of the sensor, the higher the revisit time over a given area. Microwave sensors are used in areas where the persistent cloud coverage prevents the efficient use of optical sensors.

At the global scale, NOAA-AVHRR data were extensively tested in the 1990s, most commonly from a multitemporal comparison of NDVI images [PER 99]. More recently, other global burnt area datasets were derived from SPOT VEGETATION or ENVISAT-ATSR-2 [GRE 03, SIM 04]. Although these data provide gross estimates of burnt areas at the global level, the lack of extensive validation and agreement between them limit their use at regional or national levels. Pereira [PER 99] showed that the accuracy of the burnt area maps with AVHRR data was about 80% for large fires in the European Mediterranean region. The methods were considered suitable only for fires larger than 1000 ha, and reliable for fires larger than 2000 ha.

Since 2000, the launch of satellites with moderate or high revisiting time that allow a global coverage allowed for the development of various products of global burn areas (L3JRC, MODIS MCD45A1, GBS, Globscar, GBA2000, GFEDv3, GLobcarbon) as detailed in [MOU 14]. This is particularly the case for the free MODIS data of medium spatial resolution that provided a definite impulse for the use of remote sensing for burn mapping at the regional and global scales [ROY 08, BOS 10, GIG 06, GIG 10]. MODIS images are not only of better spatial resolution but also have a better radiometry and a higher spectral information quality for burnt area mapping [BOS 10]. They are also acquired with enough revisit time because of the simultaneous operation of both the TERRA and AQUA satellites that can obtain up to four images per day. At the global scale, the MODIS program has released a standard product on burned areas that is based on a multitemporal change detection approach that analyzes differences between modeled and actual reflectance by taking into account bidirectional reflectance distribution function corrections. The resulting products reached a high agreement level with the MODIS active fire product [ROY 08].

At the regional scale, as for the active fire detection, it is the MODIS burn products that are used in the operational systems [SAN 12]. Full MODIS mosaics are received and processed twice daily by the EFFIS team to provide near-real time monitoring of active wildfires and map burnt areas. EFFIS is thus able to provide accurate information of fire impacts and support to decision and policy making. The use of higher spatial resolution imagery from the Advanced Wide Field Sensor (AWIFS) for regional coverage in Europe was recently tested, but the limited revisit time of the sensor is not compensated by the advantage of using high spatial imagery. It did not enhance the standard rapid damage assessment module of EFFIS that is based on MODIS imagery. Figure 3.9 compares the extent of burnt area mapped from MODIS and AWIFS imagery in the case of the 2007 wildfire in the Peloponnese province (Greece).

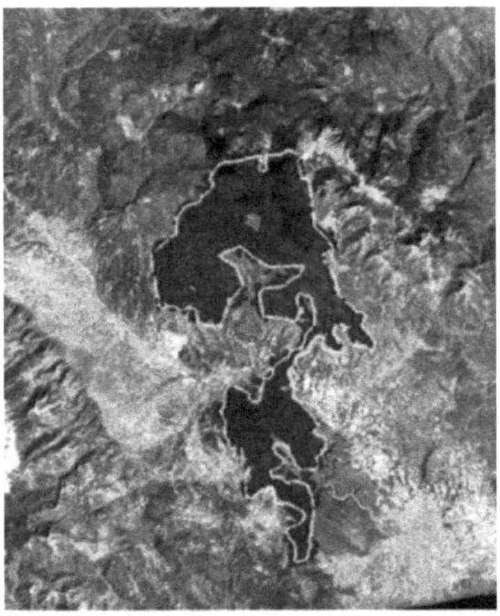

Figure 3.9. *Perimeter of burnt areas obtained from MODIS (yellow) and AWIFS (blue) in one of the large fires in the Peloponnese province, Greece in 2007 (adapted from [SAN 12]). For a color version of this figure, see www.iste.co.uk/baghdadi/6.zip*

Other sensors of similar spatial and spectral resolutions to those of MODIS, such as the ENVISAT-MERIS, were recently used for a similar production of global burn area products (e.g. ESA FIRE CCI4) [ALO 15]. Similarly, the recent near-real time availability of high spatial resolution Landsat images provide new means for detailed burnt scar mapping at regional to global scales. The required computational needs may be provided by Google Earth Engine, which recently published the first 30 m resolution maps of global forest change for the 2000–2012 period[5].

Given the great number of sensors and resulting burnt area products, the Global Wildfire Information System (GWIS) was established by the Group on Earth Observations and the Global Observation of Forest Cover Fire Implementation Team (GOFC Fire IT)[6]. GWIS has the aim to be a warehouse for information provided by different sources at regional to global scales in order to facilitate the product intercomparison and their accessibility across the world. The variability of burnt area estimations between products remain important with a known underestimation of these products with regard to the local analysis done with high-resolution images such as Landsat, because of the difficulty of these sensors to map the small fires. Randerson *et al.* [RAN 12] estimated that between 24 and 54% of burnt areas due to small fires can be added to the MODIS areas by coupling these products with active fire detection products.

At national to local scales, the wide variety of medium to high-resolution images having pixel size ranging from 10 to 30 m allows for accurate mapping of burnt areas. However, as mentioned above, the increase in spatial resolution is often accompanied by a decrease in revisit time of the sensor, which prevents the acquisition of this imagery for large areas. High-spatial burnt area mapping has been performed with Landsat-TM imagery complemented in some cases by SPOT or ASTER imagery, such as in the study of Goodwin and Alett [GOO 14], which uses an image series of Landsat-TM, which were acquired between 1986 and 2013 over a region of Australia that has experienced a variety of savannahs fires between 1997 and 2010. Various indices computed from the original spectral bands and many different techniques were used to enhance the mapping of burnt areas

4 http://www.esa-fire-cci.org/.

5 https://earthengine.google.org.

6 ftp://ftp.earthobservations.org/GEO-X/GEO-X_10_GEO%202012-2015%20Work%20Plan%20Update.pdf.

[PER 99, GIT 12]. However, this was, in most cases, limited to the mapping of burnt areas at local and subnational scale. An exception to this is the case of Portugal, where an operational system able of processing Landsat-TM scenes for mapping burnt areas was set up [OLI 12].

High-resolution images are critical to the management of wildfires because they provide a great amount of spatial details, needed for an accurate analysis of fire damage and for the planning of restoration measures. Rapid coverage of critical fire events is also possible because of the availability of a great variety of high and very high spatial resolution imagery, such as those provided by QUICKBIRD, IKONOS, FORMOSAT, EARLYBIRD and RAPIDEYE. Data provision for critical fire events has been supported by the International Space Charter, an agreement between space agencies, which allows for the rapid provision of available imagery from a series of satellites, including RADARSAT, ERS, ENVISAT, SPOT, IRS, SAC-C, NOAA, LANDSAT, ALOS and DMC supporting crisis management. Furthermore, programs such as the European Copernicus provide the means for the national civil protection services to activate the request for data acquisition and processing[7].

So far, burnt area mapping has been mainly carried out with optical imagery, but single-polarized SAR images were also tested over various forest ecosystems [BOU 02, GIM 04]. While burn scars are visible in optical imagery because of changes in vegetation condition and structure, it is the change in moisture content of the burnt surface with respect to the unburned area that allows for the detection of burnt areas using SAR images. Burnt areas usually have a higher moisture content than unburned areas, and thus a higher backscatter. This is particularly the case for boreal forests over discontinuous permafrost where frost water melts with the fire effect. Thus, wet burnt areas appear with a brighter tone than the surrounding non-affected area. More recently, RADARSAT-2 polarimetric SAR images have been tested to map fire scars. When mapping 2 month-old burns over a rough terrain in the Calabria peninsula in Southern Italy, with Radarsat-2 polarimetric SAR images acquired at 32.4°–34° incidence angle and an east-looking direction, Leblon [LEB 11] showed that the Freeman–Durden decomposition images were better than the Cloude–Pottier decomposition

7 http://emergency.copernicus.eu/mapping/.

images. This difference can be explained by the fact that the Freeman–Durden decomposition implicitly considers the intensity information, which is not the case of the Cloude–Pottier that only relies on the polarization state. It is also with a Freeman–Durden-based unsupervised classifier applied to a RADARSAT-2 polarimetric SAR image with a 39° incidence angle and a west-facing direction that Goodenough *et al.* [GOO 11] were able to retrieve 85% of the area related to 8 year-old fire scars delineated over a SPOT-5 image in the case of a flat area in Northeast China. However, they retrieved 84% of of the area related to 9 year-old fire scars over a rough area in Alberta when applying a Cloude–Pottier-based unsupervised classifier to a RADARSAT-2 polarimetric SAR image with a 30° incidence angle and an east-facing direction. This classifier was shown to be less influenced by topographic variations.

3.5. Perspectives

This chapter described the use of optical, thermal infrared and radar images for prefire and postfire condition monitoring. Prefire conditions are discussed with a particular emphasis on fuel moisture monitoring. So far, there is no operational system that uses remote sensing for mapping fuel moisture. Optical, thermal infrared, single-polarized and polarimetric SAR images have been tested. While synergisms between optical and thermal infrared were successfully implemented in fuel moisture monitoring and while there were some trials of coupling optical and thermal infrared remote sensing data with single-polarized SAR images, synergisms with polarimetric SAR images are still to be developed. Indeed, SAR images have been shown to be highly complementary to optical and thermal infrared images. SAR images allow for detecting surface status by providing information on the scattering mechanisms that are linked to surface roughness (and thus to the presence or absence of vegetation as well as to the vegetation type) and to the fuel moisture. Optical images are associated with the reflective properties of the surface that are linked to presence or absence of vegetation, the vegetation type and the surface moisture. With the future development of new SAR sensors such as the RADARSAT-3 SAR satellite constellation, such synergism with SAR images will be of importance as the revisit period of SAR sensors (daily acquisition) will be the same as to the ones of optical and thermal infrared sensors such as MODIS.

Much of the aforementioned remote sensing studies for prefire condition management are based on the estimation of fuel moisture or other indices through empirical relationships. Further research is needed to develop analytical methods. There are also other prefire condition parameters required by an operational fire danger system, such as wind parameters, which should be estimated. These are more difficult to map using remote sensing data and additional ground-based information will be required to effectively monitor fire danger. In this regard, *in situ* sensing systems that integrate microsensing, computer and wireless communication technologies as described in Teillet *et al.* [TEI 01] will be useful.

Fire detection with remote sensing techniques can be considered fully operational. At the local scale, active fire detection is mainly based on the use of visible and thermal infrared cameras for the detection of thermal anomalies or smoke plumes. At regional to global scales, information on active fires is provided by geostationary satellite sensors (GOES, SEVIRI and MTSAT-2) or geosynchronous sensors (AVHRR, ATSR and MODIS). The high revisit time of the geostationary satellites provides frequent information (15–30 min) that can be needed for monitoring active fires. By contrast, geosynchronous satellites provide a lower revisit time (one to two daily passes), but provide global wildfire information, such fire regimes. Therefore, MODIS has become the standard sensor for active fire monitoring at regional to global scales because of its high spatial resolution, good fire detection capabilities and global coverage. However, differences in accuracy among the existing products still persist. Moreover, optical and thermal infrared images cannot be acquired under cloudy conditions, and SAR images should be tested for environments where cloudy conditions are common.

Burnt area mapping from remote sensing is also operational. Global burnt area datasets are derived from AVHRR, ATRS, SPOT-VEGETATION and recently from MODIS, which has become the standard for fire mapping at regional to global scales. Higher spatial resolution imagery (e.g. LANDSAT, LISS-3 and SPOT-HRV) are also used at the national or local level. For active fire detection, optical and thermal infrared image acquisition is limited under cloudy conditions, and SAR images from single-polarized sensors (RADARSAT-1) or polarimetric sensors (e.g. RADARSAT-2) have been also tested. An operational system using this new imagery is still to be developed.

For both active fire detection and burnt area mapping, agreements have been recently established among the space agencies in the frame of the International Space Charter for the provision of remote sensing data for wildfire crisis management, which enables rapid monitoring of critical fire events.

3.6. Key points

– Fires represent significant disturbances for grassland and forest ecosystems across the world and are considered in parts of the world as a natural hazard.

– The use of remote sensing for prefire condition monitoring is not yet operational. Prefire conditions are monitored using fire danger predicting systems. Several parameters of these systems, i.e. fuel load, fuel type and fuel moisture, can be estimated with remote sensing. Fuel load is mapped from optical/radar images using the same methods as the ones used for vegetation biomass. Fuel types are mapped from optical/radar images using the same methods as the ones used for vegetation types. Fuel moisture can be mapped using optical images, but there are several limitations in using these images. Thermal infrared are better, particularly when they are coupled with optical images. Thermal infrared and optical images cannot be acquired only during clear sky conditions, by contrast to SAR images. Single-polarized SAR data were well correlated to fuel moisture variables, but the relationships are influenced by the biomass and surface roughness. Polarimetric SAR images were also tested, because they allow advanced data processing, which could reduce the effect of these factors.

– Fire detection is better achieved using either mid-infrared or thermal infrared images, that allow to distinguish fires from the surface of the Earth, because fires are hotter than the Earth's surface. At the local scale, active fire detection is mainly carried out with ground-based cameras. At regional to global scales, active fires can be detected on images acquired by geostationary (GOES, SEVIRI and MTSAT-2) or geosynchronous (AVHRR, ATSR and MODIS) sensors. MODIS is currently the standard sensor for active fire monitoring at regional to global scales because of its high spatial resolution, good fire detection capabilities and global coverage. However, MODIS images cannot be acquired under cloudy conditions, and SAR images should be tested for areas under frequent cloudy conditions.

– Mapping burnt areas is one of the most successful applications of satellite remote sensing. It is mainly carried out using MODIS optical images. More recently, SAR images and polarimetric SAR images have been tested. An operational system using SAR imagery is still to be developed.

3.7. Bibliography

[ABB 07] ABBOTT K., LEBLON B., STAPLES G. *et al.*, "Fire danger monitoring in a northern boreal forest region using RADARSAT-1 imagery", *International Journal of Remote Sensing*, vol. 28, nos. 5–6, pp. 1317–1338, 2007.

[AKT 11] AKTHER M.S., HASSAN Q.K., "Remote sensing-based assessment of fire danger conditions over boreal forest", *IEEE Journal of Selected Topics in Applied Earth Observations and Remote Sensing*, vol. 4, no. 4, pp. 992–999, 2011.

[ALL 98] ALLEN R.G., PEREIRA L.S., RAES D. *et al.*, Crop evapotranspiration guidelines for computing crop water requirements, FAO Irrigation and Drainage paper, vol. 56, Rome, 1998.

[ALO 15] ALONSO-CANAS I., CHUVIECO E., "Global burned area mapping from ENVISAT MERIS and MODIS active fire data", *Remote Sensing of Environment*, vol. 163, pp. 140–152, 2015.

[AST 09] ASTM STANDARD "Direct moisture content measurement of wood and wood-base materials", *Annual book of ASTM Standards*, ASTM International, West Conshohocken, vol. 4, 2009.

[BAS 98] BASTIAANSSEN W.G.M., MENENTI M., FEDDED R.A. *et al.*, "A remote sensing surface energy balance algorithm for land (SEBAL): formulation", *Journal of Hydrology*, vol. 212–213, pp. 198–212, 1998.

[BIS 14] BISQUERT M., SÁNCHEZ J.M., CASELLES V., "Modeling fire danger in Galicia and Asturias (Spain) from MODIS images", *Remote Sensing*, vol. 6, pp. 540–554, 2014.

[BOS 10] BOSCHETTI L., ROY D., JUSTICE C. *et al.*, "Global assessment of the temporal reporting accuracy and precision of the MODIS burned area product", *International Journal of Wildland Fire*, vol. 19, pp. 705–709, 2010.

[BOU 02] BOURGEAU-CHAVEZ L.L., KASISCHKE E.S., BRUNZELL S. *et al.*, "Mapping fire scars in global boreal forests using imaging radar data", *International Journal of Remote Sensing*, vol. 23, pp. 4211–4234, 2002.

[BOU 07] BOURGEAU-CHAVEZ L.L., GARWOOD G., RIORDAN K. *et al.*, "Improving the prediction of wildfire potential in boreal Alaska with satellite imaging radar", *Polar Record*, vol. 43, no. 4, pp. 321–330, 2007.

[BOU 13a] BOURGEAU-CHAVEZ L.L., LEBLON B., BUCKLEY J. *et al.*, "Assessment of polarimetric SAR data for fuel moisture estimation: analysis of wet versus dry conditions", *International Journal of Remote Sensing*, vol. 34, no. 16, pp. 5709–5730, 2013.

[BOU 13b] BOURGEAU-CHAVEZ L.L., LEBLON B., BUCKLEY J. *et al.*, "Estimating fuel moisture using RADARSAT-2 polarimetric SAR images", *Remote Sensing of Environment*, vol. 132, pp. 71–85, 2013.

[BUG 14] BUGARIĆ M., JAKOVČEVIĆ T., STIPANIČEV D., "Adaptive estimation of visual smoke detection parameters based on spatial data and fire risk index", *Computer Vision and Image Understanding*, vol. 118, pp. 184–196, 2014.

[BUR 98] BURGAN R.E., KLAVER R.W., KLAVER J.M., "Fuel models and fire potential from satellite and surface observations", *International Journal of Wildland Fire*, vol. 8, no. 3, pp. 159–170, 1998.

[CHA 15] CHAKROUN, H., MOUILLOT, F., HAMDI, A., "Regional equivalent water thickness modeling from remote sensing across a tree cover/LAI gradient in Mediterranean forests of northern Tunisia", *Remote Sensing*, vol. 7, no. 2, pp. 1937–1961, 2015.

[CHU 03] CHUVIECO E., AGUADO I., COCERO D. *et al.*, "Design of an empirical index to estimate fuel moisture content from NOAA-AVHRR analysis in forest fire danger studies", *International Journal of Remote Sensing*, vol. 24, no. 8, pp. 1621–1637, 2003.

[CHU 10] CHUVIECO E., AGUADO I., YEBRA M. *et al.*, "Development of a framework for fire risk assessment using remote sensing and geographic information system technologies", *Ecological Modelling*, vol. 221, no. 1, pp. 46–58, 2010.

[CHU 14] CHUVIECO E., MARTINEZ S., ROMAN M.V. *et al.*, "Integration of ecological and socio economic factors to assess global vulnerability to wildfire", *Global Ecology and Biogeography*, vol. 23, no. 2, pp. 245–258, 2014.

[CLO 97] CLOUDE S.R., POTTIER E., "An entropy based classification scheme for land applications of polarimetric SAR", *IEEE Transactions Geoscience and Remote Sensing*, vol. 35, no. 1, pp. 68–78, 1997.

[DES 96] DESBOIS N., VIDAL A., "Real-time monitoring of vegetation flammability using NOAA-AVHRR thermal infrared data", *EARSeL Advanced in Remote Sensing*, vol. 4, no. 4, pp. 25–32, 1996.

[DUC 99] DUCHEMIN B., GUYON D., LAGOUARDE J.P., "Potential and limits of NOAA-AVHRR temporal composite data for phenology and water stress monitoring of temperate forest ecosystems", *International Journal of Remote Sensing*, vol. 20, no. 5, pp. 895–917, 1999.

[ELV 01] ELVIDGE C.D., IMHOFF M.L., KIMBERLY E.B. *et al.*, "Night-time lights of the world: 1994-1995", *ISPRS Journal of Photogrammetry and Remote Sensing*, vol. 56, no. 2, pp. 81–89, 2001.

[FOR 92] FORETS CANADA, Élaboration et structure de la méthode canadienne de prévision du comportement des incendies de forêt, Forêts Canada Groupe de travail sur les dangers d'incendie, Report no. ST-X-3F, vol. 68, 1992.

[FRE 14] FREEBORN P.H., WOOSTER M.J., ROBERTS G. *et al.*, "Evaluating the SEVIRI fire thermal anomaly detection algorithm across the Central African Republic using the MODIS active fire product", *Remote Sensing*, vol. 6, no. 3, pp. 1890–1917, 2014.

[FRE 98] FREEMAN A., DURDEN S.L., "A three-component scattering model for polarimetric SAR data", *IEEE Transactions Geoscience and Remote Sensing*, vol. 36, no. 3, pp. 963– 973, 1998

[GIG 03] GIGLIO L., DESCLOITRES J., JUSTICE C. *et al.*, "An enhanced contextual fire detection algorithm for MODIS", *Remote Sensing of Environment*, vol. 87, pp. 273–282, 2003.

[GIG 06] GIGLIO L., CSISZAR I., JUSTICE C., "Global distribution and seasonality of active fires as observed with the Terra and Aqua Moderate Resolution Imaging Spectroradiometer (MODIS) sensors", *Journal of Geophysical Research – Biogeosciences*, vol. 111, no. G2, pp. 1–12, 2006.

[GIG 10] GIGLIO L., RANDERSON J.T., VAN DER WERF G.R. *et al.*, "Assessing variability and long-term trends in burned area by merging multiple satellite fire products", *Biogeosciences*, vol. 7, pp. 1171–1186, 2010.

[GIM 04] GIMENO M., SAN-MIGUEL-AYANZ J., "Evaluation of RADARSAT-1 data for identification of burnt areas in Southern Europe", *Remote Sensing of Environment*, vol. 92, pp. 370–375, 2004.

[GIT 12] GITAS I.Z., MITRI G., VERAVERBEKE S. *et al.*, "Advances in remote sensing of post-fire monitoring: a review", in FATOYINBO T.E. (ed.), *Remote Sensing of Biomass: Principles and Applications*, InTech Pub., Rijeka, 2012.

[GOO 11] GOODENOUGH D.G., CHEN H., RICHARDSON A. *et al.*, "Mapping fire scars using Radarsat-2 polarimetric SAR data", *Canadian Journal of Remote Sensing*, vol. 37, no. 5, pp. 500–509, 2011.

[GOO 14] GOODWIN N.R., COLLETT L.J., "Development of an automated method for mapping fire history captured in Landsat TM and ETM+ time series across Queensland, Australia", *Remote Sensing of Environment*, vol. 148, pp. 206–221, 2014.

[GOU 91] GOUYET J.F., KING C., LE GLEAU H. *et al.*, "Apport des données satellitaires NOAA-AVHRR dans le suivi de la végétation forestière", *Proceedings of the 5th International Colloquium on Physical Measurements and Signatures in Remote Sensing*, Courchevel, France, pp. 625–629, 1991.

[GOW 94] GOWARD S.N., WARING R.H., DYE D.G. *et al.*, "Ecological remote sensing of OTTER satellite macroscale observations", *Ecological Applications*, vol. 4, no. 2, pp. 322–343, 1994.

[GRA 97] GRANGER R.J., "Comparison of surface and satellite-derived estimates of evapotranspiration using a feedback algorithm", *Proceedings of the 3rd International Workshop on Application of Remote Sensing in Hydrology*, Goddard Space Flight Center, Washington, DC, pp. 71–81, 1997.

[GRE 03] GRÉGOIRE J.M., TANSEY K., SILVA J.M.N., "The GBA2000 initiative: Developing a global burnt area database from SPOT-VEGETATION imagery", *International Journal of Remote Sensing*, vol. 24, pp. 1369–1376, 2003.

[HAO 14] HAO W.M., LARKIN N.K., "Wildland fire emissions, carbon, and climate: wildland fire detection and burned area in the United States", *Forest Ecology and Management*, vol. 317, pp. 20–25, 2014.

[HYE 13] HYER E.J., REID J.S., PRINS E.M. *et al.*, "Patterns of fire activity over Indonesia and Malaysia from polar and geostationary satellite observations", *Atmospheric Research*, vol. 122, pp. 504–519, 2013.

[JUS 13] JUSTICE C.O., ROMAN M.O., CSISZAR I. *et al.*, "Land and cryosphere products from Suomi NPP VIIRS: overview and status", *Journal of Geophysical Research: Atmospheres*, vol. 118, pp. 9753–9765, 2013.

[KEM 06] KEMPKA T., KAISER T., SOLBACH K., "Microwaves in fire detection", *Fire Safety Journal*, vol. 41, pp. 327–333, 2006.

[KEN 94] KENNEDY P.J., BELWARD A.S., GRÉGOIRE J.M., "An improved approach to fire monitoring in West Africa using AVHRR data", *International Journal of Remote Sensing*, vol. 15, no. 11., pp. 2235–2255, 1994.

[KON 14] KONG M., LEBLON B., MATHIEU R. *et al.*, "Fuel moisture estimation over savannahs using RADARSAT-2 images", *Proceedings IGARSS'14*, Québec Canada, pp. 5033–5036, 2014.

[LAW 08] LAWSON, B.D., ARMITAGE O.B., "Weather guide for the Canadian forest fire danger rating system", *Natural Resources Canada*, vol. 87, 2008.

[LEB 05] LEBLON B., "Using remote sensing for fire danger monitoring", *Natural Hazards*, vol. 35, no. 3, pp. 343–359, 2005.

[LEB 02] LEBLON B., KASISCHKE E.S., ALEXANDER M.E. *et al.*, "Fire danger monitoring using ERS-1 SAR images over northern boreal forests", *Natural Hazards*, vol. 27, no. 3, pp. 231–255, 2002.

[LEB 07] LEBLON B., FERNANDEZ-GARCIA, P.A., OLDFORD S. *et al.*, "Using NOAA-AVHRR cumulative indices for estimating fire danger codes in northern boreal forests", *International Journal of Applied Earth Observations and Geoinformation*, vol. 9, no. 3, pp. 335–342, 2007.

[LEB 11] LEBLON B., "Research and training on the use of RADARSAT-2 polarimetric SAR images in several environmental applications". *Proceedings if the 8th Advanced SAR Workshop*, Canadian Space Agency, Montreal, Canada, 2011.

[LEB 13] LEBLON B., BOURGEAU-CHAVEZ L.L., *Encyclopedia of Natural Hazards*, Springer-Verlag, Berlin, 2013.

[MER 87] MERRILL D.F., ALEXANDER, M.E., Glossary of forest fire management terms, 4th edition, National Research Council of Canada, Canadian Committee on Forest Fire Management, Ottawa, Ontario, vol. 26516, no. 91, 1987.

[MOR 97] MORAN M.S., VIDAL A., TROUFLEAU D. *et al.*, "Combining multi-frequency microwave and optical data for crop management", *Remote Sensing of Environment*, vol. 61, pp. 96–109, 1997.

[MOU 14] MOUILLOT F., SCHULTZ M.G., YUE C. *et al.*, "Ten years of global burned area products from spaceborne remote sensing: A review. Analysis of user needs and recommendations for future developments", *International Journal of Applied Earth Observation and Geoinformation*, vol. 26, pp. 64–79, 2014.

[NIE 10] NIETO H., AGUADO I., CHUVIECO E. *et al.*, "Dead fuel moisture estimation with MSG-SEVIRI data. Retrieval of meteorological data for the calculation of equilibrium moisture content", *Agricultural and Forest Meteorology*, vol. 150, pp. 861–870, 2010.

[OLD 03] OLDFORD S., LEBLON B., GALLANT L. *et al.*, "Mapping pre-fire conditions in the Northwest Territories, Canada, using NOAA-AVHRR images", *Geocarto International*, vol. 18 no. 4, pp. 21–32, 2003.

[OLD 06] OLDFORD S., LEBLON B., MACLEAN D. *et al.*, "Predicting slow drying fuel moisture codes using NOAA-AVHRR images", *International Journal of Remote Sensing*, vol. 27, no. 18, pp. 3881–3902, 2006.

[OLI 99] OLIOSO A., CHAUKI H., COURAULT D. *et al.*, "Estimation of evapotranspiration and photosynthesis by assimilation of remote sensing data into SVAT models", *Remote Sensing of Environment*, vol. 68, pp. 341–356, 1999.

[OLI 12] OLIVEIRA S., PEREIRA J.M.C., CARREIRAS J., "Fire frequency analysis in Portugal (1975–2005), using Landsat-based burnt area maps", *International Journal of Wildland Fire*, vol. 21, pp. 48–60, 2012.

[PER 99] PEREIRA J.M.C., "A comparative evaluation of NOAA/AVHRR vegetation indices for burned surface detection and mapping", *IEEE Transactions on Geoscience and Remote Sensing*, vol. 37, pp. 217–226, 1999.

[PIE 90] PIERCE L.L., RUNNING S.W., RIGGS G.A., "Remote detection of canopy water stress in coniferous forests using the NS001 Thematic Mapper Simulator and the Thermal Infrared Multispectral Scanner", *Photogrammetry Engineering and Remote Sensing*, vol. 56, no. 5, pp. 579–586, 1990.

[PRO 95] PROSPER-LAGET V., WIGNERON J.P., GUINOT J.P. *et al.*, "Utilisation du satellite NOAA pour la détection des risques d'incendies de forêts", *La Météorologie*, vol. 8, no. 10, pp. 28–38, 1995.

[RAN 12] RANDERSON J.T., CHEN Y., VAN DER WERF G.R. *et al.*, "Global burned area and biomass burning emissions from small fires", *Journal of Geophysical Research*, vol. 117, p. G04012, 2012.

[ROS 14] ROSAS-ROMERO R., "Remote detection of forest fires from video signals with classifiers based on K-SVD learned dictionaries", *Engineering Applications of Artificial Intelligence*, vol. 33, pp. 1–11, 2014.

[ROY 08] ROY D., BOSCHETTI L., JUSTICE C. *et al.*, "The collection 5 MODIS burned area product – Global evaluation by comparison with the MODIS active fire product", *Remote Sense of Environment*, vol. 112, pp. 3690–3707, 2008.

[SAA 07] SAATCHI S.S., HALLIGAN K., DESPAIN D.G. *et al.*, "Estimation of forest fuel load from radar remote sensing", *IEEE Transactions on Geoscience and Remote Sensing*, vol. 45, pp. 1726–1739. 2007.

[SAN 12] SAN-MIGUEL-AYANZ J., SCHULTE E., SCHMUCK G. *et al.*, "Comprehensive monitoring of wildfires in Europe: the European Forest Fire Information System (EFFIS)", in TIEFENBACHER J. (ed), *Approaches to Managing Disaster and Assessing Hazards, Emergencies and Disaster Impacts*, InTech Pub., Rijeka, 2012.

[SCO 14] SCOTT A.C., BOWMAN D.M.J.S., BOND W.J. *et al.*, *Fire on Earth: an Introduction*, Wiley-Blackwell, Oxford, 2014.

[SIM 04] SIMON M., PLUMMER S., FLERENS F. *et al.*, "Burnt area detection at global scale using ATSR-2: the GLOBSCAR products and their qualification", *Journal of Geophysical Research*, vol. 109, no. D14S02, pp. 1–16, 2004.

[STO 89] STOCKS B.J., LAWSON B.D., ALEXANDER M.E. *et al.*, "The Canadian Forest Fire Danger Rating System: an overview", *The Forestry Chronicle*, vol. 65, no. 6, pp. 450–457, 1989.

[STR 01] STRICKLAND G., LEBLON B., GALLANT L. *et al.*, "Monitoring fire danger of northern boreal forests from optical and thermal infrared NOAA-AVHRR images", *Proceedings of the 23th Canadian Remote Sensing Symposium*, Ste-Foy, Canada, pp. 667–676, 2001.

[TEI 01] TEILLET P.M., DUDELZAK A.E., PULTZ T.J. *et al.*, "A framework for in-situ sensor measurement assimilation in remote sensing applications" *Proceedings of the 23th Canadian Remote Sensing Symposium*, Ste-Foy, Canada, pp. 111–118, 2001.

[TRO 97] TROUFLEAU D., VIDAL A., BEAUDOIN A. *et al.*, "Optical-microwave synergy for estimating surface sensible heat flux over a semi-arid rangeland", *Remote Sensing Reviews*, vol. 15, pp. 113–132, 1997.

[TOU 92] TOUZI R., GOZE S., LE TOAN T. *et al.*, "Polarimetric discriminators for SAR images", *IEEE Transactions on Geoscience and Remote Sensing*, vol. 30, no. 5, pp. 973–98, 1992.

[VAN 87] VAN WAGNER C., Development and structure of the Canadian Forest Fire Weather Index, Forestry Technical Report 35, Government of Canada, Canadian Forestry Service, Petawawa National Forestry Institute, Ottawa, Ontario, vol. 37, 1987.

[VAN 10] VAN DER WERF G.R., RANDERSON J.T., GIGLIO L. *et al.*, "Global fire emissions and the contribution of deforestation, savanna, forest, agriculture and peat fires (1997-2009)", *Atmospheric Chemistry and Physics*, vol. 10, pp. 11707–11735, 2010.

[VAN 11] VAN ZYL J.J., ARII M., KIM Y., "Model-based decomposition of polarimetric SAR covariance matrices constrained for nonnegative eigenvalues", *IEEE Transactions Geoscience and Remote Sensing*, vol. 49, no. 9, pp. 3452–3459, 2011.

[VID 94] VIDAL A., PINGLO F., DURAND H. *et al.*, "Evaluation of a temporal fire risk index in Mediterranean forests from NOAA thermal IR", *Remote Sensing of Environment*, vol. 49, no. 3, pp. 296–303, 1994.

[WOO 12] WOOSTER M.J., XU W., NIGHTINGALE T., "Sentinel-3 SLSTR active fire detection and FRP product: Pre-launch algorithm development and performance evaluation using MODIS and ASTER datasets", *Remote Sensing of Environment*, vol. 120, pp. 236–254, 2012.

[YEB 13] YEBRA M., DENNISON P.E., CHUVIECO E. *et al.*, "A global review of remote sensing of live fuel moisture content for fire danger assessment: Moving towards operational products", *Remote Sensing of Environment*, vol. 136, pp. 455–468, 2013.

4

Characterization of Industrial Plumes

4.1. Introduction

Since the Industrial Revolution in the 19th Century, human activity has increased significantly, all the while being accompanied by an increase in environmental pollution (of water, soil and atmosphere), which has not stopped growing since then. According to the most recent report from the Intergovernmental Panel on Climate Change [GIE 13], it has been confirmed with a certainty of 95% that "human activity is the principal cause of the warming that has been observed" since the beginning of the 20th Century.

This chapter concentrates on air pollution resulting from human activity, which has a direct influence not only on global warming but also on human health (air quality). As discussed in section 4.2, a wide variety of gases and aerosols make up the anthropogenic plumes emitted by various industries, primarily the energy industry, the iron and steel industry, transportation and agriculture. As a result, human activity causes the emission of large amounts of carbon dioxide (CO_2), an important factor in the greenhouse effect, into the atmosphere. Its concentration has increased from approximately 280 ppm in 1750 to 399 ppm in 2014 (http://co2now.org/current-co2/co2-now/). On the other hand, aerosols, including those of anthropogenic origin, have a more limited effect on warming [LET 03] depending on whether their radiative behavior is diffusant or absorbent. It is estimated that approximately 2,370 Tg of aerosols are emitted into the atmosphere every

---

Chapter written by Pierre-Yves FOUCHER and Xavier BRIOTTET.

year, with the contribution of anthropogenic aerosols being in the order of 17%. This percentage is sufficient to affect the climate as well as to have an effect on health.

Different techniques are used to detect and characterize plumes of atmospheric waste with industrial origin: they can be from ground or remote measurements, sampling or by remote sensing.

This chapter focuses on the study of aerosols and gases using passive remote sensing with the aid of an airborne or spectral imaging space-based system. It is organized in the following way: after recalling the main characteristics of the plumes (section 4.2), their optical properties are introduced and examples for two different kinds of behavior are given in section 4.3. Finally, the methods of detection and characterization of plumes using remote sensing for aerosols and gases are described in sections 4.4 and 4.5, respectively.

The principal radiometric quantities used in this chapter are introduced in [BRI 16].

4.2. Principal characteristics of anthropogenic plumes

4.2.1. *Composition of plumes*

The composition of plumes emitted by factories primarily depends on the industrial sector, the industrial processes applied and the operating phase (for example start-up, routine, shutdown or maintenance). Industries produce a wide variety of pollutants (gas or aerosols) discharged into the atmosphere.

Among the main gaseous constituents of waste, we find gases naturally present in the atmosphere such as water vapor (H_2O), carbon dioxide (CO_2), methane (CH_4), nitrous oxide (N_2O) and rarer gases such as sulfur dioxide (SO_2), nitrogen dioxide (NO_2), hydrogen chloride (HCl), hydrogen fluoride (HF) or volatile organic compounds (VOCs) as well as purely anthropogenic gases such as chlorofluorocarbons (CFCs) and hydrochlorofluorocarbons (HCFCs).

Aerosols are fine particles, solid or liquid, suspended in a gaseous or liquid environment. They are mostly present in the lower layers of the atmosphere (the troposphere), and are primarily found in the boundary layer (between 0 and 2 km). They can come from natural sources (for example dust raised by the wind, ocean spray, particles resulting from burning of biomass or volcanic emissions) or be emitted by human activity (vehicle emissions, industrial wastes or wood heating). We distinguish between primary aerosols, which are particles directly emitted into the atmosphere, and secondary aerosols, which are the product of physicochemical interactions. In effect, these particles form from gaseous compounds (nucleation) that clump together (coagulation) or undergo modifications of their physical properties (dilution and condensation) [RIE 09].

The family of industrial aerosols covers a wide range of dimensions, from a few dozen nanometers to several hundred micrometers, and consists of a complex mixtures of particles. The abundance of these aerosols is also very variable. For example, in 1 year, refineries (cement plant, respectively) emit an average of 100–700 (500, respectively) tons/year of particles into the atmosphere, corresponding to an average emission level of 100 mg/m^3 (considerable variability depending on the type of chimney: between 10 and 3,000 mg/m^3). The discharge from a chimney can vary from 1 to 100 g/s in PM10 (particles whose size is less than 10 μm) for the same type of industry depending on the activity and processes in progress.

Moreover, there is a wide variety of particles emitted and little information about their physical properties is available. This variety is due to various chemical compositions: soot, sulfates, organic matter, metallic oxides [CHO 09, ROG 06] and different shapes.

The majority of anthropogenic sources occur close to ground level. Nevertheless, due to convection movements, these aerosols can be found both in the boundary layer and in the free troposphere. Their residence time in the atmosphere is in the order of 1 to several weeks depending on the size of the particles. The spatial distribution of these particles is directly linked to the location of the sources and meteorological phenomena, which determine the distance that aerosols can be transported as well as their vertical distribution in the atmosphere. When aerosols are eliminated from the atmosphere by dry deposition, particles near the surface are deposited by several physical mechanisms (gravity, turbulence, etc.) in the absence of precipitation, or by leaching when it rains. It follows that the spatial distribution of aerosols is highly variable. There are also particles of secondary origin, formed in the atmosphere by photochemical reaction from precursor gases.

The European Environment Agency (European Pollutant Release and Transfer Register; http://prtr.ec.europa.eu) provides a list of the main pollutants provided by thousands of industries across Europe, organized by region, industrial sector or nature of the pollutant. For example, Table 4.1 provides the information available for the year 2008 on the ArcelorMittal factory in Fos-sur-Mer (France).

Name of pollutant	Amount emitted during the year 2008
Methane (CH_4)	897 t
Carbon dioxide (CO_2)	6,460,000 t
Hydrofluorocarbons (HFCs)	148 kg
Nitrous oxide (N_2O)	123 t
Volatile organic compounds (VOCs)	452 t
Nitrogen oxides (NOx/NO_2)	6,520 t
Sulfur oxide (SOx/SO_2)	6,960 t
Hydrochlorofluorocarbons (HCFCs)	393 kg
Fluorine and inorganic compounds (such as hydrogen fluoride)	12.5 t
Arsenic and its compounds (As)	40 kg
Cadmium and its compounds (Cd)	439 kg
Chrome and its compounds (Cr)	195 kg
Copper and its compounds (Cu)	587 kg
Mercury and its compounds (Hg)	69 kg
Nickel and its compounds (Ni)	711 kg
Lead and its compounds (Pb)	4.96 t
Zinc and its compounds (Zn)	1.64 t
Benzene (C_6H_6)	18.7 t

Table 4.1. *Emissions of pollutants into the air (solid and gaseous) by the Arcelor-Mittal factory in Fos-sur-Mer in 2008 (source: European Pollutant Release and Transfer Register)*

4.2.2. *Specifics of observing plumes in an industrial environment*

Industrial environments are characterized by significant spatial variation of the scene due to the large number of materials presenting optical properties and very heterogeneous temperatures. Similarly, anthropogenic gas plumes and aerosols present significant spatial variation in their properties: strong concentration gradients and temperature of the plume, and sometimes very rapid change in the chemical and physical composition. This can lead to real difficulty in modeling local spatial distribution of this type of plume, hence the need for access to means of observation at a fine scale. There are different types of airborne and space-borne instruments for sensing the gas composition of the atmosphere.

The DOAS (differential optical absorption spectroscopy) technique makes it possible to probe the atmosphere and take measurements at very high spectral resolution (typically 1–2 nm) around very narrow absorption bands of the gases with signatures in the ultraviolet and the visible (UV/VIS), 300–600 nm, range. This technique was originally described by Platt *et al.* [PLA 94, PLA 08], and the reader can consult recent work using the Airborne Multi Axis Differential Optical Absorption Spectroscopy (AMaxDOAS) and Airborne Compact Atmospheric Mapper (ACAM) instruments. The principal is to use differential measurement to characterize the radiative impact of the atmospheric layer by spectral analysis in the UV/VIS range, where certain gases of interest for air quality such as NO_2 or SO_2 have signatures, and where certain properties of aerosols are observable. If the results are suitable for global monitoring of air quality, spatial resolution often remains in the order of square kilometer (km^2), since a single measuring point is associated with an entire portion of the landscape acquired by an airborne sensor. To fix this problem of spatialization and spatial resolution, the introduction of imaging spectrometers appears today as a viable solution for obtaining resolutions in the order of a hundred meters. The second main limitation concerns the instrument calibration, since it is necessary to know the radiative contribution of the ground and of the atmosphere (strong link between aerosols and gases).

Satellite instruments involved in characterizing atmospheric composition probe the atmosphere over the entire optical spectrum at very high spectral resolution on a pixel with kilometric ground sampling resolution to retrieve, over the line of sight path, gases of interest for air quality and greenhouse effect. The reader can refer to publications focusing on SCanning Imaging

Absorption spectroMeter for Atmospheric CHartographY (SCIAMACHY) instruments [FRA 05, BUC 05] or even Global Ozone Monitoring Experiment (GOME), Greenhouse Gases Observing Satellite (GOSAT), Orbiting Carbon Observatory (OCO), Infrared Atmospheric Sounding Interferometer (IASI) and Atmospheric Infrared Sounder (AIRS). Because of limitations related to spatial resolution, they will not be studied in this book; the characterization of anthropogenic plumes requires acquisitions at high spatial resolution.

Finally, airborne hyperspectral imagers combine a fine spectral measurement (typically 5 nm) with a large spectral range of metric resolution. The retrieval methods are inspired by the DOAS principle [THO 14] while offering a certain control of the materials present in the scene in order to spatially delimit a gaseous event or specific aerosols. Although they have less radiometric sensitivity, the contribution of spatial information and a wide spectral range are undeniable assets for the characterization of anthropogenic plumes, in particular for decoupling the effects of gases and aerosols. We will subsequently describe in more detail the characterization of the plume by this type of instrument.

4.3. Optical properties of plumes

4.3.1. *Anthropogenic aerosols*

4.3.1.1. *Introduction*

There is a very wide variety of anthropogenic aerosols of industrial origin due to very high diversity of production in the industrial sector. As a result, the composition of an anthropogenic atmospheric plume will depend on the type of industry, the type of filtering of the waste air produced, but also significant and rapid modifications resulting from interections with the atmosphere.

Each year INERIS (National Institute for Study of Industrial Environments and Risks) publishes the French Register of Polluting Emissions (iRep) that lists the data provided by the operators of industrial facilities, water purification plants and livestock production concerning their emissions into the air, water and ground, including the PM10. In 2012, the level of emission of particles with a diameter of less than 10 μm (PM10) in Metropolitan France was 270 kt (10^3 tons, see http://www.citepa.org/fr/).

Every sector contributes to emission of this pollutant. Figure 4.1 shows the contribution to emissions of this pollutant by large industrial sectors for the year 2012 in decreasing order: residential/tertiary (33%); due to combustion of wood, and to a lesser extent of coal and fuel oil; the manufacturing industry (29%), of which 35% is produced by the subsector of construction; agriculture/silviculture (20%); road transportation (14%); other transportation (non-road transportation, 2%) and finally energy conversion (2%).

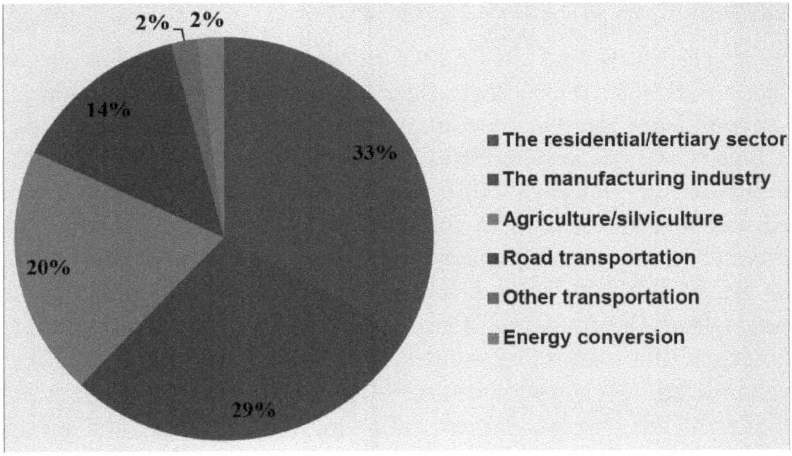

Figure 4.1. *Division Distribution of emissionsof PM10 in France by industrial sector in 2012 (source: http://www.citepa.org/fr/). For a color version of this figure, see www.iste.co.uk/baghdadi/6.zip*

Before presenting the characteristics of some types of industrial aerosols, their main parameters should be recalled. For more details on the definition of optical and radiative parameters, refer to [BRI 16].

Physiochemical analysis is generally carried out on samples collected and then characterized in a laboratory in order to determine their composition, the size distribution of the particles and their shape (for example with an electron microscope).

Knowledge of the optical index is accessible whether in databases (OPAC [HES 98]) of a large set of natural aerosols, or by laboratory measurements on collected samples. For example, Ceolato [CEO 13] has proposed a method that estimates the equivalent index of the environment by hyperspectral and polarized measurementsof the scattering diagramof the cloud of particles, as well as the concentration and the distribution of particle size in a simple diffusion regime.

From this information, their optical properties (extinction coefficients, diffusion and absorption and phase function) are deduced by using, for example, the Mie theory where the hypothesis of spherical particles is valid.

4.3.1.2. *Examples of anthropogenic aerosols*

The properties of aerosols emitted by industry are little-known. Studies have been conducted to estimate their optical and physical properties, but these have focused more on aerosols far from their sources of emission, such as in China [CHE 08], on the East Coast of the United States [RUS 99] or in France at Fos-sur-Mer [MAL 05]. Due to the distance from the source, the studied aerosols correspond to mixtures of natural and industrial aerosols where we cannot determine exactly what kind of industry they come from. Mallet *et al.* [MAL 03] have characterized aerosols collected several kilometers from the industrial sources at Fos-sur-Mer and show that these are composed of a mixture of sulfates, soot and organic matter. Studies have been conducted for aerosols emitted by a power plant in South Africa [ANN 83] and for a plume emitted following the explosion of a storage site in Buncefield, Hertfordshire, in the United Kingdom [MAT 07]. In these two cases, the aerosols studied were largely made of soot. Finally, Moffet *et al.* [MOF 07] published a study on the characterization of industrial metallic aerosols in Mexico MILAGRO (Megacity Initiative: Local And Global Research Observations) and their changes over time. They show that the concentration of lead can reach elevated levels: 1.14 $\mu g/m^3$ in PM10 and 0.76 $\mu g/m^3$ in PM2.5.

The radiative behavior of aerosols (diffusion and absorption) strongly depends on their chemical composition. The following examples give the optical properties of two types of "pure" anthropogenic aerosols: sulfate aerosols with diffusion behavior, and absorbing metallic aerosols.

4.3.1.2.1. The case of diffuse aerosols: sulfates

Table 4.2 shows the optical and physical properties of anthropogenic sulfate aerosols.

Size distribution	Mass density	Extinction coefficient	Single scattering albedo	Asymmetry factor	Hygroscopic property
Monomodal (0.1; 1.5)	1.75 g/cm^3	For a concentration of 1 µg/m^3 in a layer 100 m thick, with a moisture content of 90% (water/aerosol mixture): 4.9×10^{-6} m^{-1}	1.0	0.65	RM = RS (1 – RH/100) $^{-0,28}$ where RS is the "dry" modal radius, RM the "real" radius and RH the relative humidity in %

Table 4.2. *Physical (size distribution, mass density) and optical properties at 550 nm (extinction coefficient, single scattering albedo and asymmetry factor see, [BRI 16] for definitions) of absorbing anthropogenic sulfate aerosols. The monomodal distribution is characterized by the maximum size of distribution (here 0.1 µm) and its geometric standard deviation in µm (here 1.5 µm). The hygroscopic property column describes a model for correcting the "dry" modal radius (RS) depending on the relative humidity (RH)*

4.3.1.2.2. The case of absorbing aerosols: metal aerosols

More recently, Marris *et al.* [MAR 12] have analyzed laboratory samples of atmospheric effluents collected in the industrial area of Dunkirk (France) manufacturing iron and manganese alloys. Several steps are necessary for the manufacturing process that generates these different aerosols. Starting with a mixture of manganese and graphite, the first step is firing followed by cooling in order to transform the raw material into grains that bind to each other and are then crushed. There is a chimney for the firing phase and another for the cooling phase. The second step produces Fe–Mn alloy in a blast furnace at 1,400 °C, which is then cast. A chimney is also present in the casting unit. Sampling was carried out on these three chimneys for aerosols trapped in the filters and sampled in the environment upwind of the chimneys, and at 500 and 1,500 m downwind of the chimneys.

The size distribution of these particles taken from the filters of each chimney is shown in Figure 4.2(a). They show a monomodal distribution, centered around 5 µm for the cooling phase chimney, 1 µm for the firing phase chimney and 70 nm for the chimney of the casting phase. In the ambient air (Figure 4.2(b)), these small particles have a distribution that changes with the distance of the sample. Therefore, downwind, at 500 m, the distribution is bimodal (the first maximum at 60 nm and the other at 18 nm), which has a tendency to become monomodal with distance.

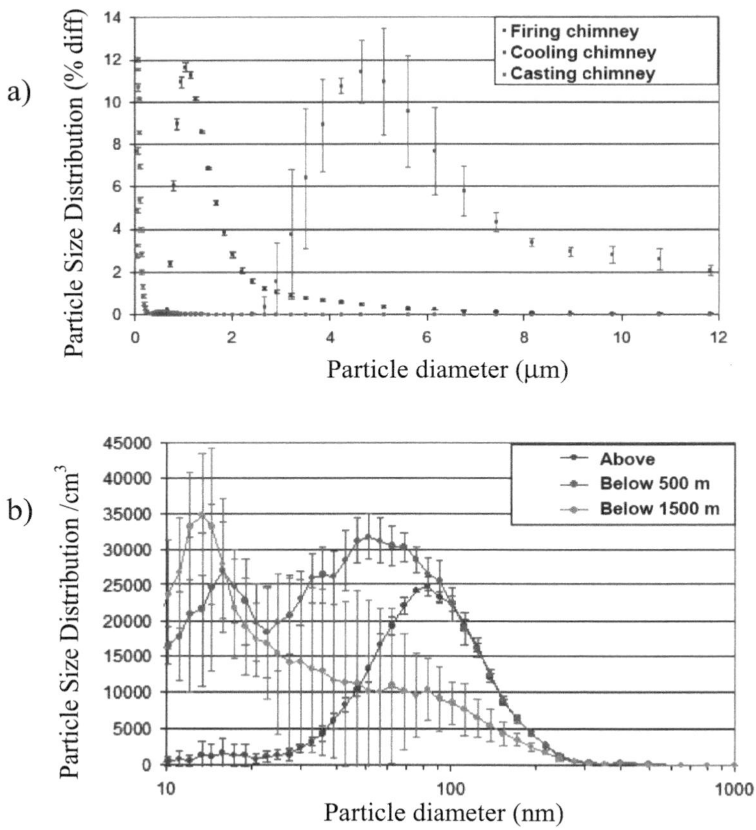

Figure 4.2. *a) Particle size distribution, in normalized volumetric concentration, of the particles collected in the filters located in the three firing, cooling and casting chimneys [MAR 12]; b) Average particle size distribution in terms of number of particles sampled in the environment upwind of the chimneys, and at 500 m and 1,500 m downwind of the chimneys [MAR 12]. For a color version of this figure, see www.iste.co.uk/baghdadi/6.zip*

The chemical analysis carried out by the Laboratory for Physico-Chemistry of the Atmosphere (LPCA) [MAR 12] is shown in Figure 4.3.

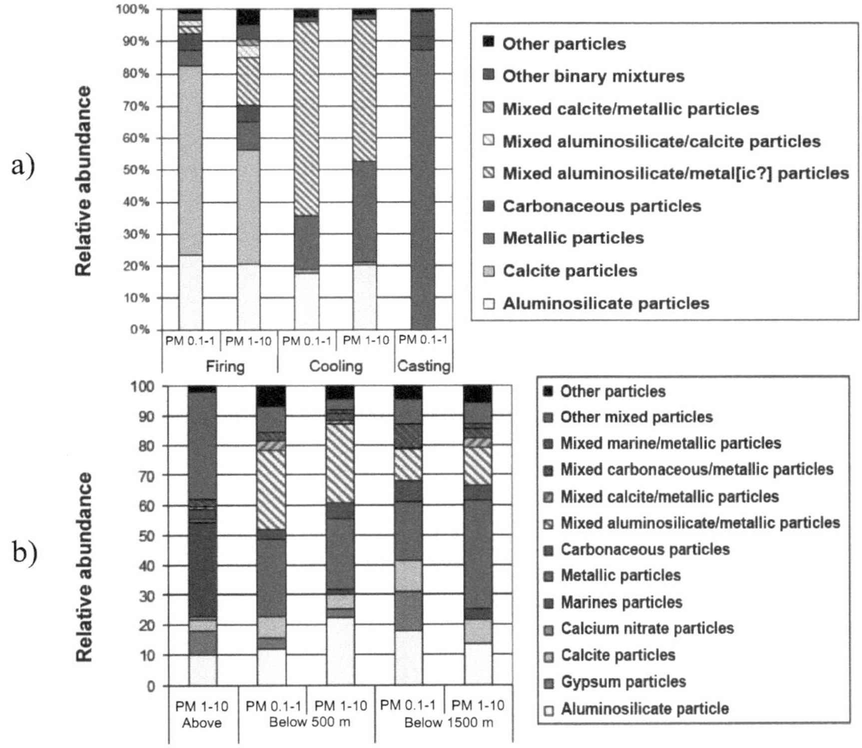

Figure 4.3. *Chemical analysis of samples taken collected in the filters (a) and in the ambient air (b). Analysis carried out by the Laboratory for Physico-chemistry Chemistry of the Atmosphere (LPCA, France), where PM represents all of the particles with diameters included in the range given in micrometers (μm).For a color version of this figure, see www.iste.co.uk/baghdadi/6.zip*

At the chimney level (Figure 4.3(a)), following each phase the ratios of the mixture change in a very significant way. For the firing phase, the composition of aerosols is essentially dominated by mixed aluminosilicate/calcite particles; for the cooling phase, by mixed aluminosilicate/metallic particles and for the casting phase, by metal particles. In the ambient air (Figure 4.3(b)), the mixture is essentially made up of metallic particles and mixed aluminosilicate/metallic particles whose relative abundance decreases with distance. We also note the emergence of a mixture of anthropogenic aerosols with marine aerosols whose presence increases with distance.

Measurements of the optical index have been carried out at different wavelengths (Table 4.3). Aerosols coming from the firing chimney are mostly made up of aluminosilicates, which are not very absorbent (null imaginary part of the optical index). However, for aerosols coming from the cooling chimney, the measured absorption is significant in terms of the presence of metal particles, aluminosilicates and metallic/aluminosilicate aggregates that strongly absorb light in the visible and near-infrared range.

Wavelength (nm)	Index of refraction of the sample collected in the firing chimney	Index of refraction of the sample collected in the cooling chimney
532	1.55	1.85 + 0.3i
633	1.55	0.90 + 0.4i

Table 4.3. *Refraction indexes [CEO 13]*

Using these data, Deschamps *et al.* [DES 13] have modeled the radiative properties of these aerosols. The extinction coefficient of the aerosols (Figure 4.4) emitted by the firing chimney is very different from that obtained for the other models due to the quasi-absence of metallic particles. The presence of these particles has the effect of increasing absorption and therefore reducing the single scattering albedo. The radiative properties of the aerosols sampled in the cooling chimney, at 500 and 1,500 m, are very similar, since this chimney is the principal emitter of aerosols.

4.3.2. *Anthropogenic gases*

The composition of the plumes emitted by the factories depends primarily on the sector of activity and industrial processes. A wide variety of pollutants are discharged into the atmosphere by industries. Some of these pollutants are presented in the following sections, grouped by family according to the current regulations, their physiochemical characteristics or their effects.

a)

b)

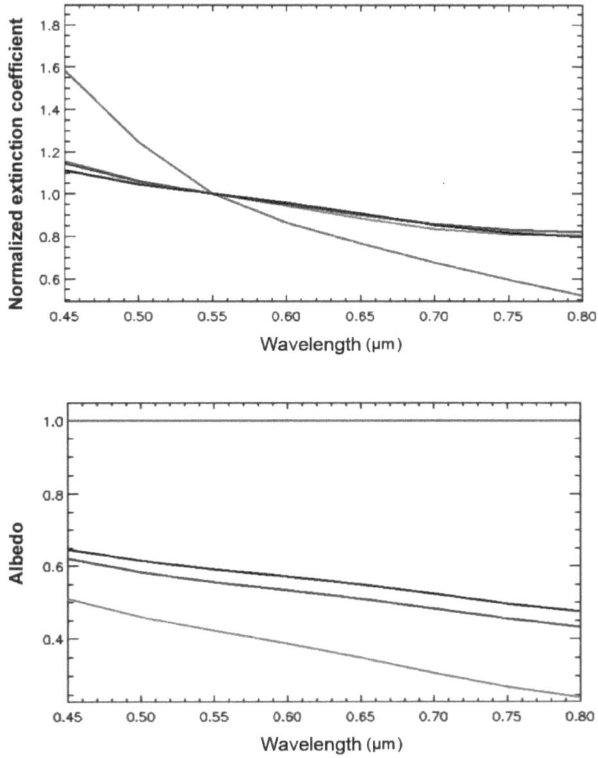

Figure 4.4. *Radiative properties obtained for mixtures of mixed aerosols emitted from the factory (red: output from the firing chimney, green: output from the cooling chimney, blue: at 500 m downwind, black: at 1,500 m downwind): a): extinction coefficients normalized by their values at 550 nm, b): single scattering albedo [DES 13]. For a color version of this figure, see www.iste.co.uk/baghdadi/6.zip*

4.3.2.1. *Gases of interest*

The emission of some of these gases is subject to European regulations. The reader can refer to the INERIS and ADEME (French Agency for Environment and Energy Management) sites to access this list. Following is the non-exhaustive list of the main gases whose observation is relevant to reports on air quality:

– *Carbon dioxide (CO_2)* is well known for its role in global warming. The combustion of fossil fuels or biomass is the main source of CO_2 emission.

This combustion is found in industry, transportation or domestic heating. This gas is also emitted naturally during forest fires, volcanic eruptions and during respiration by living beings;

– *Sulfur dioxide (SO_2)* is a gas that is primarily produced by the use of sulfurous fossil fuels (coal, fuel oil, gas oil, etc.). The production of electricity and oil refining are the sectors that emit the most SO_2. Depending on the exposure and the concentrations involved, this gas can cause health problems ranging from simple respiratory obstruction to death by respiratory failure. It is also the source of sulfate aerosols;

– *Nitrogen oxides (NO_x)* include nitrogen monoxide (NO), nitrogen dioxide (NO_2) and nitrogen tetroxide (N_2O_4). The principal sources of NO_x emission are anthropogenic, particularly the combustion of fossil fuels and emissions from diesel vehicles. They also originate from other industrial processes such as the production of nitric acid or the manufacturing of fertilizer;

– *Ammonia (NH_3)* is a pollutant usually linked to agricultural activity (organic waste from livestock production), but it can also be caused by the use of cars equipped with a catalytic converter;

– *Carbon monoxide (CO)* is emitted after incomplete combustion of combustibles or fuels. It is also found in the emissions of incinerators or car mufflers;

– *Volatile organic compounds (VOCs):* The family of VOCs includes a large number of gases with extremely variable characteristics. They consist of (according to the Council Directive of 1999/13/EC of the European Council of March 11, 1999) gases that contain at least the element carbon as well as one or several of the following elements: hydrogen, halogens, oxygen, sulfur, phosphorus, silicon or nitrogen. There are several subfamilies of VOCs:

- alkanes: propane (C_3H_8), butane (C_4H_{10}), pentane (C_5H_{12}), etc.
- alkenes: butene (C_4H_8), pentene (C_5H_{10}), etc.
- dienes and terpenes: butadine (C_4H_6), hexadien (C_6H_7), etc.
- oxygen compounds: aldehydes, ketones, esters, alcohol, etc.
- mono or polycyclic aromatics: benzene (C_6H_6), toluene (C_7H_8), etc.

Among the aromatics, polycyclic aromatic hydrocarbons (PAH) are particularly monitored because of their chemical and toxicological characteristics. They are often toxic, mutagenic and/or carcinogenic. There are various sources of VOC emissions. They can involve the phenomena of combustion, evaporation of solvents (present in paint, dyes, glues, removers and cosmetics), evaporation of organic matter (issued from fuels, for example) or biological reactions. The residential and construction sectors are the activity sectors that emit the most VOCs due to the use of paint (http://www.citepa.org/fr/air-et-climat/techniques-de-reduction/mmesures-primaires-et-secondaires-de-reduction-des-emissions-de-covnm/28-categories-francais/pollution-et-climat/polluant-et-ges/aep).

4.3.2.2. *Ejection of plumes*

Depending on the type of industry, the compounds emitted will vary widely and the emissions are either channeled (chimneys), or diffuse (storage of hydrocarbon, for example). The discharge involved is traditionally somewhere between 1 g/s (case of insignificant leaks) to more than 1 kg/s for the largest chimneys (or in the case of accidents). The temperature of ejection plumes can still have strong variation: ambient air temperature for diffuse emissions, temperature below the ambient temperature for leaks with depressurization, temperature of 150–500 °C for emissions from combustion ejection chimneys (refining, cement works, incinerators and thermal power plant).

Table 4.4 provides an example of average discharge of plumes from the refineries in Europe. We invite the reader to consult reference documents on the emissions from different industries (http://eippcb.jrc.ec.europa.eu/reference/BREF/REF_BREF_2015.pdf) for more details.

Gas	Unit	Average amount	Range
C_2H_6O	kg/h	300	0–500
C_2H_6	kg/h	1,000	0–2,000
C_2H_4	kg/h	200	0–500
C_3H_6	kg/h	150	0–300
NH_3	kg/h	350	0–500
C_6H_6	kg/h	NA	0–2

Table 4.4. *Example of average discharge at the level of refinery storage in Europe*

The relationship between the ejection parameters and the concentrations in the plume is strongly dependent on the meteorological conditions and the configuration of the scene. An example of one of the simplest models is represented by a Gaussian function in the form:

$$C_g(x,y,z) = \frac{Q_g}{2.\pi.\sigma_y.\sigma_z.u} e^{-\frac{y^2}{2.\sigma_y{}^2}} \left[e^{-\frac{(z-H)^2}{2.\sigma_z{}^2}} + e^{-\frac{(z+H)^2}{2.\sigma_z{}^2}} \right] \qquad [4.1]$$

where the parameters (σ_y, σ_z) define the lateral and vertical extensions of the plume according to the meteorological conditions, and for which there are pre-calculated tables [ZAN 90]; H is the average height of the plume, which depends on the height of the chimney, the speed of the vertical ejection, the ejection temperature, and the diameter of the chimney; u is the speed of the wind and Q_g is the emission rate of gas g. There are different tools and software programs for calculating the spatial evolution of the concentration from the ejection parameters [CAR 94].

4.3.2.3. *Absorption, transmission and emission*

A gas plume will interact with electromagnetic radiation by absorption or emission according to its thermodynamic state and to the range of wavelengths. The absorption of a gas g is described by its absorptivity coefficient A_g that measures the capacity of a body to absorb the energy crossing it. This absorption is defined from the normalized effective cross-section, k_g (in ppmv^{-1}·.m^{-1}), of the ratio between the volume of gas and the volume of air c_g (in ppmv) defined by the crossed area by the path length l (in m):

$$A_g = c_g.l.k_g \qquad [4.2]$$

According to Kirchhoff's law, the emissivity ε_g of a gas is equal to its absorptivity [BRI 16]: $\varepsilon_g = A_g$.

These parameters A, ε and k depend on the wavelength, but for reasons of simplicity the latter will be systematically omitted in this chapter.

This spectral absorption is calculated for the simplest gases from databases detailing the position and the intensity of the absorption lines such as HITRAN [ROT 12] or GEISA [JAC 09]. Figure 4.5 shows the spectral normalized effective crosos-section k_g for different gases.

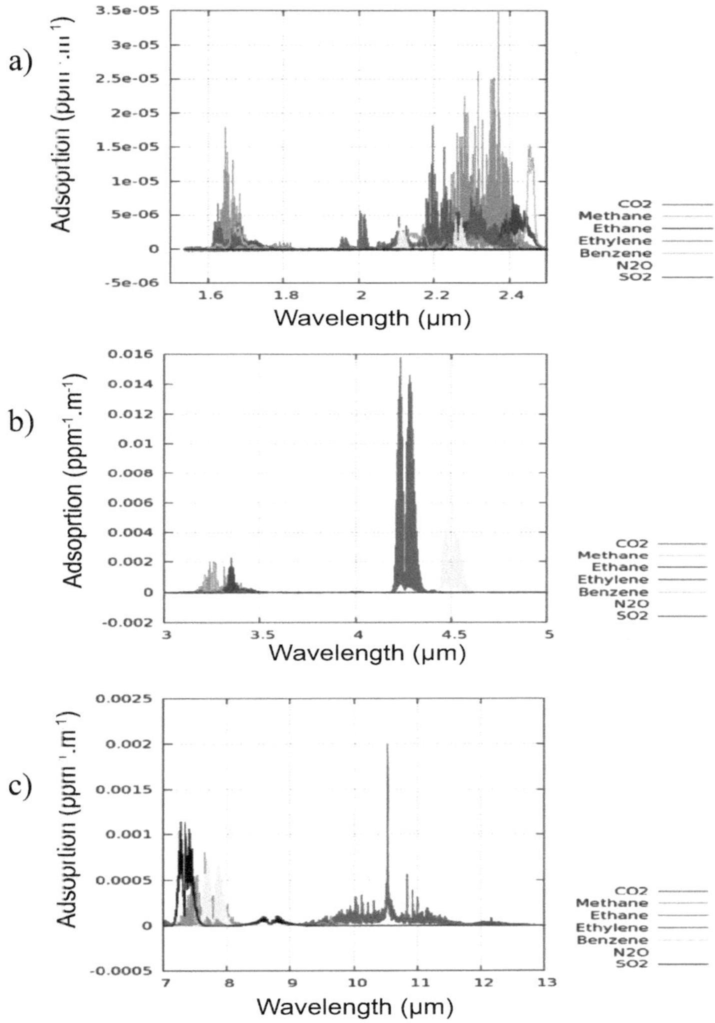

Figure 4.5. *Normalized absorptivity coefficients of different gases of interest for the three infrared atmospheric windows: a) SWIR: short wave infrared (1.5—2.5 µm); b) MWIR : mid wave infrared (3—5 µm); c) LWIR: long wave infrared (7—13 µm).For a color version of this figure, see www.iste.co.uk/baghdadi/6.zip*

The transmission t_g of a gas whose average volume mixing ratio (c_g) and length (l) is then written:

$$t_g = e^{-c_g.l.k_g} \quad [4.3]$$

Figures 4.6 and 4.7 show the transmission of different gases of interest for integrated concentrations ranging from 100 to 2,000 ppm.m according to the compounds (other figures are presented in the following paragraphs). These figures clearly illustrate that each gas has its own spectral signature and that the intensity of these signatures varies considerably between gases, for the same gas and between spectral ranges. Moreover, they also show the problem of interferences between the gases. This can occur in certain spectral ranges.

a)

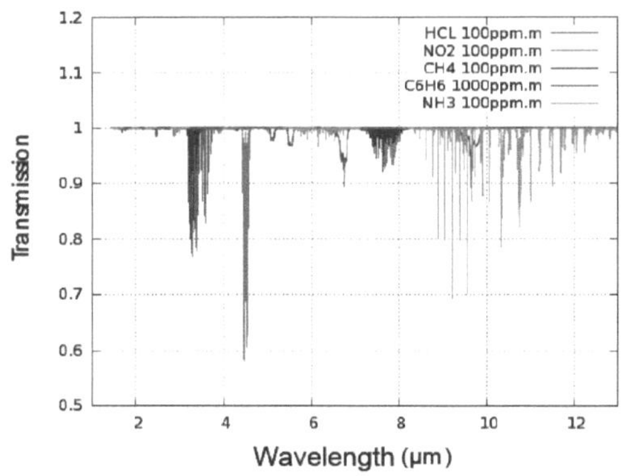

b)

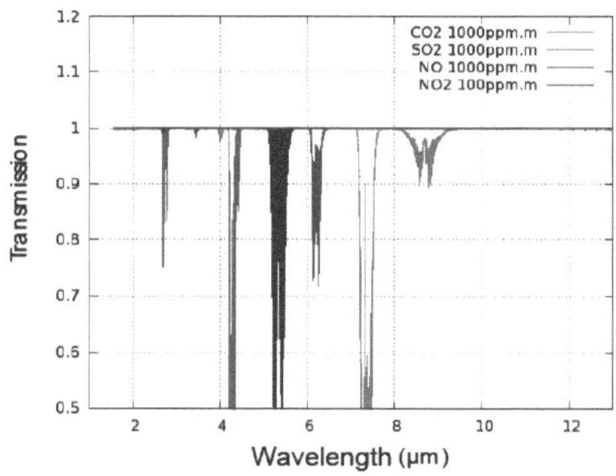

Figure 4.6. *Examples of transmission of the main gases of interest. For a color version of this figure, see www.iste.co.uk/baghdadi/6.zip*

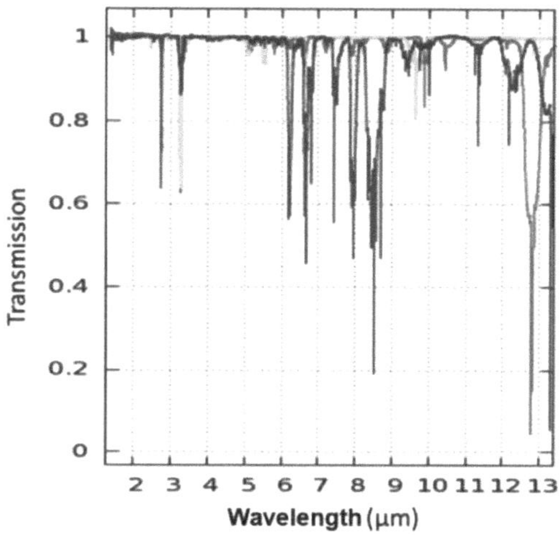

Figure 4.7. *Examples of transmission of polycyclic aromatic hydrocarbons (PAH). For a color version of this figure, see www.iste.co.uk/baghdadi/6.zip*

4.4. Characterization of aerosol plumes using remote sensing

4.4.1. *Modeling of the spectral signature of aerosols in the reflective domain*

The signal measured by a remote sensing instrument in the 0.4–2.5 μm range will be described by posing the following hypotheses:

– the ground surface is homogeneous, flat and characterized by a Lambertian reflectance;

– the plume is described as a flat, homogeneous layer;

– thermal emissions of gases are ignored (this hypothesis is verified when the temperature of the plume does not exceed 500 K).

In the presence of a clear sky, the expression of the total spectral radiance L_T measured by a space-based or airborne sensor viewing a surface of reflectance ρ is in the form [BRI 16]:

$$L_T = L_{atm,m} + \frac{\mu_s.E_{ground}}{\pi}\frac{t_{atm}.\rho}{1-\rho.s} \quad [4.4]$$

with E_{ground} the total incident irradiance, s the spherical albedo of the atmosphere, $L_{\text{atm},m}$ upwelling atmospheric radiance, t_{atm} on the upwelling path and μ_s the cosine of the solar zenith angle.

In the presence of a plume, the expression of the total spectral radiance $L_{T+\text{plume}}$ becomes:

$$L_{T+plume} = L_{atm,m} + L_{plume} + \frac{\mu_s.E_{ground}}{\pi}\frac{t_{atm}.t_{plume}.\rho}{1-\rho.[s+s_{plume}]} \qquad [4.5]$$

where L_{plume} is the spectral radiance due to the plume's own diffusion, t_{plume} is the total transmission of the plume (upwelling and downwelling) and s_{plume} its spherical albedo.

If the assumption of simple scattering can be applied, the spectral radiance of the plume can be written as [KAU 99]:

$$L_{plume} = \frac{E_{ground}}{4\pi}\frac{1}{\mu_s+\mu_v}t_{atm}.\omega_0.P(\Theta).e^{-\left[c_g.l.k_{ext}\left(\frac{\mu_s+\mu_v}{\mu_s.\mu_v}\right)\right]} \qquad [4.6]$$

with ω_0 the single scattering albedo, $P(\Theta)$ the phase function of the plume and μ_v the cosine of the view zenith angle.

Assuming that the plume has a very limited extension, the diffuse contribution of the plume is ignored, the global radiative impact of the plume is then written from the preceding equations and some manipulations:

$$\Delta L = L_{T+plume} - L_T \approx L_{plume} + \frac{\rho}{\pi}\left(E_{ground}\Delta t_{atm} + t_{atm}\Delta E_{ground}\right) \qquad [4.7]$$

with:

$$\Delta t_{atm} = t_{atm}\left(t_{plume} - 1\right) = c_g.l.k_{ext}.T_{atm}\left(\frac{1}{\mu_v} - \omega_0\right) \qquad [4.8]$$

and:

$$\Delta E_{ground} \approx c_g\, l\, k_{ext}E_{ground}\left(\frac{1}{\mu_s} - \omega_0\right) \qquad [4.9]$$

k_{ext} is the normalized extinction coefficient.

This formulation, proposed by Kaufman [KAU 99], shows limitations that are linked in particular to not taking into account the diffuse component of equation [4.6].

In this expression, we can see the strong dependence on ΔL with the value of ground reflectance. For ground surfaces with low reflectance, the component resulting from its own diffusion is preponderant (and positive), and the differential signal will almost always be positive. For a very reflective surface, there is then competition with the extinction term. The term ΔL is therefore generally negative. In this case, the absorbing plumes (soot in particular) will have a tendency to reduce the signal received as sensor input.

4.4.2. *Example of differential radiance ΔL*

Taking into consideration a standard *mid-latitude summer* atmosphere made up of rural aerosols, Figure 4.8 shows the total radiances as a result of the diffusion of ambient particles only (term L_{atm}). Based on this case, Figure 4.9 shows the impact on a pixel of the presence of a plume of absorbing aerosol(in the case of soot) or diffusant (in the case of sulfate aerosols), and the radiance differentials introduced by an aerosol plume.

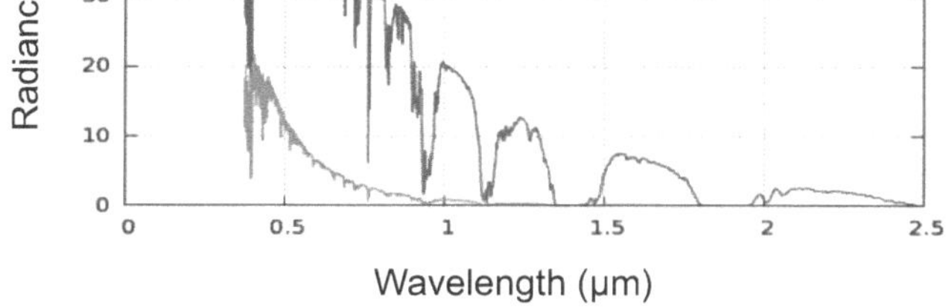

Figure 4.8. *Upwelling total and diffuse radiances for a Mid-Latitude Summer atmosphere, with rural aerosols of visibility 23 km. For a color version of this figure, see www.iste.co.uk/baghdadi/6.zip*

For a layer of 80 m thickness composed of sulfate aerosols and an optical depth of 0.04 at 550 nm ($k_{ext\_550\ nm}$ = 0.5) located above a material with homogenous reflectance equal to 0.1, the resulting spectral radiance $L_{T+plume}$ is higher than the radiance without a plume. Therefore, the radiance differential is increased to a maximum of 2 W/m^2/sr/µm, which is a gain of around 1.5%.

In the case of a layer of aerosols of fine soot, which also has a thickness of 80 m and an optical depth of 0.04–550 nm located above a material with homogenous reflectance equal to 0.1, the resulting spectral radiance $L_{T+plume}$ is less than the radiance with a plume. In this case, the radiance differential is reduced by a maximum of 3 W/m^2/sr/µm, which is a loss of around 2%.

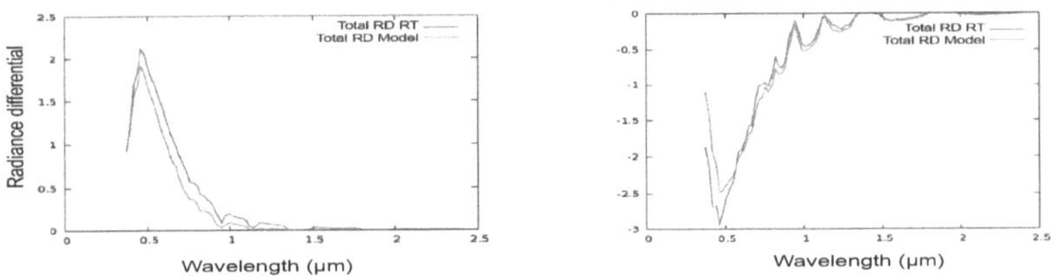

a) b)

Figure 4.9. *Radiance differential "RD" (in %) caused by the presence of a plume of sulfate aerosols; a) and soot aerosols; b). Green: example of impact modeled from the intrinsic properties of the particles. Red: exact calculation of multiscattering radiative transfer model (RT). For a color version of this figure, see www.iste.co.uk/baghdadi/6.zip*

4.4.3. *Characterization of aerosol properties*

As we mentioned earlier, aerosols have a marked radiative impact for short wavelengths (0.4 –2.5 µm): it is therefore in this range that we find the majority of methods used for characterizing them. Due to the large variety of aerosols and no trivial link between the signature of a plume and the microphysical properties of the aerosols, which make it up, pre-calculated tables are commonly used (Look-up-Table (LUT)), belonging to each type of aerosol, each type of ground and each instrument for analyzing plumes.

Therefore, Alakian *et al.* [ALA 09] developed the L-APOM (LUT – Aerosol Plume Optical Model) method to determine the concentration of particles in a biomass fire plume, as well as their average size and the proportion of soot in the mixture of aerosols from airborne hyperspectral acquisitions. Its method consists of comparing the measured radiance by each pixel to a Look-up-Table (LUT) of synthetic spectral radiances corresponding to different types of aerosols and atmospheres. The selected plume composition is the one for which aerosols minimize the bias between these two radiances. The selected spectrum defines the microphysical properties of aerosols. Spatial constraints have also been introduced to limit disparities in the properties obtained for near pixels.

4.5. Characterization of gas plume properties

4.5.1. *Introduction*

Characterizing gas plumes in industrial environments by optical remote sensing requires the use of instruments that allow the detection of spectral shape while providing a good spectral sensitivity in order to assure good characterization of the properties of the plume such as type of gas and abundance. The following sections will lead us to consider the question of detectability of a gas plume in an industrial environment and its characterization according to the different spectral ranges based on specific gases such as methane, carbon dioxide or sulfur dioxide. This section concerns the study of the detectability of a gas plume depending on the spectral range being considered. It is based on a formulation of the radiative transfer equation in order to emphasize a particular pixel's signature that is to say in a line of sight, due to the presence of a specific layer of gas between the remote sensor and the ground. For each spectral range, we define the conditions allowing the detection of a gas and tools to extract its spatial dimension.

4.5.2. *Reflective domain*

In the reflective domain (0.4–2.5 μm), the impact of a gas plume from a radiative point of view occurs mostly via the absorption of decreasing solar fluxes.

4.5.2.1. *Equation of radiative transfer in the presence of gas*

By taking equation [4.5] and considering that the scattering elements of the molecules in the plume are negligible in view ofthe natural molecular diffusion of the atmosphere (this last hypothesis is not always valid in the UV range for dense plumes), the equation above is simplified, and the corrected L'_{plume} radiances of the atmosphere become:

$$L'_{plume} = \frac{\mu_{s.}E_{ground}}{\pi}\frac{\rho.t_{plume}}{1-s.\rho} \quad [4.10]$$

$$\Delta L' = L'_{T} - L'_{T+plume} = \frac{\mu_{s.}E_{ground}}{\pi}\frac{\rho}{1-s.\rho}\left(1 - t_{plume}\right) \quad [4.11]$$

with t_{plume} the total transmission of the gas plume (upwelling, downwelling, direct and diffuse).

Therefore, by successively measuring the radiance with and without the presence of a plume, it is possible to estimate this radiance differential. The operation is carried out either from measurements acquired at the same scene but at different times, or from the recording of an image positioned on and outside of the plume.

It can be clearly seen that it is the transmission of the target gas that contains signature of the plume. The spherical atmospheric albedo being rather weak, the $\Delta L'$ signature in transmission is directly modulated by ground reflectance. For a signal-to-noise ratio and given lighting conditions, the detectability will therefore be conditional on the value of the transmission of the plume and the type of ground materials. Thus, the same plume, above a landscape with a higher reflectance, will be more easily detected:

$$\Delta L' \approx \frac{\mu_{s.}E_{s}}{\pi}\rho\left(1 - t_{plume}\right) \quad [4.12]$$

4.5.2.2. *Transmission of the target gas in the reflective domain*

The transmission t_{plume} is the product of downwelling and upwelling transmissions due only to gaseous absorption. In the case of fine layers where the absorption is far less than 1, the transmission of a plume with thickness h, concentration c_g in the direction θ along a path j (upwelling or downwelling) for a gas g can be linearized as:

$$t_{plume,j,g} = e^{-c_g \frac{h}{cos\theta} k_g} \approx 1 - c_g \frac{h}{\mu} k_g \quad [4.13]$$

with c_g (ppmv) the volume mixing ratio of the gas (ratio between the volume of gas and the volume of the air), integrated into the layer with a thickness h (the length of the path $l = h/\cos(\theta)$), k_g the spectral absorption coefficient of the gas being studied ($m^{-1}\cdot ppmv^{-1}$), θ the view zenith angle for the upwelling transmission and the solar zenith angle for the downwelling transmission and μ the cosine of the zenith angle being considered.

In the presence of several gases, the total transmission on a path j is the product of the transmission of each gas g:

$$t_{plume,j} \approx 1 - \frac{h}{cos\theta} \sum_{g=1}^{ngas} c_g k_g \quad [4.14]$$

with $n$gas the number of gases present in the layer.

We subsequently see that q_g, the integrated value in the column gas g, and the total transmission are written as:

$$t_{plume} \approx 1 - \frac{\mu_s + \mu_v}{\mu_s \mu_v} \sum_{g=1}^{ngas} q_g k_g(\lambda) \quad [4.15]$$

Figure 4.10 shows, for different gases of interest having non-null absorption in the reflective domain, the value of the upwelling transmission corresponding to an integrated value of 1,000 ppm.m, whether 1 ppm of 1,000 m thickness or 10 ppm of 100 m thickness. We can see that detectability is a function of the absorptivity of the gas. It should be noted that the case of CO_2 is distinctive, since it is naturally present in relatively strong concentrations in the atmosphere, typically 400 ppm throughout the troposphere.

4.5.2.3. *Expression of differential radiance*

Starting with the preceding differential expression [4.12] and by using the linearized expression of the transmission [4.15], the signature of a plume (in the case of a thin transmission layerclose to 1) is written as:

$$\Delta L' \approx \rho \frac{E_s}{\pi} \left(\frac{\mu_s + \mu_v}{\mu_v} \right) \sum_{g=1}^{ngas} q_g k_g \quad [4.16]$$

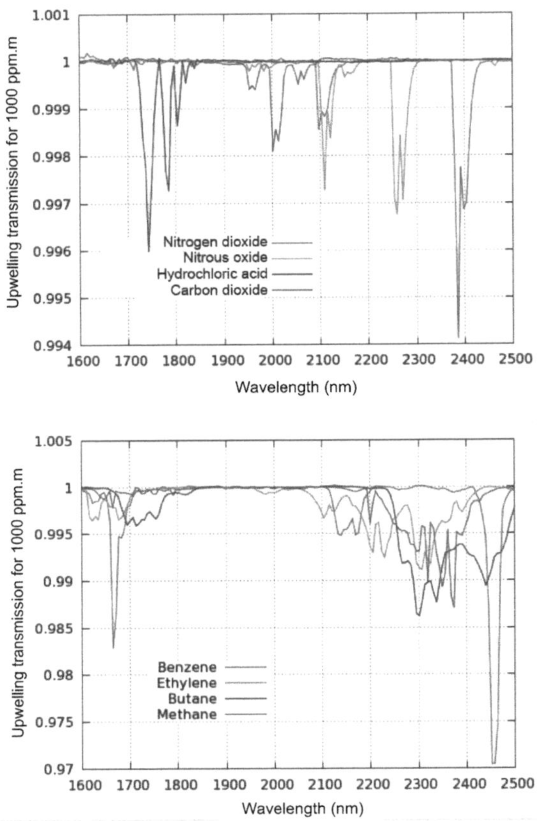

Figure 4.10. *Transmission in the reflective domain of near near-infrared for different anthropogenic gases. For a color version of this figure, see www.iste.co.uk/baghdadi/6.zip*

We recall the importance of the ground reflectance on the differential signal and the necessity of being familiar with the properties of the ground during the processes of detection and characterization.

4.5.2.4. *Illustration using a semi-synthetic example*

Figure 4.11 shows the 1,650 nm band of a hyperspectral image in radiance acquired from an urban/industrial zone. In this image, a plume of methane corresponding to a discharge of 10 g/s has been added synthetically. This plume is not observable by using only the band at 1,650 nm, but is nevertheless characteristic of a minimum of methane transmission as shown in Figures 4.10 and 4.12(b). Figure 4.12 shows two spectra characteristic of the image presented (Figure 4.12(a)) and the transmission values of methane (noted tau CH_4 in the figure) for a pixel close to the source and at 100 m from the source (Figure 4.12(b)).

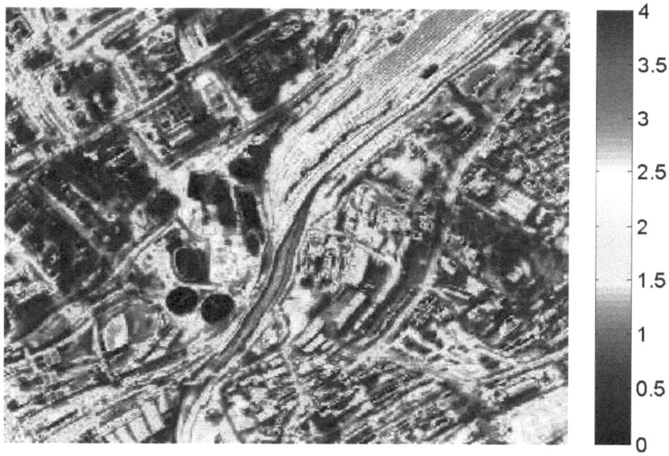

Figure 4.11. *Image of an industrial area (Refinery refinery in Provence, France) showing a plume of CH4. Radiance values (W.m^{-2}.sr^{-1}.nm^{-1}) observed by the Hyspex instrument in the 1,650 nm band corresponding to a local minimum of the transmission of methane. For a color version of this figure, see www.iste.co.uk/ baghdadi/6.zip*

Figure 4.13 shows for a pixel close to the source, the radiance reference in the sensor input (red) and the radiance with the attenuation due to the plume (green) in the case of ground with low reflectance (dark pixel) as well as close up around 1,650 and 2,350 nm. Figure 4.14 shows the radiance (Figure 4.14(a)) and the differential signal observed (Figure 4.14(b)) close to the source for the case of a pixel with high reflectance (light material) and with low reflectance (dark material on the ground), respectively. We can see that the signature of the plume even close to the source can be rather weak and disturbed by instrumental noise for low reflectance. Then, as shown in

Figure 4.14(b), a surface with a reflectance that is two times higher allows detection with an amplitude three times greater, nevertheless methane quantity is two times lower. We can see, in particular, in the last graph that the level of noise (characteristic of a hyperspectral sensor) added to the images with the plume is in the order of 0.01 in radiance.

a)

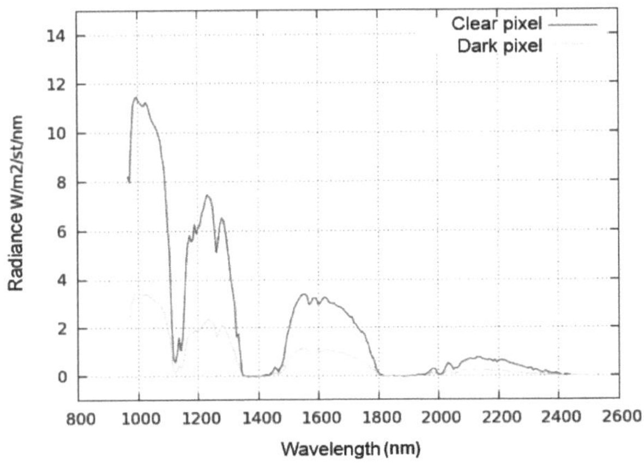

b)

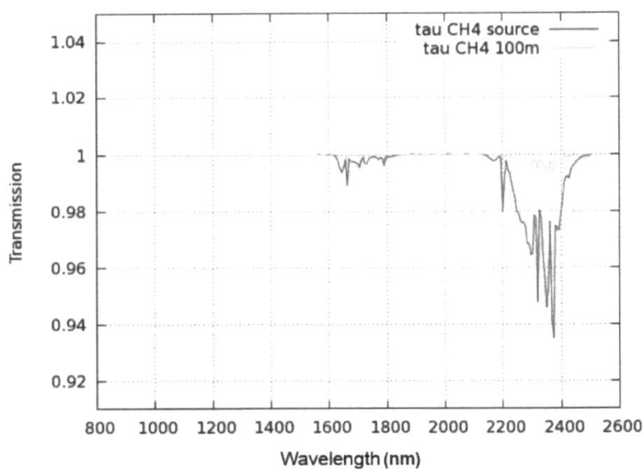

Figure 4.12. *a) Typical spectral radiance of the image area for materials with low reflectance (dark pixel), and higher reflectance (light pixel); b) Transmission (noted tau) in the methane plume close to the source (red) and at 100 m from the source (green).For a color version of this figure, see www.iste.co.uk/baghdadi/6.zip*

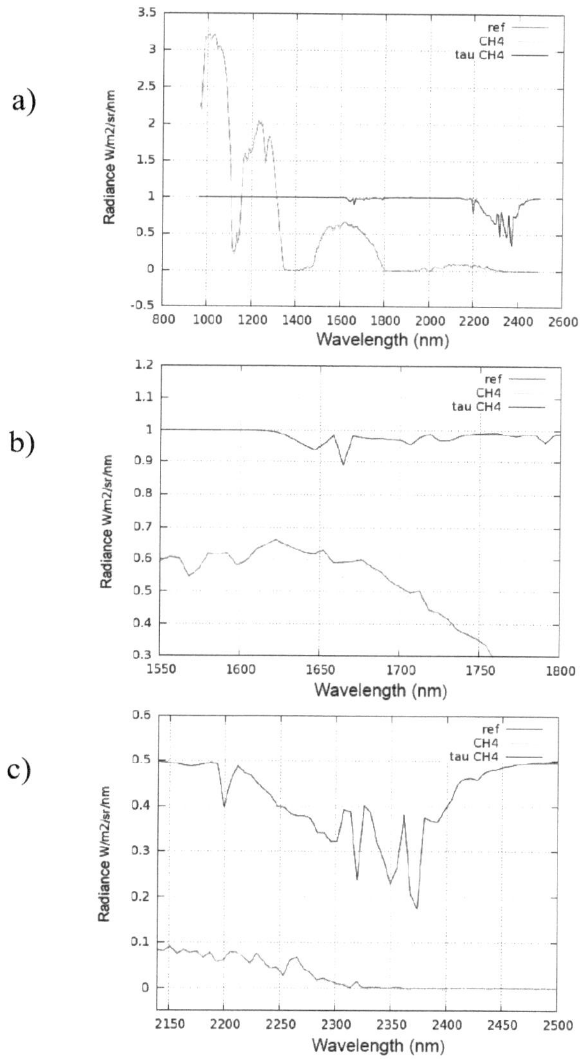

Figure 4.13. *Spectral radiance of the plume on a pixel with low reflectance; a) Standard reference spectrum close to the source in red, spectrum with the signature of methane in green and transmission of methane (tau noted CH_4) in blue; b) zoom around 1,650 nm on the first spectral absorption zone of methane, we can see a transmission peak at 0.9 for the wavelength 1,660 nm; c) Zoom around 2,350 nm on the second absorption zone of methane around 2,350 nm (the spectrum in transmission has been offset by 0.5 for the purposes of representation). So the transmission shows a different minimum of 2,320, 2,350 and 2,370 nm. For a color version of this figure, see www.iste.co.uk/baghdadi/6.zip*

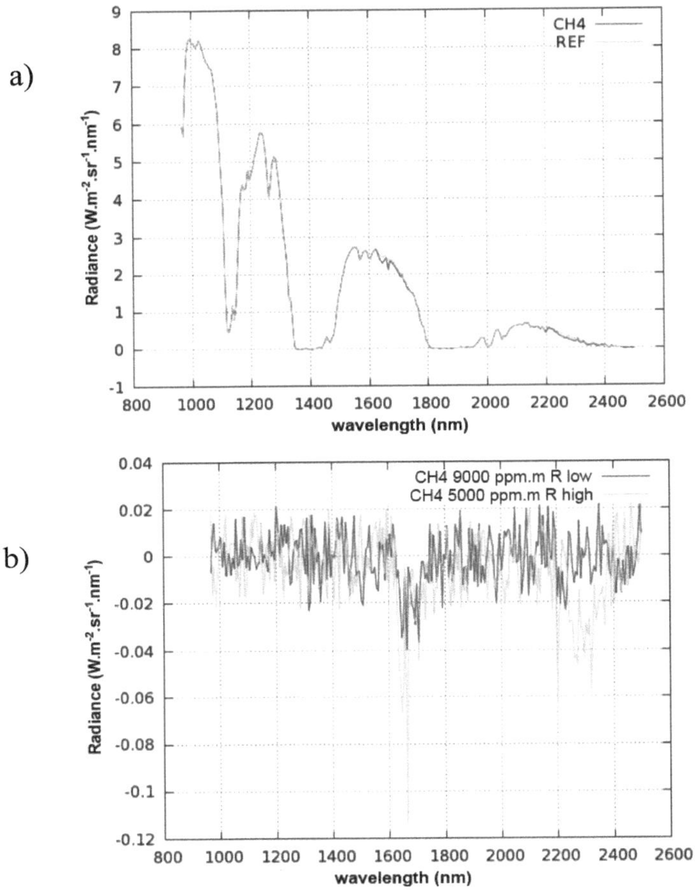

Figure 4.14. *a) Spectral radiance of the plume on a reference pixel with high reflectance close to the source (red) and containing the signature of methane (green); b) Difference in radiance: spectrum without methane – spectrum with methane, for an integrated concentration of CH_4 of 9,000 ppm.m, on a ground with low reflectance (red) and for a concentration of 5,000 ppm.m on a ground with sizeable reflectance (green). For a color version of this figure, see www.iste.co.uk/baghdadi/6.zip*

Therefore even if the band around 2,350 nm seems the most favorable since methane absorption is the highest, due to the relatively low radiance (decrease in solar irradiance) and to the degradation of the signal-to-noise ratio, the band at 1,650 nm shows detection results roughly similar.

4.5.2.5. *Process of detection/characterization*

As shown earlier, without *a priori* knowledge of the ground, detecting a plume proves to be very difficult. It is necessary to perform a preliminary step in order to estimate the background properties, that is to say the reference spectrum for each pixel. The solution commonly allowed in hyperspectral imaging is to project this image onto a new base B represented by L^B. There are several possible bases: vectors belonging to the pixels in the image (assuming that the pixels containing gas have little influence on this calculation), simulated bases from physical data of the properties of the ground, from which mean vectors coming from a classification of the image. A base B is defined by P vectors b_i; the L^B background radiance is expressed as follow:

$$L^B = \sum_{i=1}^{P} \alpha_i . b_i \qquad [4.17]$$

or in vectorial notation:

$$L^B = \overline{\alpha} . \overline{B} \qquad [4.18]$$

In the presence of a plume, it is therefore possible to write radiance as the sum of the background radiance L^B and an additive term proportional to the quantity of gas q_g:

$$\begin{aligned} L^B_{plume} &= L^B(1 - t_{plume}) = \bar{\alpha}\,\bar{B} - q_g S + b \\ S &= f(\mu_s, \mu_v, \bar{\alpha}, \bar{B}, k_g) \end{aligned} \qquad [4.19]$$

The vector S therefore corresponds to the intrinsic signature of the gas and b corresponds to the error in reconstructing the image with a plume. Assuming a Gaussian instrumental noise, characterized by its covariance matrix Σ, an optimal linear solution is therefore:

$$q_g = \frac{S.\Sigma^{-1}(L^B - \overline{\alpha}.\overline{B})^T}{S.\Sigma^{-1}.S^T} \qquad [4.20]$$

Note that in practice we carry out a pre-classification of the image and estimate α, S and Σ by the class of soil type [FUN 01], in order to obtain a detection that is rapid and homogeneous in concentration and limit the impact of the ground on the gas estimation. This classification by soil type is based either on a library of material signatures, or on the average spectrums most often represented in the image. We show applicative examples using classifications in the following sections.

The following detection filter is also commonly found in the literature:

$$f = \frac{G.\Sigma_c^{-1}}{\sqrt{G.\Sigma_c^{-1}.G^T}} \quad [4.21]$$

This filter is applied by soil class on the image in normalized vector G which is exactly proportional to the transmission of the gas to be detected. Second, in order to carry out a class-by-class quantification, the sizes α and S are to be estimated pixel by pixel to solve equation [4.20]. This process is explained in detail for applications in the thermal range below where the same methods can be applied. The reader may also refer to the work of Thorpe *et al.* [THO 13] or Deninson *et al.* [DEN 13] for a more detailed analysis of applicative cases of remote sensing of fine plumes in the reflective domain.

4.5.3. *Thermal infrared range*

In the thermal infrared range (3–12 μm), the radiative impact of a gas plume is due to its own emission and to the absorption of the flux emitted by the surface. In this range, we discuss as much about brightness temperature as radiance.

4.5.3.1. *Radiative impact of a gas plume*

4.5.3.1.1. Long-wave infrared (LWIR; 7.5–13 μm) case

In this spectral range, the impact of solar flux is ignored. The traditional expression of radiative transfer then leads to the following equation for the sensor input radiance L (W/m^2/sr/cm):

$$L_T = L_{atm,m} + \left(L_{atm,d}(1-\varepsilon_S) + \varepsilon_S L_{BB}(T_S)\right) t_{atm,m} \quad [4.22]$$

After atmospheric correction, equation [4.22] is written as:

$$L'_T = L_{atm,d}(1-\varepsilon_S) + \varepsilon_S L_{BB}(T_S) \quad [4.23]$$

L'_T is the total corrected radiance of the atmosphere, $L_{atm,m}$ is the upwelling atmospheric radiance of the atmospheric layers located between the ground and the sensor, $t_{atm,m}$ is the atmospheric transmission due to the atmosphere (except target gas) between the ground and the sensor, L_{BB} is the black body law, ε_s and T_s are emissivity and the ground temperature, respectively, and $L_{atm,d}$ is the downwelling atmospheric radianceof all the atmospheric layers.

Atmospheric correction consists of estimating the terms $L_{atm,m}$ and $t_{atm,m}$ from *a priori* knowledge of the atmosphere and/or the use of a reference zone in the image whose ground is known. We consequently suppose that these termsare constant for the entire image, and we can therefore correct them for all of the pixels in the image and similarly for pixels with or without a plume. For more details on this estimation, we refer the reader to the work of Young *et al.* [YOU 02].

In an atmospheric layer with a thickness h containing a gas g with a volumic concentration c_g (ppmv) characterized by its normalized spectral absorption $k_g\,(\mathrm{m}^{-1}\cdot\mathrm{ppmv}^{-1})$, the total absorption A_g, equal to the emissivity ε_g of the layer of gas g, is written as:

$$A_g = c_g \frac{h}{\cos\theta} k_g = \varepsilon_g \qquad [4.24]$$

θ is the zenith angle defining the direction of the considered path. The transmission in this layer underneath the incidence angle θ in the case of thin layers where absorption is far less than 1 can then be linearized and we get:

$$t_g = e^{-c_g \frac{h}{\cos\theta} k_g} = 1 - c_g \frac{h}{\cos\theta} k_g = 1 - \varepsilon_g \qquad [4.25]$$

The presence of a gas plume g modeled by a layer with a thickness h at ground level is expressed by the radiance sensor input in absorbance and in emission. The new $L_{T+\mathrm{plume}}$ radiance is then written as:

$$L_{T+panache} = L_{atm,m} + \begin{pmatrix} (\int L_{atm,d}\cdot f(t_{panache,d},\Omega).d\Omega + \Omega_{panache}.\varepsilon_{panache}.L_{CN}(T_{panache}))(1-\varepsilon_s) \\ +\varepsilon_s.L_{CN}(T_s))t_{panache,m} + \varepsilon_{panache}.L_{CN}(T_{panache}) \end{pmatrix} t_{atm,m} \qquad [4.26]$$

where t_{plume} corresponds to the transmission of the plume, $\varepsilon_{\mathrm{plume}}$ and T_{plume} its emissivity and temperature, respectively, the term Ω_{plume} corresponds to the portion of the solid angle by which the plume is viewed from the ground, and the function f characterizes the changes in the transmission depending on the incident zenith angle.

By taking the simplifying hypothesis where the plume is homogeneous with an infinite extent, and after atmospheric correction, we get:

$$L'_{T+panache} = ((L_{atm,d}.t_{panache} + \varepsilon_{panache}.L_{CN}(T_{panache})).(1-\varepsilon_s) + \varepsilon_s.L_{CN}(T_s)).t_{panache} + \varepsilon_{panache}.L_{CN}(T_{panache}) \quad [4.27]$$

where $L'_{T+\mathrm{plume}}$ is the radiance in the presence of the plume corrected for the effects of the atmosphere.

By disregarding the decreasing reflective terms $L_{atm,d}$, and by using equation [4.25], we get a simple differential expression by linearizing the emission of the plume:

$$L'_{T+plume} = L'_T + \varepsilon_{plume}\{L_{BB}(T_{plume}) - \varepsilon_s L_{BB}(T_s)\} \quad [4.28]$$

$$\Delta L' = L'_T - L'_{T+plume} = \frac{\sum_{g=1}^{ngas} q_g k_g}{\mu_v}\{L_{BB}(T_{plume}) - \varepsilon_s L_{BB}(T_s)\}$$

4.5.3.1.2. The case of mid-wave infrared (MWIR; 3–5 μm)

In this spectral range, the daytime solar influence must be taken into account. Expression [4.28] becomes:

$$\Delta L' = \left[(\varepsilon_S - 1)\frac{E_S}{\pi}(\mu_v + \mu_S) + L_{BB}(T_{plume}) - \varepsilon_S L_{BB}(T_S)\right]\frac{\sum_{g=1}^{ngas} q_g k_g}{\mu_v} \quad [4.29]$$

The term $(\varepsilon_S - 1)\frac{E_S}{\pi}(\mu_v + \mu_S)$ comes from equation [4.16], it concerns solar irradiance reflected by the surface and then absorbed by the plume.

4.5.3.2. *Illustration using synthetic hyperspectral images*

The radiative impact in the thermal range is illustrated here based on simulations of realistic scenes [IDO 14]. The first scene corresponds to the case where the plume is hotter than the background, which shows up mostly as emission. In the second, the plume is colder than the background, which shows up mostly as absorption.

4.5.3.2.1. Emission signature: the case of hot plumes

Figure 4.15 shows the absorbance of gases of interest at a resolution characteristic of a hyperspectral instrumentand Figure 4.16 shows the differential of the simulated signature for the same material of the scene with and without a plume. The emission of a hot plume causes an increase of more than 10 K in brightness temperature in the bands characteristic of the gases present in the plume. The involved concentrations correspond to discharges in the order of 100 g/s, which is the integrated concentration of the emission in the order of 1,000 ppm.m.

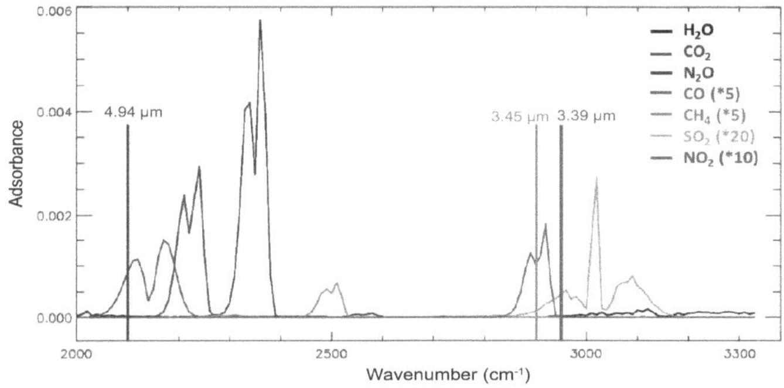

Figure 4.15. *Absorption of the different gases making up the simulated plume. For a color version of this figure, see www.iste.co.uk/baghdadi/6.zip*

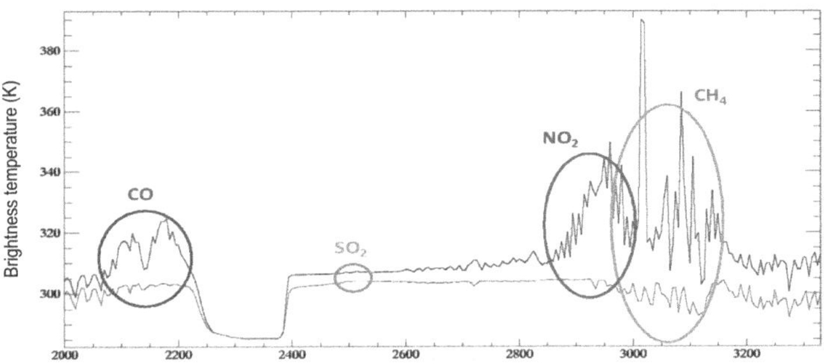

Figure 4.16. *Brightness temperature spectra of two points of the refinery scene showing the presence of gases (CO, SO_2, NO_2 and CH_4) in the MWIR spectra band. The spectrumin red(respectively, blue) represents a pixel outside of (respectively, inside the) plume. For a color version of this figure, see www.iste.co.uk/ baghdadi/6.zip*

4.5.3.2.2. Absorption signature: the case of cold plumes

Here, we are studying a plume of Freon F114 emitted with a temperature 10K cooler than the ambient air (itself already colder than the ground). Figure 4.17(a) shows the transmission of F114 in the LWIR (8–12 μm) range. We clearly see the four absorption zones of Freon at 920, 1,050, 1,140 and 1,190 cm^{-1}. Figure 4.17(b) shows the radiative brightness temperature depending on the wavelengths. We can also clearly see that the radiative brightness temperature in the presence of a plume (referred as F114) is less than that without the plume (referred as REF). A difference of 3 K in brightness temperature is shown for the 1,050 cm^{-1} absorption band.

a)

b)

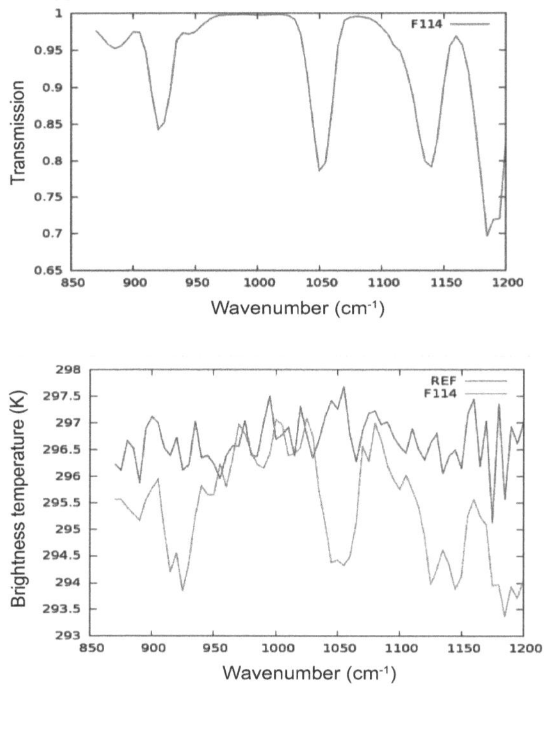

Figure 4.17. *a) Spectral transmittance of Freon F114; b) Radiative impact on radiative brightness temperature of the same F114 plume in the synthetic refinery (Freon F114 plume discharging 1 g/s and an ejection temperature of -10 K compared to the ambient environment on a homogeneous ground for the absorption peak at 1,050 cm^{-1}). For a color version of this figure, see www.iste.co.uk/baghdadi/6.zip*

4.5.3.3. *Detection methods*

The general principle is to rely on a spectral signature of the studied gas from either the absorptivity coefficients, or simulations of radiance or transmission via a radiative transfer model. Indeed, as we have previously seen, it is possible to linearize the problemby creating a "background" term to which a spectrally marked component corresponding to the studied gas is added. In the literature, several comparisons of the different existing detection tools have been carried out, the reader can consult the works of Messinger *et al.* [MES 04] in particular. Here, we present the most common tools.

4.5.3.3.1. Spectral angle (Spectral Angle Mapper (SAM))

The first step consists of subtracting the average spectrum $L_m$of the image from the radiance L_i of pixel i, and then carrying out the measurement of the angle between the spectrum obtained and the vector of the absorption coefficient of the gas being studied k_g:

$$\theta_g^{SAM}(i) = cos^{-1}\left[\frac{k_g(L_i-L_m)^T}{\|k_g\|\|L_i-L_m\|}\right] \qquad [4.30]$$

For each pixel i and each gas g, we then get an angle that tends to 0 when the gas is present. This tool, frequently used for any type of detection, is adapted for gases when the "average" spectrum of an image or a part of the image is representative of the spectrum without a plume.

4.5.3.3.2. Spectral matched filter (SMF)

This filter is already mentioned in equation [4.20] for detection in the reflective domain:

$$SMF_g(i) = \frac{{k_g}^T {\Sigma_L}^{-1}(L_i - L_m)}{{k_g}^T {\Sigma_L}^{-1} k_g} \; avec\, \Sigma_L = \frac{1}{N-1}\sum_{n=1}^{N}(L_i - L_m)(L_i - L_m)^T \qquad [4.31]$$

In practice, this is well-suited when the SAM does not provide accurate results due to spectral correlation and instrumental noise.

4.5.3.3.3. Generalized likelihood ratio test (GLRT)

This method makes it possible to work in differential mode, assuming that the radiance without a plume can be described by a base noted B (either by the image's own vectors or from spectral signatures coming from a database of optical properties). When $B = [b_1, ..., b_{Nb}]$, this base is made up of Nb vectors noted as b_i, we then define a base Z_g for each gas, corresponding to the concatenation of bases B and vector k_g. The GLR operator is therefore defined by the projection ratiofor each pixel i between base B and Z_g assuming that the error follows a normal multivariate law and that the B and K matrices are known and independent:

$$GLR_g(i) = \left[\frac{L_i^T \left(P_B \right) L_i}{L_i^T \left(P_{Z_g} \right) L_i} \right]^{p/2} \qquad P_Y = I - Y(Y^T Y)^{-1} Y^T \qquad [4.32]$$

with p the number of spectral points for L_m and P_Y the orthogonal projector on base Y. It is common to use a similar operator, the matched subspace detector(MSD), defined as follows:

$$GLR_g(i) = \left[1 + MSD_g(i) \right]^{p/2} \qquad MSD_g(i) = \left[\frac{L_i^T \left(P_B - P_{Z_g} \right) L_i}{L_i^T \left(P_{Z_g} \right) L_i} \right] \qquad [4.33]$$

4.5.3.3.4. Cluster-tuned matched filter (CTMF)

In order to further improve detection, the goal is to operate the SMF after classifying the image in order to compensate for the impact of the ground reflectance on the power of detection involved in the reflective orthermal contrast [FUN 01]. SMF is therefore applied class by class. The results are then re-agglomerated to create the detection image. For each class J, we define a specific filter for gas g:

$$q^J{}_g(i) = \left[\frac{\Sigma_J^{-1} k_g}{k_g^T \left(\Sigma_J^{-1} \right) k_g} \right] \qquad CTMF^J{}_g(i) = (q^J{}_g)^T (l_i), i \in J \qquad l_i = (L_i - L_m)/\sigma \qquad [4.34]$$

where l_i is the normalized radiance (subtracted from the average value of the entire image and divided by the standard deviation σ estimated for the entire image). Using an image of a refinery acquired in the LWIR region by the airborne infrared hyperspectral sensor HyperCam (Telops company,

Figure 4.18), we illustrate the type of results obtained after this detection step.

a)

b)

Figure 4.18. *Airborne acquisition of a refinery; a) visible image; b) infrared image (image obtained from the hyperspectral instrument HyperCam from Telops Company). For a color version of this figure, see www.iste.co.uk/baghdadi/6.zip*

In Figure 4.19(a), SMF is applied to the zone surrounded in red in Figure 4.18, where the outlet of the refinery is locatedto search for the presence of SO_2. Using the map obtained from Figure 4.19(a), a threshold is applied by keeping only the higher intensity points after having studied the histogram of the detection map (Figure 4.19(b)). The pixels whose SMF is higher than the threshold are set as 1, the others at 0: we then get a detection mask (Figure 4.19(c)). It is often then necessary to apply morphological operations to flatten this mask and/or expand it.

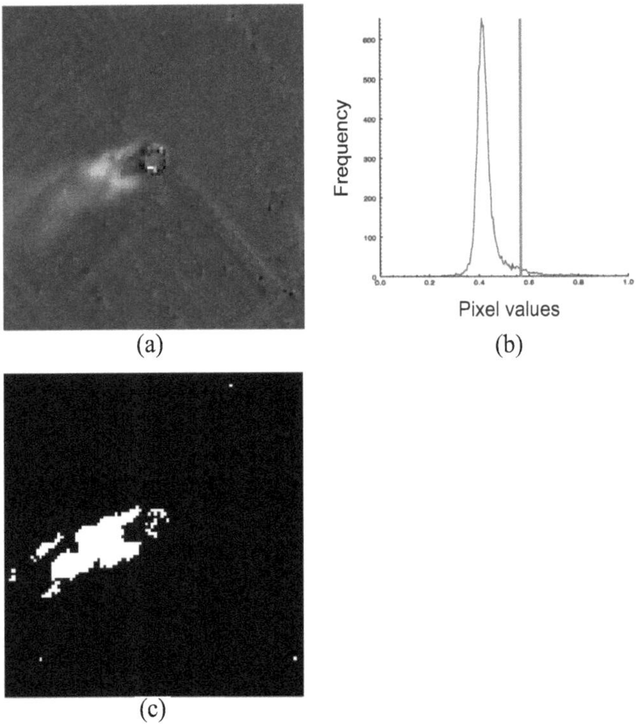

Figure 4.19. *Illustration of the approach for defining the threshold: a) image obtained after the application of the SMF method; b) histogram made from the image; a) the green line shows the threshold selected for detecting the plume, c) mask obtained with a thresholding of 0.57. For a color version of this figure, see www.iste.co.uk/baghdadi/6.zip*

4.5.3.4. *Characterization of gas properties*

From the detection map of the plume obtained gas by gas, quantifying the concentration of the plume requires the knowledge of the background parameters; the ground under the plume and the temperature of the plume.

After the detection step, the other principal steps of the characterization process are characterization of the background underneath the plume and quantification, which are explained in the following section. We also encourage the reader to consult the work of Manolakis *et al.* [MAN 14] and Golowich *et al.* [GOL 14] for a detailed review of existing methods and achievable performance.

4.5.3.4.1. Estimation and definition of the background radiance underneath the plume

"Simple" classification

After rejecting the spectral points corresponding to the absorption zones of the detected gases, a classification is applied to the image as shown in Figure 4.20. There is a mean vector associated with each class formed this way; this vector is a first approximation of the local background vector L' for all of the pixels in the class.

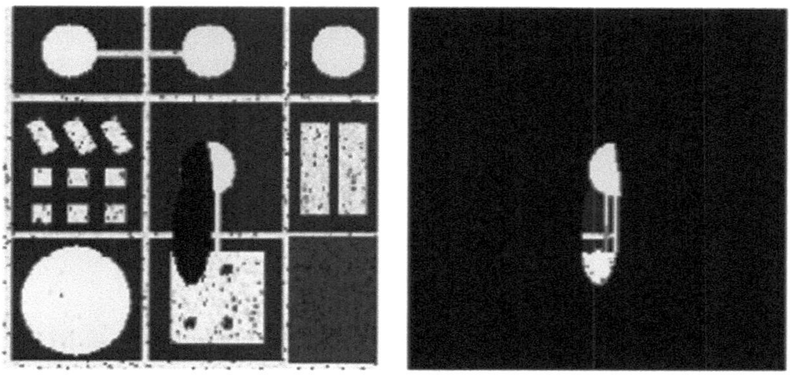

a) b)

Figure 4.20. *Illustration of a classification made outside of; a) and underneath; b) the plume in the case of a synthetic scene. After these two classifications, mapping makes it possible to associate a ""background"" with each pixel of the plume. For a color version of this figure, see www.iste.co.uk/baghdadi/6.zip*

Linear regression from a spectral base of soil properties

Another frequently used method [YOU 02] is to define a spectral base for L' vectors. There are several possible bases: vectors belonging to the image's pixels (those for which detection filters have shown no detection), simulation base from physical data of the properties of the ground and mean vectors of a pre-classification. As previously, the vector L' is written in base B: $L'=\overline{\alpha}.\overline{B}$. The idea is then to determine the parameters α and q of the gas concentration simultaneously by traditional linear resolution:

$$L'_p = \alpha\, B + q_g K + E \;;\; L'_p = \beta\, A \;;\; A = \begin{pmatrix} B & 0 \\ 0 & K \end{pmatrix};$$
$$\beta = \left[\alpha_1, \dots, \alpha_{p,} q_1, \dots, q_{ngas}\right] \qquad [4.35]$$

with:

$$\beta = \left(A^T \Sigma_L^{-1} A\right)^{-1} \left(A^T \Sigma_L^{-1} L_m\right) \qquad [4.36]$$

The vector β therefore corresponds to the parameters of the ground and to the concentrations of the gases being studied, matrix A is the block diagonal, block B corresponds to the base of ground materials and block K corresponds to the base defined by the spectral absorption of the gases being studied. The operation therefore consists of resolving the linear problem by assuming Gaussian instrumental noise, characterized by its covariance matrix Σ_L.

Methods of extrapolating the absorption bands of the background radiance from the radiance spectrum containing gas outside the absorption bands have also been studied with interesting results [NIU 14, MAR 04].

4.5.3.4.2. Differential inversion

Generalized linear problem: the case of a plume made up of a single homogeneous layer

From the preceding equations of the normalized differential, the linear vectorial form in the case of a monolayer gas plume near the surface is as follows:

$$L_m - L_{sp} = \Delta L = C.K + \mathrm{E} \qquad [4.37]$$

where E is the error vector to be minimized, C the targeting vector containing the concentrations of different gases present in the plume, K the matrix containing the spectra of signatures calculated from spectral absorptivity of the gases present and the properties of the ground, L_m the spectrum measured and L_{sp} the spectrum without a plume estimated by class or pixel to pixel. The exact solution to the problem is as follows:

$$C=\left(K^T \Sigma_L^{-1} K\right)^{-1}\left(K^T \Sigma_L^{-1} \Delta L\right) \ avec \ \Sigma_L=\frac{1}{N-1}\sum_{n=1}^{N}(L_n-L_m)^T(L_n-L_m) \quad [4.38]$$

This principle was applied to the image of the refinery on the SO_2 mask selected; the map obtained in ppm.m of SO_2 is presented in Figure 4.21.

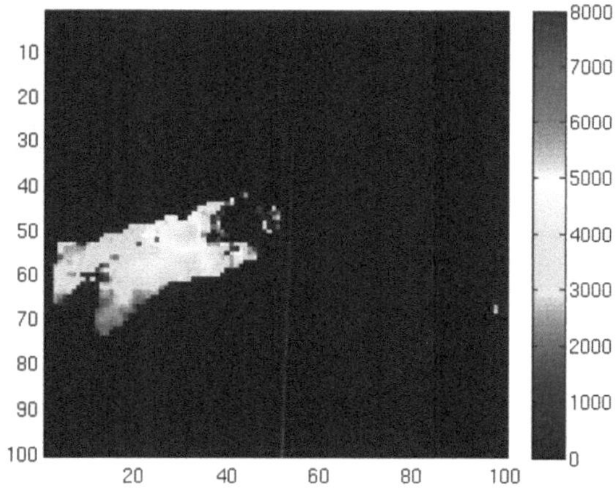

Figure 4.21. *Image after linear inversion applied to the estimated radiance differential on the plume detected on the image of the refinery. The values are in ppm.m of SO2.For a color version of this figure, see www.iste.co.uk/baghdadi/6.zip*

Iterative quantification method

The general principle of this method is to assume a multivariate distribution of the error and to consider each parameter as a probability density focused on an average value and characterized by a dispersion connected to a standard deviation. Moreover, we assume *a priori* knowledge of X noted as Xa associated likewise with an *a priori* probability. Therefore, minimizing the error results in minimizing the following function:

$$\chi^2=\left((L-F(X))^T \Sigma_L^{-1}(L-F(X))+(X-Xa)^T \Sigma_X^{-1}(X-Xa)\right) \quad [4.39]$$

where Σ_L^{-1} is the inverse of the covariance matrix of the data observed L, and Σ_X^{-1} is the inverse of the covariance matrix of the parameter vector.

Several methods have been proposed for solving this problem:

– Iterative regression methods using the partial derivatives of the F function (Jacobin), such as the Levenberg–Marquardt method;

– Monte Carlo type methods (connected or not with Markov chains), although such methods are very time-consuming.

Figure 4.22 shows the contribution of an iterative method [IDO 14] based on physical modeling of the scene, making it possible to take into account vertical discretization of the plume, in particular the temperature profile. After the iterative method, the blue spectrum obtained is closer to the spectrum obtained from inverse parameters using the results of the linear inversion. Nevertheless, these methods remain very time-consuming.

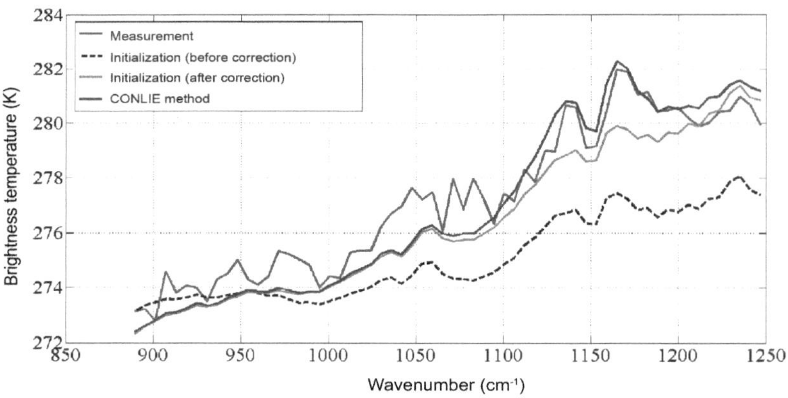

Figure 4.22. *Comparison of spectra during the inversion process. The spectrum coming obtained from Telops data on the SO_2 plume is in red; the spectrum obtained from inverse parameters by linearizing the problem is in black. The green spectrum stemming arising from the linear inversion (black) and correction of the continuum (slope) with respect to the measurement is here used initially for the iterative method. The final spectrum obtained after minimization of the CONLIE iterative method [IDO 14] is illustrated in blue. For a color version of this figure, see www.iste.co.uk/ baghdadi/6.zip*

4.6. Conclusions and outlook

Characterizing the aerosols and gases present in industrial plumes requires the use ofthe entire optical region. The current methods allow for the detection and characterization of the plumes. Their principal limitations are as follows:

– a lack of physical data for aerosols characteristic of different types of industries;

– an uncertainty in estimating the backgrounds underneath the plume and a signal-to-noise ratio of the instrument, which is not high enough for the characterization of gases;

– the difficulty of simultaneously implementing hyperspectral instruments covering the entire optical region.

Nevertheless, significant methodological efforts have been set up in the communityin order to better characterize aerosol particles. Moreover, new hyperspectral instruments (SYSIPHE [ROU 15]) of better radiometric quality are in the process of being developed or in the acceptance phase. Finally, it is becoming more and more important for the industrial and the scientific community to provide airborne instruments able to simultaneously cover the entire optical region.

4.7. Key points

– The characterization of gases and anthropogenic aerosols using remote sensing in airborne passive imaging is often a secondary effect compared to the signatures of the ground or of the atmosphere itself.

– "Differential" modeling makes it possible to remove the background signal and reveal the spectral characteristics of gases and aerosols: diffusion, absorption and emission.

– The main effects of anthropogenic aerosols are observable up to wavelengths of 1.5–2 µm, while gases have signatures spread over the entire electromagnetic spectrum.

– For aerosols, reversal methods are limited by the lack of knowledge about anthropogenic aerosols, of which there are a very wide variety linked to types of industry, and by an uncertainty involving knowledge of the background below the plume.

– For gases, detection methods are robust, particularly in the thermal domain, but inverse quantification methods are primarily limited by an uncertainty about ground properties underneath the plume as well as by the temperature differential between the plume and the atmosphere.

4.8. Bibliography

[AKI 03] AKIMOTO H., "Global air quality and pollution", *Science*, vol. 302, pp. 1716–1719, 2003.

[ALA 09] ALAKIAN A., MARION R., BRIOTTET X., "Retrieval of microphysical and optical properties in aerosol plumes with hyperspectral imagery: L-APOM method", *Remote Sensing of Environment*, vol. 113, pp. 781–793, 2009.

[ANN 83] ANNEGARN H., LESLIE A., WINCHESTER J. *et al.*, "Particle size and temporal characteristics of aerosol composition near coal-fired electric power plants of the Eastern Transvaal", *Aerosol Science and Technology*, vol. 2, pp. 489–498, 1983.

[BER 89] BERK A., BERNSTEIN L.S., ROBERTSON D.C., MODTRAN: a moderate resolution model for LOWTRAN 7, Final report, GL-TR-89-0122, AFGL, Hanscom AFB, MA, p. 42, 1989.

[BRI 16] BRIOTTET X., "Radiometry in the optical domain", in BAGHDADI N., ZRIBI M. (eds), *Optical Remote Sensing of Land Surfaces*, ISTE Press Ltd, London and Elsevier Ltd, Oxford, 2016.

[BUC 05] BUCHWITZ M., DE BEEK R., NOËL S. *et al.*, "Carbon monoxide, methane and carbon dioxide columns retrieved from SCIAMACHY by WFM-DOAS: year 2003 initial data set", *Atmospheric Chemistry and Physics*, vol. 5, pp. 3313–3329, 2005.

[CAR 94] CARRUTHERS D.J., HOLROYD R.J., HUNT J.C.R. *et al.*, "UK-ADMS: a new approach to modelling dispersion in the earth's atmospheric boundary layer", *Journal of Wind Engineering & Industrial Aerodynamics*, vol. 52, pp. 139–153, 1994.

[CEO 13] CEOLATO R., Data fusion system for hyperspectral, polarimetric, and angular scattering. Application to optical diagnostic of dense and complex media, PhD Thesis, University of Toulouse, 8 November 2013.

[CHE 08] CHENG Y., EICHLER W.A., SU H. *et al.*, "Aerosol optical properties and related chemical apportionment at Xinken in Pearl River Delta of China", *Atmospheric Environment*, vol. 42, pp. 6351–6372, 2008.

[CHO 09] CHOËL M., DEBOUDT K., FLAMENT P., "Development of time-resolved description of aerosol properties at the particle scale during an episode of industrial pollution plume", *Water, Air, & Soil Pollution*, vol. 209, pp. 93–107, 2009.

[DEN 13] DENNISON P.E., THORPE A.K., PARDYJAK E.R. *et al.*, "High spatial resolution mapping of elevated atmospheric carbon dioxide using airborne imaging spectroscopy: radiative transfer modeling and power plant plume detection", *Remote Sensing of Environment*, vol. 139, pp. 116–129, 2013.

[DES 13] DESCHAMPS A., MARION R., BRIOTTET X. *et al.*, "Simultaneous retrieval of CO2 and aerosols in a plume from hyperspectral imagery: application to the characterization of forest fire smoke using AVIRIS data", *International Journal of Remote Sensing*, vol. 34, no. 19, pp. 6837–6864, 2013.

[FRA 05] FRANKENBERG C., PLATT U., WAGNER T., "Iterative maximum *a posteriori* (IMAP)-DOAS for retrieval of strongly absorbing trace gases: model studies for CH4 and CO_2 retrieval from near infrared spectra of SCIAMACHY onboard ENVISAT", *Atmospheric Chemistry and Physics*, vol. 5, pp. 9–22, 2005.

[FUN 01] FUNK C.C., THEILER J., ROBERTS D.A. *et al.*, "Clustering to improve matched filter detection of weak gas plumes in hyperspectral thermal imagery", *IEEE Transactions on Geoscience and Remote Sensing*, vol. 39, pp. 1410–1420, 2001.

[GIE 13] GIEC 2013, "Contribution du groupe de travail I au cinquième rapport d'évaluation de groupe d'experts intergouvernemental sur l'évolution du climat", Intergovernmental Panel on Climate Change, 2013.

[GOL 14] GOLOWICH S., MANOLAKIS D., "Cramer-Rao bounds for long-wave infrared gaseous plume quantification", *Optical Engineering*, vol. 53, no. 2, p. 021109, 2014.

[HES 98] HESS M., KOEPKE P., SCHULT I., "Optical properties of aerosols and clouds: the software package OPAC", *Bulletin of the American Meteorological Society*, vol. 79, no. 5, pp. 831–844, 1998.

[IDO 14] IDOUGHI R., FOUCHER P.Y., POUTIER L. *et al.*, "A non-linear optimal estimator for plume concentration retrieval using airborne hyperspectral measurement", *6th Workshop on Hyperspectral Image and Signal Processing: Evolution in Remote Sensing*, Lausanne, Switzerland, June 2014.

[JAC 09] JACQUINET-HUSSON N., CREPEAU L., ARMANTE R. *et al.*, "The 2009 edition of the GEISA spectroscopic database", *Journal of Quantitative Spectroscopy and Radiative Transfer*, vol. 112, pp. 2395–2445, 2009.

[KAU 02] KAUFMAN Y.J., TANRÉ D., BOUCHER O., "A satellite view of aerosols in the climate system", *Nature*, vol. 419, pp. 215–223, 2002.

[KIE 95] KIEHL J.T., RODHE H., "Modeling geographical and seasonal forcing due to aerosols", CHARLSON R.J., HEINTZENBERG J.J. (eds), *Aerosol Forcing of Climate*, Wiley and Sons Ltd, New York, pp. 281–296, 1995.

[LET 03] LE TREUT H., "Les scénarios globaux de changement climatique et leurs incertitudes", *Comptes Rendus Geoscience*, vol. 335, nos. 6–7, pp. 525–533, 2003.

[MAL 03] MALLET M., ROGER J., DESPIAU S. *et al.*, "Microphysical and optical properties of aerosol particles in urban zone during ESCOMPTE", *Atmospheric Research*, vol. 69, pp. 73–97, 2003.

[MAL 05] MALLET M., VAN DINGENEN R., ROGER J. *et al.*, "*In situ* airborne measurements of aerosol optical properties during photochemical pollution events", *Journal of Geophysical Research*, vol. 110, pp. D03205, 2005.

[MAN 14] MANOLAKIS D., GOLOWICH S., DIPIETRO R., "Long-wave infrared hyperspectral remote sensing of chemical clouds", *IEEE Signal Processing Magazine*, vol. 120, pp. 1053–5888, 2014.

[MAR 04] MARION R., MICHEL R., FAYE C., "Measuring trace gases in plumes from hyperspectral remotely-sensed data", *IEEE Transactions on Geoscience and Remote Sensing*, vol. 42, no. 4, pp. 854–864, 2004.

[MAR 12] MARRIS H., DEBOUDT K., AUGUSTIN P. *et al.*, "Fast changes in chemical composition and size distribution of fine particles during the near-field transport of industrial plumes", *Science of the Total Environment*, vols. 427–428, pp. 126–138, 2012.

[MAT 07] MATHER T., HARRISON R., TSANEV V. *et al.*, "Observations of the plume generated by the December 2005 oil depot explosions and prolonged fire at Buncefield (Hertfordshire, UK) and associated atmospheric changes", *Proceedings of the Royal Society*, vol. 463, pp. 1153–1177, 2007.

[MES 04] MESSINGER D.W., "Gaseous plume detection in hyperspectral images: a Comparison of methods", SHEN S.S., LEWIS P.E. (eds), *Algorithms and Technologies for Multispectral, Hyperspectral, and Ultraspectral Imagery X*, *Proceedings of SPIE*, vol. 5425, pp. 592–603, 2004.

[MIE 05] MIESCH C., POUTIER L., ACHARDV. *et al.*, "Direct and inverse radiative transfer solutions for visible and near infrared hyperspectral imagery", *IEEE Transactions on Geoscience and Remote Sensing*, vol. 43, no. 7, pp. 1552–1562, 2005.

[MOF 08] MOFFET R., DESYATERIK Y., HOPKINS R. *et al.*, "Characterization of aerosols containing Zn, Pb, and Cl from an industrial region of Mexico City", *Environmental Science & Technology*, vol. 42, pp. 7091–7097, 2008.

[NIU 14] NIU S., GOLOWICH S.E., INGLE V.K.. *et al.*, "New approach to remote gas-phase chemical quantification: selected-band algorithm", *Optical Engineering*, vol. 53, no. 2, p. 021111, 2014.

[PLA 94] PLATT U., "Differential optical absorption spectroscopy (DOAS)", in SIGRIST M.W. (ed.), *Air Monitoring by Spectroscopic Techniques*, John Wiley, 1994.

[PLA 08] PLATT U., STUTZ J., *Differential Optical Absorption Spectroscopy: Principles and Applications*, Springer-Verlag, Berlin, Heidelberg, 2008.

[RIE 09] RIEMER N., WEST M., ZAVERI R. *et al.*, "Simulating the evolution of soot mixing state with a particle-resolved aerosol model", *Journal of Geophysical Research*, vol. 114, p. D09202, 2009.

[ROG 06] ROGER J.C., MALLET M., DUBUISSON P. *et al.*, "A synergetic approach for estimating the local direct aerosol forcing: application to an urban zone during the Expérience sur Site pour Contraindre les Modèles de Pollution et de Transport d'Émission (ESCOMPTE) experiment", *Journal of Geophysical Research*, vol. 111, D13208, 2006.

[ROT 13] ROTHMAN L.S., GORDON I.E., BABIKOV Y. *et al.*, "The HITRAN 2012 molecular spectroscopic database", *Journal of Quantitative Spectroscopy and Radiative Transfer*, vol. 130, pp. 4–50, 2013.

[ROU 15] ROUSSET-ROUVIÈREL., "SYSIPHE system: a state of the art airborne hyperspectral imaging system. Initial results from the first airborne campaign", EARSeL 2015, Luxembourg, 14–16 April 2015.

[RUS 99] RUSSEL P., HOBBS P., STOWE L., "Aerosol properties and radiative effects in the United States East Coast haze plume: an overview of the troposheric aerosol radiative forcing observational experiment (TARFOX)", *Journal of Geophysical Research*, vol. 104, no. D2, pp. 2213–2222, 1999.

[SHA 04] SHARPE S.W, JOHNSON T.J., SAMS R.L. *et al.*, "Gas-phase databases for quantitative infrared spectrometry", *Applied Spectrometry*, vol. 58, pp. 1452–1461, 2004.

[THO 13] THORPE A.K., ROBERTS D.A., BRADLEY E.S. *et al.*, "High resolution mapping of methane emissions from marine and terrestrial sources using a cluster-tuned matched filter technique and imaging spectrometry", *Remote Sensing of Environment*, vol. 134, pp. 305–318, 2013.

[THO 14] THORPE A.K., FRANKENBERG C., ROBERTS D.A., "Retrieval techniques for airborne imaging of methane concentrations using high spatial and moderate spectral resolution: application to AVIRIS", *Atmospheric Measurement Techniques*, vol. 7, pp. 491–506, 2014.

[YOU 02] YOUNG S.J., "Detection and quantification of gases in industrial-stack plumes using thermal-infrared hyperspectral imaging", *Aerospace Corporation Technical Report*, vol. 1, p. 8407, 2002.

[ZAN 90] ZANNETTI P., "Gaussian models", Chapter 7 in ZANNETTI P. (ed), *Air Polution Modeling: Theories, Computational Methods and Available Software*, Computational Mechanics Publications, Southampton, Boston and Van Nostrand Reinhold, New York, 1990.

5

Monitoring of Earth Surface Motion and Geomorphologic Processes by Optical Image Correlation

5.1. Introduction

The detection and quantification of motion of the Earth's surface are essential for the study and understanding of many geomorphological and geological processes such as tectonics, earthquakes, landslides and ice glaciers, to name a few [DEL 04, KAA 02, LEP 07a, SCA 92]. The direct measurement of earth displacement and deformation falls into larger discipline of geodesy, which also includes studies of the shape, movement and gravitational field of the Earth. Traditional surveying methods to measure surface motion comprise the use of theodolites, electronic distance meters, total stations and nowadays typically global navigation satellite systems (GNSS). While such methods can provide very precise point-wise measurements, aerial and satellite remote sensing techniques are generally more suitable for measuring dense motion fields of larger and/or inaccessible areas.

Today, the most important remote sensing techniques for motion measurements in the geosciences are light detection and ranging (LiDAR; [CHA 16]), interferometric Synthetic Aperture Radar (SAR) (InSAR;

Chapter written by André STUMPF, Jean-Philippe MALET, Anne PUISSANT and Julien TRAVELLETTI.

[FER 16]) and photogrammetric image techniques. In particular, InSAR has proven to be a very precise technique for measuring slow movements in the range of several millimeters per month along the line of sight of the satellite [COL 06]. LiDAR provides very precise three-dimensional (3D) models of the terrain and can be used to recover all components of the surface motion where multitemporal surveys are available. Optical image matching techniques, which are discussed in this chapter, are typically used to recover the two horizontal components of the 3D motion with magnitudes reaching from a fraction of the image pixel size to several dozen pixels. If available, multitemporal stereo-images can also be used to reconstruct the topography at different dates and infer the vertical component of the motion.

The earliest documented measurements of surface displacement from aerial photographs date back to the 1940s and 1950s when glacier surface velocities were determined by visually tracking surface features from multitemporal aerial photographs [HOF 58, RIC 46]. The first studies started to exploit multitemporal satellite images in the 1980s in order to measure glacier velocities in Antarctica [LUC 86] still relied on visual tracking of persistent surface features.

The first technical descriptions and implementations of automatic digital image matching techniques can be traced back to the 1960s when such techniques were primarily conceived for image coregistration and photogrammetric reconstructions [GRU 12]. The first documented geoscientific study in motion field estimation was probably a study conducted by Leese *et al.* [LEE 71] on cloud motion observed from a geostationary satellite. Initially, the application of digital image correlation (DIC) was considered too computationally expensive but its use quickly became widespread as faster computers and algorithms became available [BER 76]. Although during the 1980s image correlation techniques were already commonly used in laboratory experiments for material science [PET 82] and fluid mechanics [ADR 84], their application for horizontal displacement measurements of the earth surface was not demonstrated until the 1990s [CRI 91].

Today, image matching techniques have become an indispensable tool for measuring surface displacement and deformation, and many studies have demonstrated their value for applications in seismo-tectonics [LEP 07a, VAN 00], glacier flow [DEH 15, KAA 00, SCH 08], monitoring of rock glaciers [WAN 06], landslides [DEL 04, LAC 15, STU 14a], sand dune migration [BRI 12, NEC 09], ocean wave velocities [DEM 12], river surface velocities [BEL 14] and surface motion related to volcanic activity [DEM 07].

This chapter first summarizes the available image data sources and the necessary preprocessing steps such as coregistration and orthorectification. In a subsequent section, the theoretical concepts behind DIC have been reviewed. In addition, the advantages and limitations of several alternative implementations are discussed. While many of the theoretical aspects apply to different imaging modalities such as radar, LiDAR and optical, the main focus is on the processing of optical images. Postprocessing techniques that can be applied to increase the robustness and accuracy of the measurements are outlined. The final sections review a number of recent applications as well as current limitations and active areas of research.

The term “image correlation” is used here to refer mainly to the correlation of optical images for measurements of horizontal surface displacement and deformation. It should be noted that image correlation and other matching techniques (feature-based, relational) are also crucial for the photogrammetric reconstruction of 3D surfaces from two or multiple images [GRU 12], image coregistration [ZIT 03] or video-based river gauging [MUS 11], which have not be treated explicitly in this chapter.

5.2. Sources of image data

While a comprehensive review of the vast variety of optical image datasets is beyond the scope of this chapter, we provide a general overview of common sources grouped in four categories including medium- to high-resolution optical satellites, very high resolution optical satellites, aerial photography and terrestrial images. An overview of the spatio-temporal resolution of selected satellite systems is given in Figure 5.1(a). Medium- to high-resolution optical satellites include systems that deliver images at resolutions between 5 and 30 m pixel resolution. The most prominent satellite missions in this category are the Landsat series (most recently Landsat-8, 15 m panchromatic), SPOT1–5 (10–2.5 m panchromatic), ASTER (15 m panchromatic), the Indian Remote Sensing satellite series

(5.8 m panchromatic) and very recently Sentinel-2 (10 m panchromatic). It is important to note that among the listed satellites, Landsat and Sentinel-2 are the most accessible since imagery can be downloaded at no cost from the databases of the respective space agencies. Similarly, ASTER and SPOT5 images can be obtained without major restrictions at comparatively low processing charges or for free through dedicated programs for non-commercial use (e.g. SPOT5-Take5, SPOT World Heritage program). The time lag between two acquisitions of ASTER and Landsat is constant at approximately 16 days. While the nominal repeat-pass cycle of the latest Indian Resource Sat-2 is 5 days for its panchromatic, Linear Imaging Self Scanner (LISS IV) camera exploitable imagery usual requires longer time lags (2–3 weeks) if narrow incidence angles are to be avoided. While other platforms such as RapidEye (5 m) also fall into this category, they have – to the best of our knowledge – not been exploited for motion measurements. This is also true for the KH-9 Hexagon satellite series, which have proven useful for the reconstruction of glacier mass loss when using stereo-photogrammetric methods [PIE 13], whereas the large time lags between individual acquisitions appear to be a challenging hurdle for multitemporal image matching.

Considering a sub-pixel accuracy, which in practice typically does not exceed one-fifth of a pixel (section 5.42.), the targeted displacement usually has to exceed 1 to several meters to yield distinguishable signals when using medium- to high-resolution satellites. In gerenal, the accumulated displacement between two dates should exceed the measurement error which in turn depends to a large degree on the image resolution. This relationship is depicted in Figure 5.1(b).

Since the launch of Ikonos-2 in 1999, the fleet of very high resolution optical satellites has greatly increased and is now capable of very short revisit times and imaging at sub-meter resolution (Figure 5.1(a)). Very high spatial resolution (VHSR) satellites are operated by national space agencies (Pléiades 1A and 1B, SPOT 6/7) or private satellite operators (e.g. Geoeye-1, WorldView-2) and images can be purchased from their respective resellers. While recent studies have demonstrated that, in particular, gravitational mass-movements can be monitored at decimeter accuracies [LAC 15, STU 14a], it should be considered that the required images still bear significant costs, especially if stereo and tasked acquisition are needed. Though prices have reduced significantly in the past decade, this is still a hindering factor for operational monitoring over larger areas.

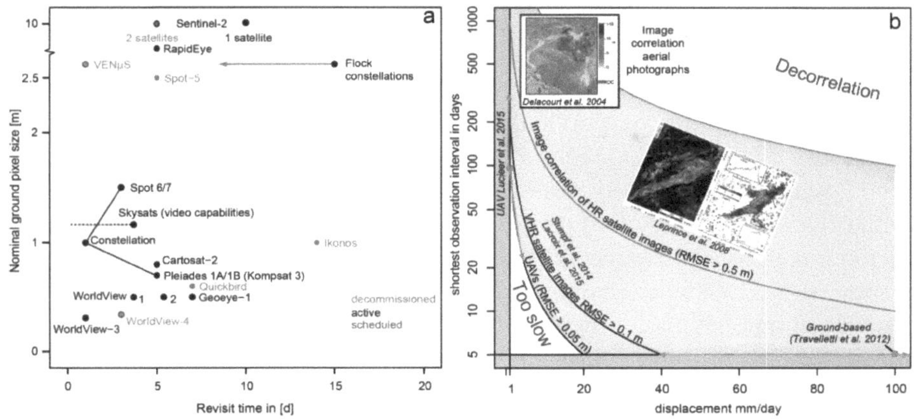

Figure 5.1. *a) Overview of the selected optical satellite systems (mainly very high spatial resolution [VHSR]) showing their revisit time and ground pixel size; b) The feasibility to accurately measure surface motion depends on a combination of factors including the displacement rates, the intervals between observations and the measurement errors. Faster displacement rates will typically require shorter observation periods since significant surface changes will lead to decorrelation among the image pairs. Slower displacements rates typically demand imaging systems with higher spatial resolutions or longer observation intervals to yield accumulated displacements that exceed the measurement error. The curves and case studies indicate the suitability of different approaches using satellite images, classical aerial photographs and aerial images acquired with unmanned aerial vehicles (UAVs) or ground-based imaging. This figure is based on several studies on the observation of landslides. Mass movements such as glaciers can reach displacements up to several meters without leading to decorrelation. For a color version of this figure, see www.iste.co.uk/baghdadi/6.zip*

In many countries, classical aerial photography has been conducted since the 1960s and earlier, typically at sub-meter resolution. While the time intervals between such surveys typically exceed several years, they constitute valuable archives to reconstruct historical seismic events [HOL 12, MIC 06] or slow-moving mass-movements [CAS 03]. Classical aerial surveys often comprise stereo-images and therefore also provide the possibility of reconstructing the surface and achieving precise orthorectification. For detailed studies of smaller study sites, aerial photography with unmanned aerial vehicles (UAVs) has become a cost-effective alternative and has been shown to allow very detailed displacement measurements of mass movements several times per year [IMM 14, TUR 15]. This is also a result of the availability of commercial and open-source software suites that make the generation of digital surface models (DSM) and orthophotographs from large image collections straightforward.

For the monitoring of ground motion at specific sites, DIC of terrestrial images has been shown to yield not only historical records of glacier evolution [KAU 12], but is also becoming more frequently used for near-real-time monitoring of gravitational mass-movements such as landslides [TRA 12] and glaciers [MES 15]. A great advantage of such techniques is their high spatio-temporal resolution, which can be achieved at low costs with off-the-shelf single-lens reflex (SLR) cameras. To employ such systems in mountainous environments, the stability of the control mechanisms and power supplies under harsh environmental conditions requires special considerations. Where wireless networks are available, images can be transferred directly for near-real-time monitoring or otherwise must be stored on site. If the observed motion comprises a strong vertical component, frequent updates of the digital elevation model (DEM) used for orthorectification become necessary (section 5.3).

5.3. Preprocessing

A crucial prerequisite for successful measurements of surface displacement and deformation with DIC is an accurate coregistration of all images including the correction of topographic effects (orthorectification). Space agencies and private data providers often distribute orthorectified processing level imagery but the quality of the geometric corrections is mostly insufficient for direct displacement measurements. Whenever possible, it is preferable to start from non-orthorectified images comprising complete sensor models. Delivered sensor models establish the acquisition geometry for each image, either by describing all of the physical elements of the imaging system (rigorous sensor model) or through complex polynomial functions that relates ground and image coordinates without explicitly describing physical components (rational polynomial function). The sensor model enables a fine coregistration and the use of a site-specific DEM, which might provide greater precision than global datasets such as Shuttle Radar Topography Mission (SRTM). While a detailed description of the coregistration and orthorectification is beyond the scope of this chapter, a general outline is given in Figure 5.2.

The original sensor models are typically not precise enough to achieve a coregistration of multitemporal images at sub-pixel accuracy. To correct systematic errors (predominately translation and rotation), feature-based matching techniques (e.g. Scale-Invariant Feature Transform (SIFT),

[LOW 04]) can be used to detect tie points among the original non-orthorectified images (Figure 5.2(a)). This implicitly assumes that certain areas depicted in the images have remained stable over time and care must be taken to assure that the tie points are not located on surfaces that have moved. Bundle adjustment is used to minimize the residual re-projection errors of the tie points and correct the bias in the original sensor models. The residuals are evaluated through a projection of the image points in ground geometry and a subsequent re-projection into image space according to the sensor model. A least-square procedure is typically used to estimate additional terms that are added to the sensor model (e.g. simple translation, translation and rotation, or an affine transformation) and minimize the residual re-projection error [GRO 03]. If available, ground control points (GCPs) can be used in this process to assure geolocation accuracy but are not strictly needed for precise coregistrations [STU 14a].

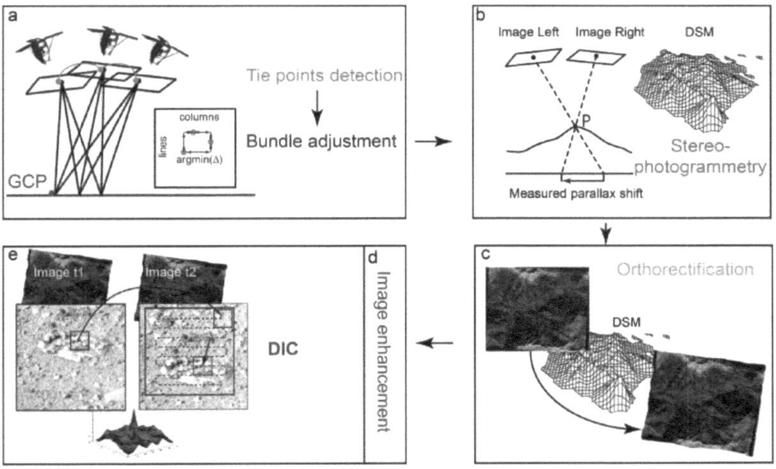

Figure 5.2. *Overview of a typical preprocessing chain for DIC-based displacement measurements comprising; a) tie-point detection and bundle adjustment including the optional use of ground control points (GCPs) for the adjustment and validation, b) stereo-photogrammetric surface reconstruction, c) orthorectification, d) image enhancement and; e) the image correlation step. For a color version of this figure, see www.iste.co.uk/baghdadi/6.zip*

Variations in the satellite's viewing angle between two acquisitions yield an apparent shift of the observed objects, known as parallax shift (Figure 5.2(b)), which is directly related to the topographic height of the terrain. Especially in mountainous environments, this effect has to be corrected before measurement of the actual ground displacement. If the image datasets comprise stereo-pairs, they can be exploited to generate DSMs that can be used in a subsequent step for the orthorectification of the images (Figure 5.2(c)). Other DEM sources such as national databases (digital terrain models [DTMs][1]), global and freely available DSMs derived from SRTM, ASTER and Advanced Land Observing Satellite / Panchromatic Remote-sensing Instrument for Stereo Mapping (ALOS PRISM), or regional LiDAR scans (preferable DSMs) can be used for the orthorectification. When using external sources of topographic data with a resolution that approaches the uncertainty of the sensor model, the use of GCPs should be considered to assure the coregistration of the topographic dataset with the image projections.

During orthorectification, the position of each pixel is corrected for the geometric effects of the terrain and the resulting orthoimage depicts the landscape as it would appear if each location had been imaged exactly from above the location. This step requires a resampling of the image and care must be taken to use resampling schemes that avoid aliasing effects, which might bias the motion detected through image correlation [ING 07, LEP 07a]. Commonly used resampling methods such as nearest neighbor, bilinear or bicubic interpolation are fast but introduce aliasing. They should be combined with a low-pass filter or replaced by sinc interpolation to avoid biases in subsequent displacement measurements. Recent initiatives by space agencies to open data archives are facilitating access to readily orthorectified images. It has been reported that the coregistration of Landsat-8 images can be sufficiently precise for direct application of DIC without further preprocessing [AVO 14]. For archived data from previous Landsat missions, coregistration errors can easily exceed 50 m but can be reduced using image matching and a second-order polynomial to model the miss-registration among the images [DEH 15]. In this case, care must be taken to not consider areas with actual surface motion when estimating the image transformation.

The use of image enhancement techniques before the actual image correlation step (Figure 5.2(d) and (e)) can, in some cases, help to increase the robustness and accuracy of the displacement measurements. A commonly used approach when dealing with multiband images is the use of

1 DIC measurements are often limited to areas without significant canopies where DSMs and DTMs correspond closely, since matching becomes increasingly difficult with denser vegetation.

principal component analysis (PCA), which enables to concentrate and exploit information from several different bands and may help to compensate radiometric variations in time series [DEH 15]. Though displacement measurements from cross-correlation of images from different satellites are still rather rare, its feasibility has been demonstrated using a regression analysis to radiometrically align image bands from ASTER and SPOT [NEC 09]. Another technique to reduce the impact of radiometric differences is to compute gradient images before the actual correlation step, which is closely related to the original optical flow formulation of the image matching problem. This is often achieved using a Sobel filter, and the correlation is then carried out on the magnitude of the gradient [TRA 12] or the orientation of the gradient [FIT 02, HEI 12].

5.4. Principles of optical image correlation

As mentioned previously, image matching techniques are important for many applications and can be classified as intensity-based matching, feature-based matching and relational matching techniques. Here, we will focus on intensity-based matching techniques that are also commonly termed "image correlation" and are the most commonly used approach to solve the following problem.

Let I_1 and I_2 be two perfectly coregistered gray scale images of the earth surface taken from the same distance at time t_1 and t_2, respectively. The intensity values at given image location $P = [x\ y]^T$ can be denoted as $I_1(x, y)$ and $I_2(x, y)$, respectively. The objective of image matching is to find for any given point P_1 in I_1 the corresponding point location $P_2 = [x + \Delta x\ \ y + \Delta y\ ]^T$ in I_2, where Δx and Δy are the two components of a simple translation[2]. Typically, Δx denotes the translation along the lines and Δy along the columns of the images. Elementary to the problem of finding corresponding points is the need for a formal description of similarity among the two images that can be exploited to implement a search algorithm. A straightforward approach to this can be taken by making the assumption that the brightness values of the two images remain constant so that the

2 Many processes such as glaciers, landslides or earthquake generate complex deformation patterns that also involve rotation and shear. However, in practice, locally piecewise translation is often a very good approximation that allows decomposing more complex motion patterns.

intensity in the first image will correspond to the intensity in the second image after the offset:

$$I_1(x,y) = I_2(x + \Delta x, y + \Delta y) \quad [5.1]$$

For small displacements (less than 1 pixel), the right-hand side of the above equation can be approximated using first-order Taylor approximation:

$$I_2(x + \Delta x, y + \Delta y) \approx I_2(x,y) + \frac{\partial I}{dx}\Delta x + \frac{\partial I}{dy}\Delta y \quad [5.2]$$

where $\frac{\partial I}{dx}$ and $\frac{\partial I}{dy}$ denote the intensity gradients along the x and y axes, respectively. Together this leads to the expression:

$$\left(I_2(x,y) - I_1(x,y)\right) + \frac{\partial I}{dx}\Delta x + \frac{\partial I}{dy}\Delta y \approx 0 \quad [5.3]$$

This is the basic formulation of the optical flow problem, which is a differential matching technique. With some additional assumptions such as that the motion field is globally smooth [HOR 81] or constant within the size of a certain interrogation window [LUC 81], the equation can be solved for the displacement vector $d = [\Delta x \;\; \Delta y\,]^T$. In recent years, numerous enhanced algorithms for solving the optical flow problem have been proposed including faster and more robust methods that can handle larger displacements and determine the displacement at sub-pixel precision (http://vision.middlebury.edu/flow/eval, [BAK 11]). Nevertheless, optical flow techniques are generally more suitable to quantify small displacements among images with little noise and small brightness variations and are predominantly used for near-real-time processing of video streams for motion detection and 3D reconstruction.

Region-based matching techniques are generally considered more suitable for remote sensing images since they are more robust to noise and small changes of the surface aspect resulting from the larger time intervals between two image acquisitions or changes in the sensor geometry. The basic idea for region-based matching techniques is illustrated in Figure 5.3. A template from the first image with a constant neighborhood size $(2\omega_x + 1) \times (2\omega_y + 1)$ is compared iteratively to the second image to evaluate the function $s(\Delta x, \Delta y)$ for all possible values of Δx and Δy. Prior knowledge on the maximum expected displacement ($\Delta x_{\max}$ and $\Delta y_{\max}$) can be used to define a search window size, which reduces the likelihood of false matches and the computational load. This can also be seen as a sliding window operation, where the template is moved over the search window to compute $s(\Delta x, \Delta y)$ for every possible displacement vector $d = [\Delta x \ \ \Delta y\,]^T$.

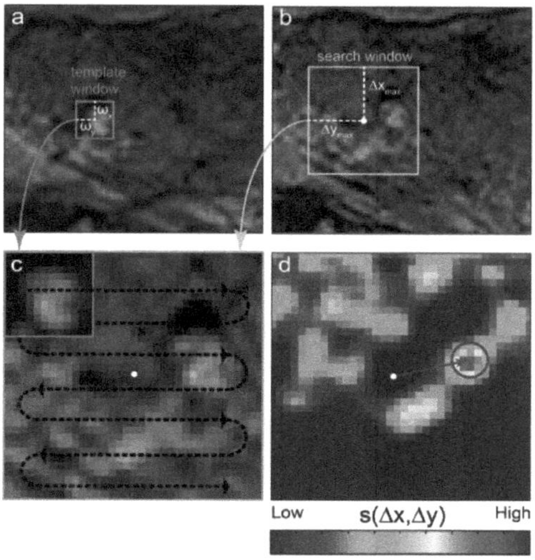

Figure 5.3. *Schematic illustration of a region-based image matching algorithm using normalized cross-correlation (NCC): a) A template is extracted from the first image (the template size is defined by ω_x and ω_y); b) To reduce the computational load, the search is typically centered at the estimated position of the template in the image (white point) and constrained to a search space of the size Δx_{max} by Δy_{max}; c) For each location in the search space, the similarity to the template is evaluated in a sliding window fashion; d) The global maximum in the resulting similarity map (here using NCC) typically marks the position of the template in the second image from which the displacement vector can be calculated. For a color version of this figure, see www.iste.co.uk/baghdadi/6.zip*

The computed function is commonly called a distance measure when smaller values of the function indicate greater similarity. Alternatively, it is called a similarity measure, when greater values of the function indicate greater similarity. A classical distance measure that can be computed very efficiently is the sum of squared difference (SSD):

$$SSD(\Delta x, \Delta y) = \sum_{x-\omega_x}^{x+\omega_x} \sum_{y-\omega_y}^{y+\omega_y} (I_1(x,y) - I_2(x+\Delta x, y+\Delta y))^2 \qquad [5.4]$$

If the intensity values of the patches being compared are exactly the same, the SSD equals 0. The global minimum of the SSD matrix signals the most likely position of the search template within the search window. For the illustration in Figure 5.3, the normalized cross-correlation (NCC) – one of the most popular similarity measures in region-based matching – was used. In this case, the global maximum of the resulting correlation matrix marks the most likely position of the template from I_1 in the image I_2. In this context, it is important to note that even if the template pattern is not present in the search window, there will still be a global minima or maxima. It is therefore important to also define a minimum threshold for a correct match. In practice, the matching is repeated with templates for all $P_1 = [x\ y]^T$ to obtain a dense displacement map for the entire scene.

5.4.1. *Similarity–distance measures and search strategies*

In the previous section, the concept of distance and similarity measures was introduced and we also discussed a simple search strategy that enables us to limit the number of possible matches that need to be tested. While a multitude of similarity–distance measures have been proposed in the scientific literature, only a few of them are commonly used in geoscientific studies and reviewed here in greater detail. We have already introduced SSD that falls into the large group of difference-based measures. SSD is robust to global offsets of the image intensities but it suffers from local offsets and gain changes (multiplicative changes of the intensities). This can be compensated by subtracting the mean from the two templates leading to the zero-mean sum of squared difference (ZSSD). Correlation-based similarity measures (from which DIC derives its name) are inherently robust to both intensity offsets and gain changes and have proven very robust for many practical applications. In particular, the NCC (also known as Pearson's r or zero-mean normalized cross-correlation) is widely used for motion measurement. To denote the NCC, we change the notation slightly to now only consider the template window $w_1(x - \Delta x, y - \Delta y)$ and a section of the search window with the same size being $w_2(x, y)$:

$$NCC = \frac{\sum_{xy}(w_2(x,y) - \overline{w_2}).(w_1(x-\Delta x, y-\Delta y) - \overline{w_1})}{\sqrt{\sum_{xy}(w_2(x,y) - \overline{w_2})^2 . \sum_{xy}(w_1(x-\Delta x, y-\Delta y) - \overline{w_1})^2}} \quad [5.5]$$

Here $\overline{w_1}$ denotes the mean of the template patch and $\overline{w_2}$ denotes the mean of the search patch. The normalization by the variance and mean makes the NCC robust against radiometric offsets and gain changes.

It is known that for applications where the motion field comprises significant discontinuities, correlation-based similarity measures are sensitive to local outliers and thus often smooth out discontinuities [HIR 09]. Non-parametric similarity measures such as the rank filter or census filter can be better suited in such cases and have shown better performance in stereo-photogrammetric applications where the motion field is a result of the camera movement and exploited to recover depth information [HIR 09]. In the context of displacement measurements in a remote sensing context, however, motion fields are mostly smooth and other factors such as noise and geometric distortion are likely to be more important.

The traditional similarity measure, which is considered for the coregistration of multimodal images, is mutual Information (MI). MI measures the statistical dependency between two signals and is therefore very robust against strong radiometric changes and changes in the signal characteristics resulting from imaging devices [WOO 15]. Compared to NCC, MI requires relatively large window sizes to obtain stable estimates of the local image statistics. It is often used for coregistration of optical and SAR images, stereo-photogrammetry or displacements measurements from polarimetric SAR data but less commonly for matching of optical remote sensing images.

A somewhat different approach to evaluating the difference among two images patches is the use of phase correlation that is based on the Fourier shift theorem stating that a shift in the image domain is equivalent to a phase shift in the frequency domain. Let us consider again two similar image patches $w_2(x, y)$ and $w_1(x - \Delta x, y - \Delta y)$ with a relative displacement of $d = [\Delta x \ \Delta y]^T$ and denote the Fourier transforms with W_1 and W_2. The Fourier shift theorem states that:

$$W_2(\xi_x, \xi_y) = W_1(\xi_x, \xi_y)e^{-i(\xi_x \Delta x + \xi_y \Delta y)} \quad [5.6]$$

where ξ_x and ξ_y denote the frequency components of the columns and rows of the image. The normalized cross-power spectrum of the two image patches is defined as:

$$\frac{W_1(\xi_x,\xi_y)W_2{}^*(\xi_x,\xi_y)}{|W_1(\xi_x,\xi_y)W_2{}^*(\xi_x,\xi_y)|} = e^{i(\xi_x \Delta x + \xi_y \Delta y)} \quad [5.7]$$

where $W_2{}^*$denotes the complex conjugate of W_2. As a result of the Fourier shift theorem, the phase of the cross-power spectrum will be equivalent to the phase difference between the two image patches. By applying the inverse Fourier transform, we can then obtain a 2D function with a sharp peak centered at $[\Delta x \;\; \Delta y\,]^T$:

$$F^{-1}\{e^{i(\xi_x \Delta x+\xi_y \Delta y)}\} = \delta(x + \Delta x, y + \Delta y) \qquad [5.8]$$

One advantage of this technique is fast computation, since the discrete Fourier transformation can be computed very efficiently using the fast Fourier transform (FFT) algorithm. This is particularly true if the templates that should be matched are rather large. The technique is also very robust to noise affecting only certain spatial frequencies, which is typically the case for remote sensing images where noise is present mainly at high spatial frequencies. Only if all frequency bands are equally affected by white noise, the sharp peak tends to be blurred. A disadvantage of Fourier-based methods can be encountered when the images depict periodically repeating patterns (e.g. waves) leading to several ambiguous peaks in the delta function.

5.4.2. *Sub-pixel precision*

All matching techniques discussed so far provide only integer pixel values for $[\Delta x \;\; \Delta y\,]^T$and their precision is thus limited by the resolution of the remote sensing image. However, several different methods are available to estimate the displacement at sub-pixel precision.

If phase correlation is used, the peak of the delta function typically spreads over several pixels. It can be shown both theoretically and in practice that the shape of the peak can be approximated very well with a 2D sinc function. The maxima of the fitted function then mark the match location with sub-pixel precision [FOR 02].

Leprince *et al.* [LEP 07a] noted that resampling of the images before matching (e.g. orthorectification) and other image defects can bias the peak location, and advocate reliance on another phase correlation technique where the spatial shift is estimated directly from the normalized cross-spectrum. In this case, the 2D phase ramp in the normalized cross-spectrum is iteratively approximated with a theoretical function. The slope of the best fitting function is directly related to the phase shift and hence the displacement $[\Delta x \; \Delta y]^T$. The advantage of this approach is that it allows estimating the shift directly at a theoretical sub-pixel precision in the order of 1/50th of a pixel.

For region-based matching techniques that use cost function in the spatial domain (e.g. NCC), three basic strategies to achieve sub-pixel precision can be distinguished. Those are an interpolation of the input images, a parametric curve fit (similar to the peak fitting for phase correlation described previously), or the interpolation of the similarity function. The three approaches are illustrated for the 1D case in Figure 5.4. Empirical studies show that when NCC is used as a similarity measure, upsampling the input images seems to be more accurate than interpolating the similarity function or parametric peak fitting [DEB 1]. A disadvantage of this approach is the strong increase in computational complexity, since the matching has to be performed on upsampled input images that are significantly larger than the originals.

It is important to note that sub-pixel matching techniques can theoretically achieve precision of 1/50th of a pixel; however, in practice, due to noise, geometric distortions and residual registration errors, it is very difficult to obtain accuracies better than one-fifth of a pixel [LAC 15, STU 14a].

5.4.3. *Search strategies*

It has been already discussed in this chapter that prior knowledge on the maximum displacement can be very useful in reducing the size of the search window in the second image; this approach can greatly reduce the computational complexity and risk of false matches (section 5.4). Another commonly used search scheme with similar objectives is a coarse-to-fine strategy often also referred to as hierarchical cross-correlation. The underlying idea is to generate several coarser resolution (e.g. Laplacian image pyramids) representations of the original images and start the correlation at the coarsest level. The displacement measured at the coarser level is then propagated as an initial guess to the next finer level and can be used to guide the matching (e.g. constrain the size of the search window). Advantages of this technique are that no initial guess regarding the maximum displacement is needed and a great reduction in the computational load since the initial estimates can be computed very fast at coarse resolution and will limit the search space at a finer resolutions [ANA 89]. The matching of larger salient image features at a coarser resolution also allows considering information from multiple spatial scales and reduces the probability of false matches. Nevertheless, hierarchical strategy still includes the risk that false matches at coarse levels are propagated to higher resolutions. To avoid such issues, back-tracking and consistency checks among the various levels should be used in hierarchical matching schemes [ZIT 03].

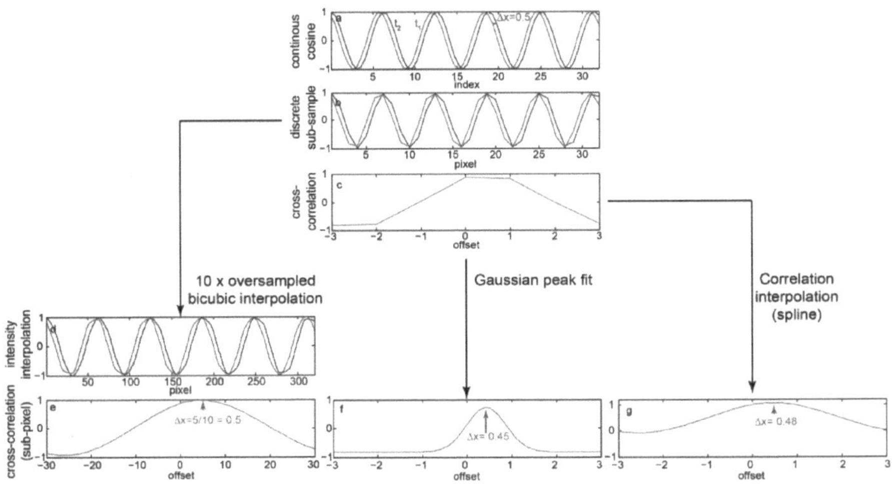

Figure 5.4. *One-dimensional (1D) illustration of three approaches used to achieve sub-pixel accuracy: a) Two continuous cosine functions with an offset of 0.5 representing the underlying signals; b) Discrete samples of the two functions at integer (pixel) intervals representing the sampling of the signal by an optical sensor; c) The cross-correlation peak at pixel resolution indicating an offset of 0 pixel; d) The first approach is oversampling of the discrete input signal to 10-fold higher resolution, which leads to e) a better localization of the correlation peak at 0.5 pixel; f) The second approach is to fit a Gaussian distribution to the cross-correlation function and allow a good approximation of the correlation peak; g) The third alternative is interpolation of the correlation function to localize the maximum at sub-pixel precision. For a color version of this figure, see www.iste.co.uk/baghdadi/6.zip*

Closely related to the use of coarse-to-fine strategies is the use of adaptive template sizes and shapes. Such techniques aim to address one of the short-comings of fixed square windows, which assume only a simple translation and do not take into account rotation or shear deformation. With sufficiently small template sizes, it is often possible to approximate more complex deformations comprising rotation and shear with a series of translations, whereas the signal-to-noise ratio (SNR) typically decreases when using smaller templates making the matching less robust. To find an optimal template size, it has been proposed to iteratively test multiple sizes and optimize the size with respect to the SNR [DEB 12a]. More complex deformations can also be modeled directly during the matching with a least-square fit of an affine deformation of the template [DEB 12b]. Such techniques, however, remain computationally very expensive and are not yet used very frequently, and dense piecewise translations measured with small fixed template sizes (e.g. 5 × 5 pixels) are generally a good approximation of more complex deformation patterns.

Though optimized versions of algorithms for image correlation in the spatial and frequency domain have been developed [LEW 95], it still remains very costly to compute dense motion fields for every pixel in large satellite images. While coarse-to-fine search strategies can alleviate the computation time, sub-sampling is another commonly employed strategy. If displacement vectors are computed, for example, only for every 10 × 10 pixels, the number of operations can be reduced by a factor of 100, which greatly speeds up the computation. If required, the resulting vector fields are then interpolated to the full image resolution. Since for small neighborhoods, the images' gray values and displacement fields are very similar (autocorrelated), the loss of information due to sub-sampling is often rather small and justified by the speed-up in computation.

A strategy that exploits the spatial autocorrelation of the motion field to increase the robustness is spatial regularization. In this case, a second term is added to the similarity measure, which takes into account deviation of motion vectors among surrounding pixels favoring matches that lead to a smooth displacement fields and rejecting matches that would lead to spikes in the displacement field [PIE 06]. Such techniques typically comprise a weighting factor that allows giving more or less importance to the similarity measure or spatial smoothness, respectively. Spatial regularization is of particular interest when using coarse-to-fine strategies, since it significantly reduces the number of outliers and prevents their propagation from coarser to finer levels.

5.4.4. *Postprocessing*

Despite great advances in image matching and geometric corrections, displacement fields resulting from DIC often still comprise significant amounts of mismatches and artifacts from geometric distortions that require special attention before a thematic interpretation of the measurements.

A common artifact resulting from the architecture of modern pushbroom satellites are stripe-like artifacts in the along-track and across directions of the satellite path (lines and columns of the images). Since such artifacts have a rather particular spatial frequency, neighboring pixels on stable ground (e.g. along a profile) can be used to compensate. Errors in the DEM used for orthorectification are another common source of geometric errors, which can lead to apparent displacements over stable areas and errors in the measurements of real surface motion. The magnitude of the measurement error (e_d) is thereby directly related to the incidence angles at which the two satellite images have been recorded (θ_1, θ_2) and the error in the topographic model e_h (Figure 5.5):

$$e_d = e_h . (tan(\theta_1) - tan(\theta_2)) \qquad [5.9]$$

For a convergence angle (the angle between the incidence angles) of 20° and $e_h = 6$ m, for example, this would lead to an error of more than 2 m in the measured displacement. This illustrates that such artifacts can be reduced by selecting image pairs with relatively small base lines and hence small convergence angles. This implies that the DEM should ideally be updated for each date if a strong vertical component of the motion is to be expected. While in practice this is often still not feasible, stereo-pairs or tri-stereo acquisitions from VHSR optical satellites are being used more and more frequently to update topographic information in DIC processing chains [LAC 15, STU 14a]. Systematic errors in the DEM (e.g. dependent on aspect and slope) can also be taken into account by fitting regression models over stable terrain [SCH 08], whereas random errors require longer time series that may allow averaging their contribution out [DEH 15] or detecting errors due to inconsistencies in the directional variance of the displacement field [STU 15a].

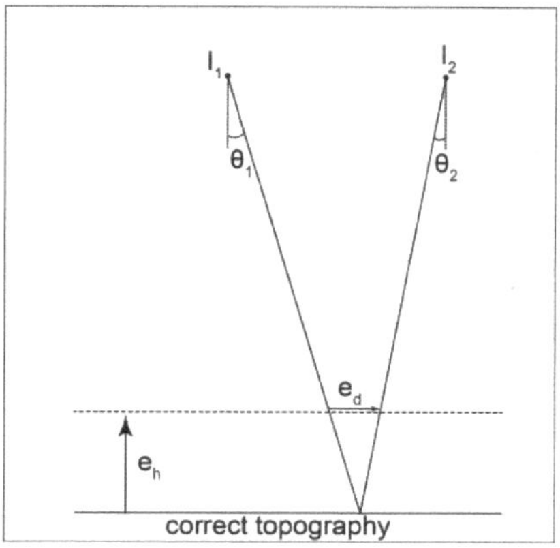

Figure 5.5. *Relationship between incidence angles (θ_1, θ_2) at which two images (I_1, I_2) have been recorded, DEM error (e_h) and the resulting error in the measured displacement (e_d) for a stable point*

Another postprocessing option is to filter measurements with low confidence according to the employed distance/similarity measure or spatial filtering to replace outliers that are not consistent with a smooth motion field [LEP 07a]. In particular, for gravitational mass-movements, it can be generally assumed that the direction of the movement will closely follow the gradient of the maximum slope, which can be exploited to remove vectors that significantly deviate from the direction of the slope [SCH 08, STU 14a]. Vegetated surfaces are typically associated with significant errors in DEMs derived from photogrammetric techniques and pose challenges for image matching techniques in general. Multispectral remote sensing images can be exploited to generate vegetation masks to discard such areas and reduce the amount of false positive matches [STU 14a].

5.4.5. *Algorithm implementation*

Many of the algorithms and techniques outlined earlier are readily implemented in software packages and are frequently used in geoscientific studies. One of the earliest tools that became available to the geoscientific community is IMCORR, which implements the technique described in Scambos *et al.* [SCA 92], which is still often used for applications such as tracking and glacier flow [GLA 11]. The source code for IMCORR can be downloaded from the website of the U.S. National Snow and Ice

Datacenter (http://nsidc.org/data/velmap/imcorr.html). A more efficient Python implementation based on IMCORR has been introduced in [FAH 16] but is not yet available publicly at the time of writing.

The probably most widely used tool for DIC-based motion measurements is COSI-Corr [AYO 15, LEP 07b], which is widely used for measurements of coseismic slip on fault lines. The software implements an advanced sub-pixel phase correlation algorithm as well as NCC-based matching, comprises several modules for geometric corrections of satellite and aerial images, and tools for postprocessing and visualization. The software is available as an executable plugin for ENVI with a graphical user interface from the website mentioned by the authors. A somewhat similar approach has also been proposed in [GON 10].

The image correlation software Correlation Image Analysis (CIAS) [HEI 12, KAA 00] has been developed and used mainly for the monitoring of glaciers and periglacial landforms. It currently includes two algorithms; an NCC-based correlator and an implementation of orientation correlation [FIT 02]. All algorithms are implemented in IDL and can be downloaded in executable form from the Web site mentioned by the authors.

More recently, the MicMac library – originally written for aerial and terrestrial photogrammetry – has been extended for the measurements of horizontal displacement fields at sub-pixel precision [ROS 15b]. It implements a hierarchical image correlation technique combining NCC-based matching and spatial regularization and has shown promising results for measurements of coseismic slip [ROS 15b], landslide motion and dune migration [STU 14b]. Source code or precompiled binaries can be downloaded as part of MicMac open-source library that is hosted at IGN France (http://logiciels.ign.fr/?Telechargement,20).

Extensions of pairwise matching to multiple pairwise matching have been proposed recently [JEO 16, STU 15b] to increase redundancy of the measurements and thereby the robustness against outliers. In principle, such techniques can be used as an extension of any image correlation algorithm. Other tools include, for example, the hierarchical multiscale image correlator described in [TRA 12] for landslide monitoring or the ImGRAFT toolbox for glacier monitoring [MES 15]. Many further algorithm implementations have been developed by various research groups for mechanical strain analysis. Such tools are not commonly used for remote sensing applications but might be potentially interesting for some users. A comparison of seven different approaches has been represented in [BOR 09] and further algorithms with publicly available implementations have been described in [GUI 08] and [BLA 15]. Comparisons of some common implementations for different applications can be found in [HEI 12] and [STU 14b].

5.5. Applications

5.5.1. *Application to active tectonics*

The assessment of seismic hazards depends to a large degree on knowledge of the slip distribution and recurrence time of major earthquakes. Considerable uncertainties arise from the fact that it is often unclear if a particular fault has in the past experienced ruptures that have activated the entire fault or a multitude of smaller events distributed over segments of the fault. DIC has proven to be a unique tool for measuring the coseismic surface displacement that occurs during large earthquakes. Since DIC of optical images is only sensitive to the horizontal motion component, studies mainly focus on fault lines with a strong horizontal component (mainly strike-slip or thrust faults) and events whose depth and magnitude allow significant displacement at the surface. The magnitude of coseismic slip is typically also investigated in the field and often measured from the offset of landscape features such as roads, channels or gullies, or through repeated mobile Differential Global Positioning System (dGPS) and permanent dGPS surveys. The number of dGPS points and landscape features is typically limited to a few points, and field surveys are generally time-consuming. In this context, DIC has the advantage that it provides dense motion fields over large areas and theoretically very soon after a given event (depending on local satellite orbits and cloud cover).

Also InSAR techniques [PIN 16] are frequently used to assess the magnitude of coseismic motion. While InSAR is generally more accurate than DIC, it is mostly limited to measurements along the line-of-sight of the satellite and less sensitive to horizontal motion toward the north and south. Strong motion near the fault line can often lead to a loss of coherence and, consequently, a lack of measurements. In such cases, DIC is an important complementary tool for quantifying the horizontal components of the motion near the fault.

Below, we provide a brief review of a selection of some historical earthquakes that have been analyzed with DIC and are summarized in Figure 5.6. Figure 5.6(a) displays the motion field of the Kashmir earthquake that occurred in 2005. With a total death toll of approximately 80,000 people, it was the most devastating earthquake in the Himalayas during the 20th Century. The study revealed a relatively simple fault geometry and continuous rupture. The analysis underlined that faults outside the main Himalayan thrust belt can rupture over longer distances than previously thought and must be considered in seismic hazard assessment for the region [AVO 06].

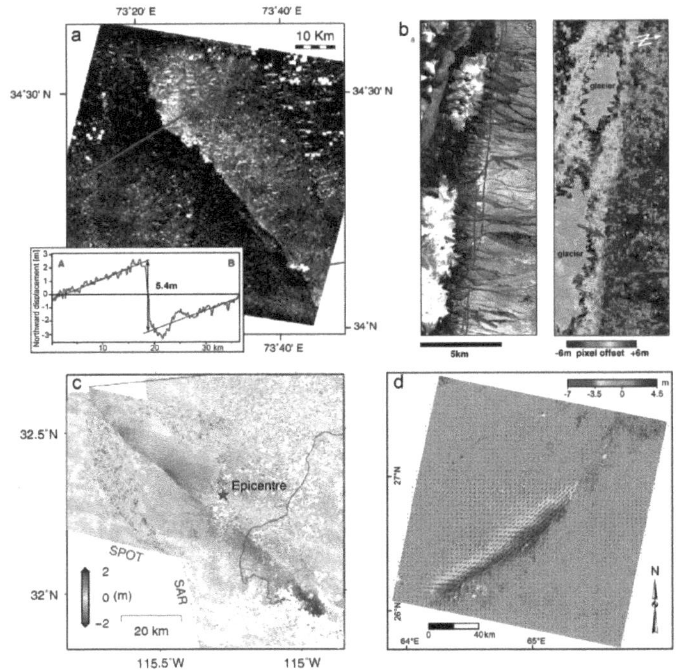

Figure 5.6. *Horizontal coseismic surface displacement measurements for selected earthquakes: a) North–south coseismic displacement from the Kashmir earthquake (2005) measured from a pair of ASTER images [AVO 06]; b) East–west coseismic displacement from the Kokoxili earthquake (2001) derived from cross-correlation of SPOT images [KLI 06]; c) North–south (red: northward, blue: southward) component of coseismic displacement from the El Mayor-Cucapah earthquake (2010) derived from cross-correlation of SPOT and SAR amplitude images [WEI 11]; d) East–west magnitude and motion vectors of the slip caused by the Balochistan earthquake (2013) derived from cross-correlation of two Landsat-8 pairs [AVO 14]. For a color version of this figure, see www.iste.co.uk/baghdadi/6.zip*

Figure 5.6(b) depicts a small section of the fault rupture of the Kokoxili earthquake that occurred in Tibet in 2001. The study showed a strong geometric control of geometric barriers along the fault line, a strong segmentation along many fault segments and an abrupt end of the rupture at a geometric barrier. It could be deduced that the segmentation of the fault has important implications for the propagation of the rupture and the likelihood for the generation of future earthquakes along neighboring faults [KLI 06].

In Figure 5.6(c), we can clearly identify the fault trace of a predominantly strike-slip fault that ruptured during the El Mayor-Cucapah earthquake in 2010. The horizontal surface displacement suggested a simple strike-slip geometry. However, the integration of evidence from various sources into a finite-fault source model showed that the sub-surface fault geometry was rather more complex and that the event started from normal fault rupture, which in turn triggered strike-slip motion in two opposite directions along the fault rupture [WEI 11].

Figure 5.6(d) depicts the coseismic surface displacement caused by the Balochistan earthquake in 2013 with a very smooth fault rupture without major geometric segmentation. The strong strike-slip component of the rupture is somewhat at odds with the relatively shallow dip angle of the fault geometry and the strong rotation of the slip vector along the fault [AVO 14]. This has led to an ongoing discussion which includes the possibility that the fault mechanism may alter between strike slip and dip-slip from one earthquake to another.

5.5.2. *Applications to ice- and rock glaciers*

Ice- and rock glaciers are of crucial importance in regional hydrological cycles and are sensitive to changes in the global climate. Glacier retreat may therefore lead to seasonal reductions of fresh water supplies that will affect the downstream environment and local population. Glacier fluctuations can also induce natural hazards such as outbursts of glacial lakes, avalanches and debris flows. Especially for fast-moving glaciers with displacement rates of up to several meters per day and frequent changes of their surface features, it remains very challenging to use techniques such as InSAR or *in situ* techniques such as permanent dGPS. As a result, DIC is often the only feasible approach to quantify the glacier velocities over larger areas and with comprehensive spatial coverage. Furthermore, remote sensing archives such as the Landsat database can be exploited to gather historical information on the glacier extents and velocities up to the present day and allow studying the long-term response of glaciers to environmental changes.

The outcomes of a recent study on the monitoring of the Greenland ice sheet outlet glaciers from Landsat images covering the period of 1999–2012 are presented in Figure 5.7(a). In accordance with previous investigations, the study found the many of the outlet glaciers have undergone acceleration during this period. It, however, also highlighted that there are also many glaciers whose surface velocity does not show a clear trend or have even slowed down during the past decade. Long-term and seasonal trends are generally very inhomogeneous in space and time and can be influenced by thinning of the ice sheet and structural changes in the glacier connectivity. This emphasizes the importance of considering such variables when relating changes in the glacier velocities to climate change [ROS 15a].

Figure 5.7(b) presents an average displacement rate for glaciers in the Karakoram mountains obtained through the analysis of a subset of the Landsat archive spanning from 1999 to 2001. The exploitation of long time series is not only useful to decipher long-term trend but also helps to increase the spatial coverage and reduces the measurement uncertainty [DEH 15]. Similar studies have shown a great variability in the glacier response to climate change depending on the debris covers, topography and precipitation regime [SCH 11].

While glaciers can easily reach displacement rates of hundreds of meters per year, rock glaciers feature significantly slower displacement rates making cross-correlation of aerial photographs an ideal tool to study the long-term behavior. Figure 5.7(c) shows the motion field of the Muragl rock glacier in Switzerland, which has an average speed of about 0.5 m/year. The study showed that most of the rock glacier's displacement field is relatively smooth, while some significant shear occurs on inactive side lobes [KAA 02].

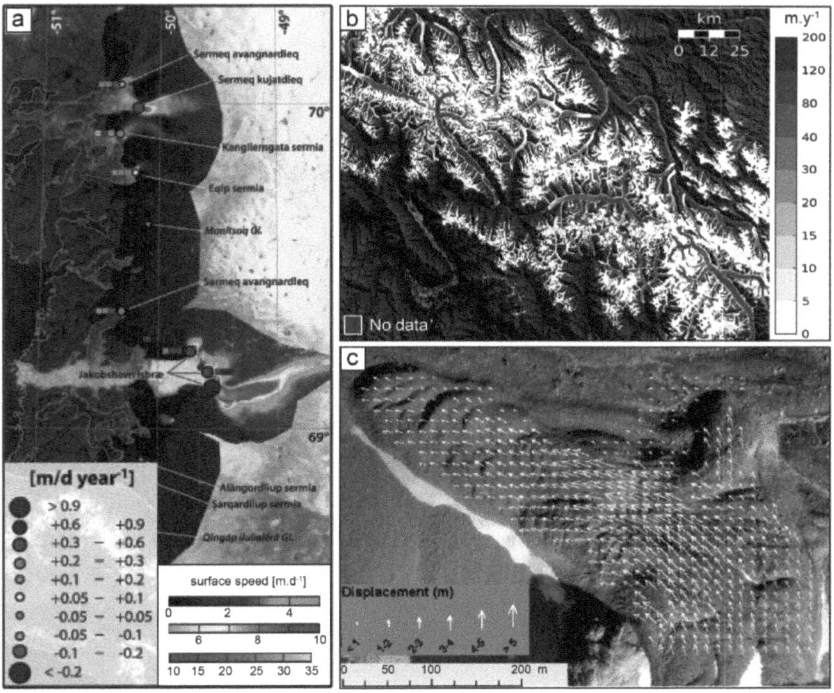

Figure 5.7. *Examples of surface displacement measurements on glaciers and rock glaciers: a) Median displacement rates (2010–2012) and changes in displacement rates for individual outlet glaciers (circles) of Central West Greenland ice sheet derived from the correlation of Landsat images from the period 1999–2012. Color-coded squares indicate the estimates of the flow-velocity trend for the three periods 1999–2003, 2004–2007 and 2008–2012 [ROS 15a]; b) Annual velocity of the Karakoram glaciers (Pakistan, India and China) for the year 2000 derived from cross-correlation of multiple Landsat images [DEH 15]; c) Surface displacement of the Muragl rock glacier (Switzerland) for the period 1981–1994 measured by cross-correlation of aerial images [DEB 11]. For a color version of this figure, see www.iste.co.uk/baghdadi/6.zip*

5.5.3. *Applications to landslides*

Landslides are a major natural hazard with socio-economic impacts worldwide and geomorphic agents that can dominate the long-term denudation rates in mountainous environments. The reduction of landslide hazard not only depends on spatio-temporal information about the frequency and distribution of landslides, and their predisposing and triggering factors, but also on the quantification and understanding of landslide kinematics. Similar to ice- and rock glaciers, landslides often occur in mountainous terrain where access is difficult and *in situ* measurements with dGPS, extensometers or similar devices are limited to few selected points. Spaceborne SAR interferometry has been used in a number of landslide investigations but provides limited spatial coverage due to layover, foreshortening and shadowing effects and, in particular, variable displacement rates in space and time often render phase unwrapping difficult. DIC of optical images is still used less frequently for landslide than for glacier flow or coseismic displacement but due to the outlined limitations of other techniques and the increasing availability of VHSR optical images, it is becoming an important tool to quantify the horizontal components of landslide motion.

Figure 5.8(a) shows the displacement field of the La Valette landslide, which is the largest earthflow in the French Alps. The largest motion has been observed at the main scarp exceeding 1 m over the 2 months between the acquisitions of the two Pléiades VHSR satellite images. The spatial distribution of the motion indicates a significant accumulation of material in the transit zone of the landslide. The study demonstrated that VHSR satellite images can be exploited to observe long-term trends and seasonal variations of the motion field with decimeter accuracy [STU 14a].

Slow-moving landslides (up to several meters per month) such as earthflows are a very common phenomenon in clay-rich rock formations and their kinematics can be very variable depending on local and regional hydro-meteorological conditions. Figure 5.8(b) shows the displacement field of the La Valette landslide (same as Figure 5.8(a)) measured through correlation of two aerial photographs with a time lag of 22 months. The spatial pattern of the motion is very similar to the observations obtained from VHSR satellite images (up to 15 m at the main scarp, accumulation in the transit zone of the landslide) and indicates a relative constant activity of the landslide over longer time periods [TRA 11].

Partial results of a further investigation on the relationship of landslide motion and seismic shaking are presented in Figure 5.8(c). The study applied DIC on VHSR Pléiades satellite images to quantify the displacement rates of nine slow-moving landslides in Colca Valley (Peru) before and after a nearby Mw6.0 earthquake at shallow depth. Comparison with point-wise dGPS measurements corroborates an accuracy of approximately 15 cm. It also showed a contrast of pre- and postseismic landslide motion, which is more pronounced for landslides closer to the epicenter, which suggests a causal relationship between shaking intensity and landslide acceleration in the valley [LAC 15].

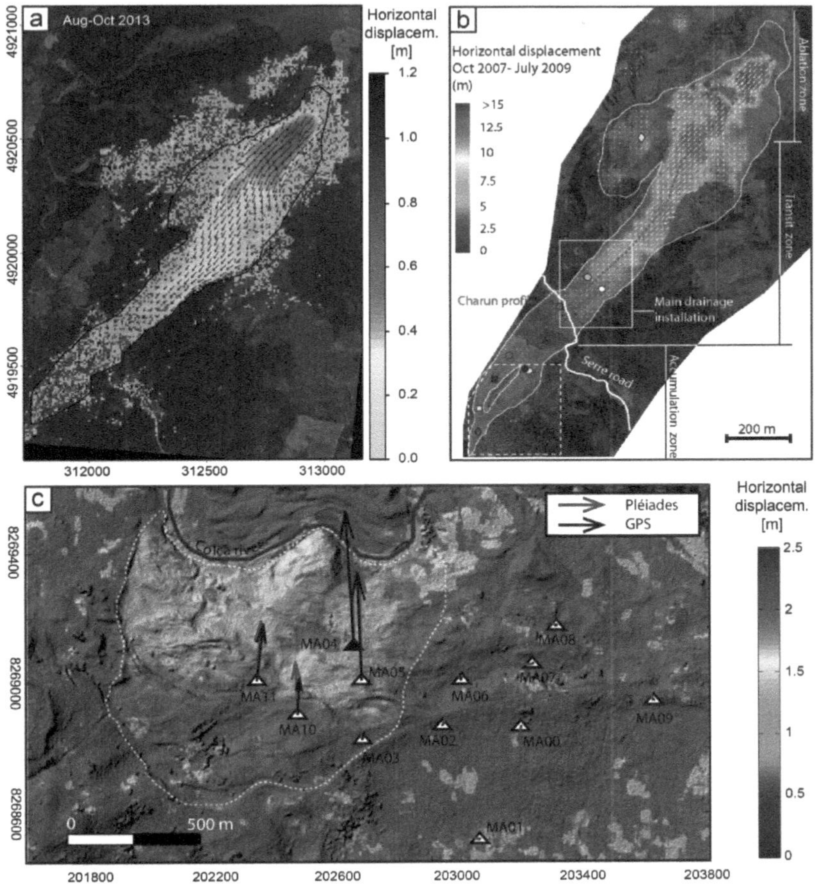

Figure 5.8. *Examples for surface displacement measurements on landslides: a) Surface velocities of the La Valette landslide (France) for August to October 2013 derived from cross-correlation of Pléiades satellite images [STU 14a]; b) Surface velocities of the La Valette landslide from October 2007 till July 2009 derived from cross-correlation of aerial photographs [TRA 11]; c) Surface displacement of the Maca landslide (Peru) between March and July 2013, derived from Pléiades satellite images. White and black triangles show the positions of repeated (March and September 2013) and permanent GPS measurements, respectively [LAC 15]. For a color version of this figure, see www.iste.co.uk/baghdadi/6.zip*

5.5.4. *Quantification of dune migration*

The formation and migration of dunes are processes that can be encountered in wide range of environmental conditions comprising terrestrial deserts and coast lines, submarine dunes on the continental shelf and extraterrestrial environments such as Mars and Venus. Despite the differences in environmental conditions and driving forces, they share common dynamical mechanism over a wide range of spatial scales [CLA 06]. The dynamics and scaling laws of dune formation are frequently investigated experimentally, whereas for observations of their long-term dynamics, remote sensing is often the only information source. Here, we briefly review two studies that employed image correlation on satellite images and submarine DEMS for the quantification of dune migration.

Figure 5.9(a) presents displacement vectors for a series of fast-moving barchan dunes in northern Chad obtained through correlation of SPOT, Landsat and ASTER images. The study showed that the celerity of the dunes remained relatively constant for a time period of 45 years (1965–2010). This suggests that the windiness in the central Sahara has changed less than 0.13% and that during this time period anthropogenic climate change did not have major impacts on the wind regime in the region [VER 12].

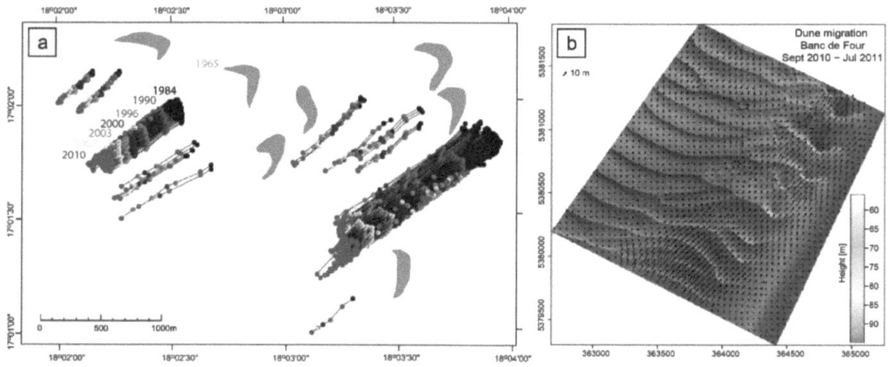

Figure 5.9. *Examples for surface displacement measurements on sand dunes migration: a) Long-term motion vectors of the world's largest and fastest barchan dunes in northern Chad. The displacement was measured combining SPOT, Landsat and ASTER images. Gray areas mark the footprints of the dunes in an older analog Corona (KH-4) spy satellite image [VER 12]; b) Two-dimensional motion field of giant submarine sand dunes at the Banc de Four (offshore western Brittany, France) measured from cross-correlation of multibeam echo sounding surface models (based on data described in [FRA 13]). For a color version of this figure, see www.iste.co.uk/baghdadi/6.zip*

Figure 5.9(b) depicts the dense motion field of a giant submarine dune formation on the Banc de Four (offshore western Brittany, France). Unlike most applications presented in this chapter, DIC was not applied on optical data but on DEMs of the seafloor acquired with a time lag of 10 months through multibeam echo sounding. The derived motion field shows two converse migration patterns with a maximum displacement of more than 20 m. The converse motion between the northwestern dune field and the eastern dune field indicates spatially variable asymmetries of the tidal currents during flood and ebb conditions [FRA 13].

5.6. Current limitations and perspectives

After more than 20 years of progress in the development and application of DIC, it is now a commonly used tool to measure horizontal surface motion related to a multitude of earth surface processes. Several factors have contributed to this trend including new correlation algorithms that deliver sub-pixel precision as well as the enhanced availability of satellite images with greater geometric accuracy and better spatial, temporal and radiometric resolution.

Despite this progress, DIC-based measurements can still comprise significant uncertainties that users should bear in mind when applying such techniques. The theoretical precision of sub-pixel correlation algorithms can reach 1/50th of the image resolution, whereas in real-world applications, radiometric noise, illumination changes and surface deformation often impose a limit one-fifth of a pixel even under ideal conditions. Further limitations can arise from geometrical inaccuracies of the input data including imperfect sensor models, coregistration residuals and orthorectification errors linked to the DEM quality, but also from study site characteristics such as dense vegetation cover, moving cast shadows, apparent movement of specular reflectance features, and mismatches due to low contrast or strong surface changes. In particular, areas with rugged topography and dense vegetation cover still pose challenges for DIC-based measurement. Quantities such as correlation coefficients or the SNR are generally good indicators for determining areas where the quality of the matching is poor but carry little information regarding the accuracy of the measurements. Given the variety of influential factors, it remains generally difficult to estimate measurement uncertainties *a priori*, and an error assessment on known stable terrain and (if available) ground truth data should be considered from case study to case study.

Despite the increasing fleet of optical satellites, it can often still be difficult to obtain image time-series with dense temporal sampling or shortly after major events (e.g. earthquakes) from the same satellite due to cloud cover or conflicting imaging schedules for VHSR steerable platforms. Furthermore, the planned lifetime for many platforms is often in the range of a half decade, which complicates the construction of long-term time series. Notable exceptions are long-term high-resolution satellite missions such as Landsat, SPOT and the recently launched Sentinel-2 mission with a repeat pass cycle of 10 days (5 days with two satellites in 2016) and a spatial resolution of up to 10 m. In particular, Sentinel-2 and Landsat-8 significantly enhance the spatio-temporal sampling and are further improving the capabilities to monitor and analyze short-term and long-term earth surface motion. At the same time, there is a multitude of VHR satellites and aerial platforms operating with variable incidence angles, spatial resolutions and radiometric characteristics. Geometric correction routines and DIC techniques are still often not sufficiently robust to enable matching despite geometric and radiometric differences. Further research on the combination of images from different satellite missions could enable a more systematic exploitation of images from different optical sensors and the construction of longer and temporally denser time-series. While the elimination of outliers with spatial filters or thresholds on the SNR is a common approach, there is still room for improvement regarding the use of multitemporal information to enhance the robustness of DIC. Some examples for the use of multiple pairwise matching to improve the robustness and reduce the uncertainties of DIC-based measurement have been presented recently in the context of the monitoring of mass movements such as glaciers and landslides [DEH 15, JEO 16, STU 15b]. However, those approaches remain limited to the exploitation of images from one single satellite mission.

Measurements of horizontal motion provide valuable constraints on the process dynamics and forcing but only depict two components of the 3D motion and often fall short to fully capture the complexity of gravitational or coseismic motion. While this is an inherent limitation, when analyzing 2D images further processing techniques and data sources such as InSAR,

LiDAR and stereo-photogrammetry can be exploited to retrieve full 3D motion fields. SAR interferometry, for example, can be used to estimate the vertical component of the motion and complement 2D displacement measurements derived from optical DIC [VAL 15]. Direct sources for 3D data in the form of 3D point clouds or gridded DEMs comprise the stereo-photogrammetric processing of satellite [BER 14, STU 14a], UAV [LUC 14] and terrestrial [STU 15c] images, but also multitemporal aerial laser and terrestrial laser scans [TRA 14]. An advantage of such datasets is that 3D displacements can be derived directly and, in particular, LiDAR offers the unique ability to penetrate vegetation and measure the terrain motion in forested areas where radar and optical imaging often fall short. A common approach to derive 3D motion vectors from 3D point clouds is the use of 3D matching techniques such as iterative closest point (ICP) introduced by [BES 92] or least squares 3D surface matching (LSSM) described first by Gruen and Akca [GRU 05]. Such techniques work well, in particular, for piecewise matching of two 3D point clouds if there is little noise and the general surface shape is well preserved over time. Applications include, for example, measurements of 3D coseismic motion close in the near-field of the fault line, where InSAR methods often suffer from decorrelation (Figure 5.10(a)). A disadvantage of such techniques, however, is their sensitivity to noise and lack of convergence when the surface has undergone significant deformation. In this case, the interpolation of raster grid images from the point clouds and matching on derived gradient images can be more robust for reconstructing the full 3D motion. Figure 5.10(b) provides a corresponding example where terrestrial multitemporal stereo imaging was used for multiple surface reconstructions and the derivation of the 3D motion field of a rock cliff. Similarly, 3D motion can also be retrieved from multitemporal satellite stereo-pairs. A promising development in this direction has recently been presented by [LEP 14] showing the possibility of reconstructing 3D motion directly from the images without the need for the intermediate generation of surface models. It should, however, be noted that height measurements from stereo-images are generally affected by greater uncertainties than the horizontal components derived from orthoimages. Those uncertainties typically amount to two times the pixel resolution and larger, depending on the surface texture and slope [BER 14, STU 14a]. As multitemporal point clouds and stereo-images become more commonly available, further research is still needed to compare the performance of the outlined methods for different scenarios.

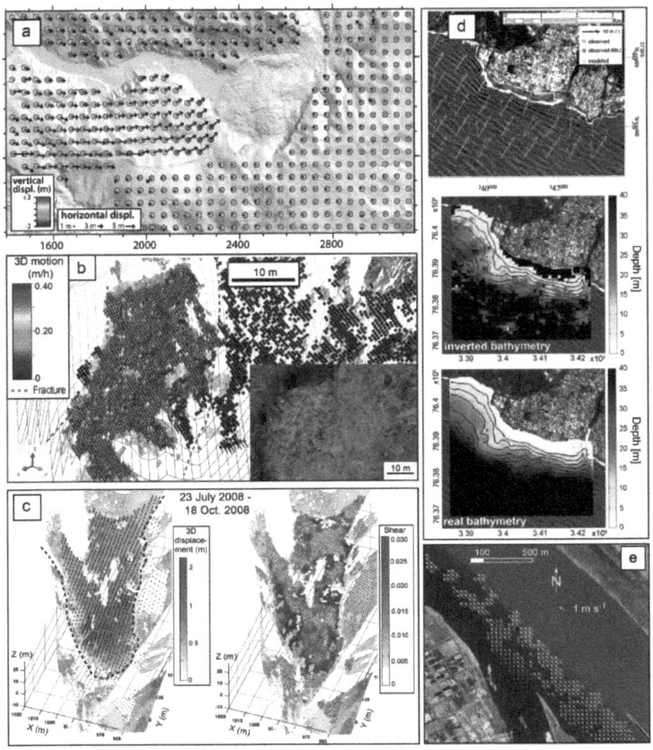

Figure 5.10. *A selection of innovative techniques that exploit non-classical data sources to derive 3D displacement, physical parameters and near-real-time motion: a) Near-field 3D displacement for the 2008 Iwate–Miyagi earthquake (Japan) obtained from ICP-based matching of pre- and postevent laser point clouds [NIS 14]; b) Three-dimensional displacement field derived from image-based correlation of stereo-photogrammetric point cloud series. The perspective view shows the Rosselin cliff (Valais, Switzerland) 1 hour before a major collapse. The inset shows a picture of the cliff taken by one of the stereo-cameras at a distance of 80 m; c) Three-dimensional displacement at the toe of the Super Sauze landslide (France; July 2008 to October 2009) derived from image-based correlation of terrestrial laser scans and the corresponding shear strain field [TRA 14]; d) Ocean wave velocities at the coast of La Reunion Island (February 6, 2010) observed through cross-correlation of nearly simultaneous panchromatic and multispectral images ($\Delta t = 2.04\ s$) acquired by SPOT-5 and a comparison against predictions from a swell propagation model [DEM 12]. The measured velocities have been used to invert the depth of the seafloor along the coast, which compares relatively well with reference bathymetric measurements [POU 14]; e) Suspended sediment velocities at the inlet of the Tennessee River into the Ohio River (USA) derived from cross-correlation of aerial photographs acquired with a time lag of 30 sec [KAA 14]. For a color version of this figure, see www.iste.co.uk/baghdadi/6.zip*

The use of DIC in earth science is often not only the first step to understanding general dynamics of tectonic and geomorphic processes but also targets a deeper understanding of the underlying mechanical processes and driving forces. The integration of displacement measurements with physical models is now commonly used for the modeling of the geometry and slip distribution of tectonic faults [AVO 06, AVO 14]. Similarly in glaciology, inverse models are routinely used to infer variables such as ice thickness and bed topography of glaciers [MCN 12] or to constrain the rheological properties of ice shelves from remotely sensed surface velocities [KHA 07]. So far, only a few studies have explored the possibility of deriving kinematic and mechanical parameters for landslides from remotely sensed motion fields including surface deformation and mass flow [BRU 06], surface strain (Figure 5.10(c)) as well as the slip surface geometry and rheological parameters [BOO 13]. Those few examples show the great potential of DIC as a tool to constrain physical and statistical models and eventually enhance our understanding of the underlying mechanics as well as short- and long-term predictions. While a comprehensive review of such inverse problems is beyond the scope of this chapter, it can be stated that the variety of factors that can influence the motion (geometry, temperature, heterogenous rheology, etc.) typical require initial guesses of the estimated quantities and precautions against overfitting that will ultimately depend on the particular phenomena and study site. For landslides, for example, further research is still needed to evaluate the reliability of different modeling approaches and inverted parameters since the landslide kinematics may range from rigid block sliding to complex flow-like behavior. In particular, the availability of reference measurements (e.g. laboratory tests, boreholes) that correspond with the timing of the image acquisition remains a major challenge for the model validation. Finally, it should also be noted that the use of prior knowledge and physical models can be used in turn to improve DIC measurements. This can achieved with simple heuristics on the direction and velocities of the movement [JEO 16, ROSE 15], but may also include dynamic models to regularize remotely sensed displacement fields [MAK 16].

While all above-mentioned examples for the application of DIC involve time lags of several weeks to years, new satellite missions such as Sentinel-2, VENμS or Dove are reducing the repeat pass-cycles to 2–5 days. The Skysat satellites even possess video imaging capabilities for small areas. In this context, DIC has the potential to become an important tool for near-real-time monitoring. It will therefore be interesting to consider techniques from experimental mechanics that often address continuous measurements and approaches that benefit from dense temporal sampling through temporal regularization [BES 12]. Many aerial and orbital imaging systems also acquire images nearly simultaneously as a side effect of the system design [KAA 14]. A notable example for the use of such image sequences is presented in Figure 5.10(d). Coastal wave velocities are measured from panchromatic and multispectral SPOT-5 image bands that are recorded with a short time delay. The derived motion field is subsequently used to invert the water depth, which shows a good agreement with bathymetric reference data [POU 14]. Image sequences with such short time lags offer many further potential applications such as velocity measurements of sediments and floating objects in rivers (Figure 5.10(e)) or the movement of cars and ships.

Throughout this chapter, we have emphasized several times that the fleet of optical earth observation satellites is continuously increasing. This has several benefits such as an enhanced spatio-temporal sampling, free data access (e.g. Sentinel-2, Landsat-8) and declining prices for images from commercial providers. Nevertheless, the increasing volume and variety of datasets also poses challenges regarding efficient and versatile processing. Easy-to-use tools such as COSI-Corr have already contributed greatly to more widespread use of DIC in the geo-planetary sciences, and new open-source libraries [ROS 15b] offer new possibilities of developing streamlined processing chains from stereo-images to the motion fields. However, to take full advantage of the large quantity of historical and daily incoming data for long-term monitoring and rapid response, it will also be important to implement such processing chains for high-performance and cloud-based infrastructures.

5.7. Conclusions

This chapter has reviewed the use of DIC for measurements of earth surface motion focusing mainly on optical images including historical and theoretical aspects, software implementations, potential data sources and applications as well as current limitations and prospects. It shows that DIC is a versatile and mature technology that is today frequently employed in diverse geoscientific studies to measure 2D horizontal motion fields. Commonly used image sources are, in particular, HSR and VHSR satellite images that, due to numerous recent launches of new platforms such as

Pléiades and Sentinel-2, are becoming available at increasingly better temporal frequencies and lower costs. This trend is complemented by a more widespread use of UAVs, terrestrial imaging, as well as aerial and terrestrial laser scans that can provide valuable additional information regarding the vertical component of the 3D surface motion. After several decades of research on the use of DIC, its limitations and potential are well understood and several commercial and open-source software solutions for DIC are available to end user from various domains.

Nevertheless, there are still major challenges regarding the use of DIC for operational monitoring and rapid response after major events (e.g. earthquakes). While bottlenecks regarding data accessibility are less and less relevant, the end-to-end processing of optical images often still comprises several complex steps (e.g. DSM generation, orthorectification, coregistration, DIC and postprocessing) that need to rely on multiple tools. This hinders the set-up of fully integrated processing chains, which would be necessary to automatically process the ever-increasing amounts of optical remote sensing data and put into place near-real-time monitoring facilities that can deliver standardized products shortly after the image acquisition. Related aspects that will require further attention to fully leverage the potential of DIC are the integration of imagery from different sensor platforms, long time series and 3D information with physical models of the investigated process.

5.8. Key points

Digital image correlation is an image processing technique that allows automatic detection of corresponding points between two images with a precision that is even greater than the actual resolution of the images. It consequently allows measurement of horizontal surface motion from two or more images of the same area acquired at different time points. It is frequently used to study the velocity of glaciers, landslides and sand dunes or to quantify the surface motion of earthquakes from aerial or satellite images. It typically provides two horizontal components of the surface motion and is therefore complementary to line-of-sight measurements from InSAR. DIC is typically less precise than InSAR measurements but less limited by the magnitude of the displacement. With the increasing fleet of optical earth observation satellites and available image archives, it has become a very useful technique for studying climatic and tectonic processes that drive surface motion over large, otherwise often inaccessible areas.

5.9. Acknowledgments

This work was jointly supported by the European Commission through the EC-FP7 Large-scale Integrating Project, *Safeland: Living with landslide risk in Europe* (Contract 226479), and by the Council of Europe through the European and Mediterranean Major Hazards Agreement Project, *Development of cost-effective ground-based and remote -monitoring system for detecting debris flow/landslide initiation*. The authors are thankful to the three anonymous reviewers who helped us to improve the quality of this Chapter.

5.10. Bibliography

[ADR 84] ADRIAN R.J., YAO C.S., "Development of pulsed laser velocimetry (PLV) for measurement of turbulent flow", in PATTERSON G., ZAKIN J.L. (eds), *Eighth Biennial Symposium on Turbulence*, Rolla, MO, pp. 170–186, 1984.

[ANA 89] ANANDAN P., "A computational framework and an algorithm for the measurement of visual motion", *International Journal of Computer Vision*, vol. 2, no. 3, pp. 283–310, 1989.

[AVO 04] AVOUAC J.P., AYOUB F., WEI S. *et al.*, "The 2013, Mw 7.7 Balochistan earthquake, energetic strike-slip reactivation of a thrust fault", *Earth and Planetary Science Letters*, vol. 391, pp. 128–134, 2004.

[AVO 06] AVOUAC J.P., AYOUB F., LEPRINCE S. *et al.*, "The 2005, Mw 7.6 Kashmir earthquake: sub-pixel correlation of ASTER images and seismic waveforms analysis", *Earth and Planetary Science Letters*, vol. 249, no. 3, pp. 514–528, 2006.

[AYO 15] AYOUB F., LEPRINCE S., AVOUAC J.P., User's guide to COSI-CORR co-registration of optically sensed images and correlation, California Institute of Technology, p. 49, 2015.

[BAK 11] BAKER S., SCHARSTEIN D., LEWIS J. *et al.*, "A database and evaluation methodology for optical flow", *International Journal of Computer Vision*, vol. 92, no. 1, pp. 1–31, 2011.

[BEL 14] BELTAOS S., KÄÄB A., "Estimating river discharge during ice breakup from near-simultaneous satellite imagery", *Cold Regions Science and Technology*, vol. 98, pp. 35–46, 2014.

[BER 76] BERNSTEIN R., "Digital image processing of earth observation sensor data", *IBM Journal of Research and Development*, vol. 20, no. 1, pp. 40–57, 1976.

[BER 14] BERTHIER E., VINCENT C., MAGNÚSSON E. *et al.*, "Glacier topography and elevation changes derived from Pléiades sub-meter stereo images", *Cryosphere*, vol. 8, no. 6, pp. 2275–2291, 2014.

[BES 92] BESL P., MCKAY N., "A method for registration of 3-D shapes", *IEEE Transactions on Pattern Analysis and Machine Intelligence*, vol. 14, pp. 239–256, 1992.

[BES 12] BESNARD G., LECLERC H., HILD F. *et al.*, "Analysis of image series through global digital image correlation", *Journal of Strain Analysis for Engineering Design*, vol. 47, no. 4, pp. 214–228, 2012.

[BLA 15] BLABER J., ADAIR B., ANTONIOU A., "Ncorr: open-source 2D digital image correlation Matlab software", *Experimental Mechanics*, vol. 55, no. 6, pp. 1105–1122, 2015.

[BOO 13] BOOTH A.M., LAMB M.P., AVOUAC J.P. *et al.*, "Landslide velocity, thickness, and rheology from remote sensing: La Clapière landslide, France", *Geophysical Research Letters*, vol. 40, no. 16, pp. 4299–4304, 2013.

[BOR 09] BORNERT M., BRÉMAND F., DOUMALIN P. *et al.*, "Assessment of digital image correlation measurement errors: methodology and results", *Experimental Mechanics*, vol. 49, no. 3, pp. 353–370, 2009.

[BRI 12] BRIDGES N., AYOUB F., AVOUAC J. *et al.*, "Earth-like sand fluxes on Mars", *Nature*, vol. 485, no. 7398, pp. 339–342, 2012.

[BRU 03] BRÜCKL E., BRUNNER F.K., KRAUS K., "Kinematics of a deep-seated landslide derived from photogrammetric, GPS and geophysical data", *Engineering Geology*, vol. 88, nos. 3–4, pp. 149–159, 2003.

[CAS 03] CASSON B., DELACOURT C., BARATOUX D. *et al.*, "Seventeen years of the "La Clapière" landslide evolution analysed from ortho-rectified aerial photographs", *Engineering Geology*, vol. 68, nos. 1–2, pp. 123–139, 2003.

[CHA 16] CHAZETTE P., TOTEMS J., HESPEL L. *et al.*, "Principle and physics of the LiDAR measurement", in BAGHDADI N., ZRIBI M. (eds), *Optical Remote Sensing of Land Surfaces*, ISTE Press Ltd, London and Elsevier Ltd, Oxford, 2016

[CLA 06] CLAUDIN P., ANDREOTTI B., "A scaling law for aeolian dunes on Mars, Venus, Earth, and for subaqueous ripples", *Earth and Planetary Science Letters*, vol. 252, no. 1, pp. 30-44, 2006.

[COL 06] COLESANTI C., WASOWSKI J., "Investigating landslides with space-borne synthetic aperture radar (SAR) interferometry", *Engineering Geology*, vol. 88, nos. 3–4, pp. 173-199, 2006.

[CRI 91] CRIPPEN R.E., BLOM R.G., "Measurement of subresolution terrain displacements using spot panchromatic imagery", *International Geoscience and Remote Sensing Symposium*, IGARSS'91, pp. 1667–1670, 1991.

[DEB 11] DEBELLA-GILO M., KÄÄB A., "Sub-pixel precision image matching for measuring surface displacements on mass movements using normalized cross-correlation", *Remote Sensing of Environment*, vol. 115, no. 1, pp. 130–142, 2011.

[DEB 12a] DEBELLA-GILO M., KÄÄB A., "Locally adaptive template sizes for matching repeat images of Earth surface mass movements", *ISPRS Journal of Photogrammetry and Remote Sensing*, vol. 69, pp. 10–28, 2012.

[DEB 12b] DEBELLA-GILO M., KÄÄB A., "Measurement of surface displacement and deformation of mass movements using least squares matching of repeat high resolution satellite and aerial images", *Remote Sensing*, vol. 4, no. 1, pp. 43–67, 2012.

[DEH 15] DEHECQ A., GOURMELEN N., TROUVE E., "Deriving large-scale glacier velocities from a complete satellite archive: application to the Pamir–Karakoram–Himalaya", *Remote Sensing of Environment*, vol. 162, pp. 55–66, 2015.

[DEL 04] DELACOURT C., ALLEMAND P., CASSON B. *et al.*, "Velocity field of the "La Clapière" landslide measured by the correlation of aerial and QuickBird satellite images", *Geophysical Research Letters*, vol. 31, no. 15, p. L15619, 2004.

[DEM 07] DE MICHELE M., BRIOLE P., "Deformation between 1989 and 1997 at Piton de la Fournaise volcano retrieved from correlation of panchromatic airborne images", *Geophysical Journal International*, vol. 169, no. 1, pp. 357–364, 2007.

[DEM 12] DE MICHELE M., LEPRINCE S., THIÉBOT J. *et al.*, "Direct measurement of ocean waves velocity field from a single SPOT-5 dataset", *Remote Sensing of Environment*, vol. 119, pp. 266–271, 2012.

[FAH 16] FAHNESTOCK M., SCAMBOS T., MOON T. *et al.*, "Rapid large-area mapping of ice flow using Landsat 8", *Remote Sensing of Environment*, vol. 119, pp. 266–271, 2016.

[FER 16] FERRO-FAMIL L., POTTIER E., "Coherent diversity mode SAR imaging: SAR polarimetry, interferometry and tomography", in BAGHDADI N., ZRIBI M. (eds), *Microwave Remote Sensing of Land Surfaces*, ISTE Press Ltd, London and Elsevier Ltd, Oxford, 2016.

[FIT 02] FITCH A.J., KADYROV A., CHRISTMAS W.J. *et al.*, "Orientation correlation", *British Machine Vision Conference*, pp. 133–142, 2002.

[FOR 02] FOROOSH H., ZERUBIA J., BERTHOD M., "Extension of phase correlation to subpixel registration", *IEEE Transactions in Image Processing*, vol. 11, no. 3, pp. 188–200, 2002.

[FRA 13] FRANZETTI M., LE ROY P., DELACOURT C. *et al.*, "Giant dune morphologies and dynamics in a deep continental shelf environment: example of the banc du four (Western Brittany, France)", *Marine Geology*, vol. 346, pp. 17–30, 2013.

[GLA 11] GLASSER N., SCAMBOS T., BOHLANDER J. *et al.*, "From ice-shelf tributary to tidewater glacier: continued rapid recession, acceleration and thinning of Röhss Glacier following the 1995 collapse of the Prince Gustav Ice Shelf, Antarctic Peninsula", *Journal of Glaciology*, vol. 57, no. 203, pp. 397–406, 2011.

[GON 10] GONZALEZ P.J., CHINI M., STRAMONDO S. *et al.*, "Coseismic horizontal offsets and fault-trace mapping using phase correlation of IRS satellite images: the 1999 Izmit (Turkey) earthquake", *IEEE Transactions on Geoscience and Remote Sensing*, vol. 48, no. 5, pp. 2242–2250, 2010.

[GRO 03] GRODECKI J., DIAL G., "Block adjustment of high-resolution satellite images described by rational polynomials", *Photgrammetric Engineering & Remote Sensing*, vol. 69, no. 1, pp. 59–68, 2003.

[GRU 05] GRUEN A., AKCA D., "Least squares 3D surface and curve matching", *ISPRS Journal of Photogrammetry and Remote Sensing*, vol. 59, no. 3, pp. 151–174, 2005.

[GRU 12] GRUEN A., "Development and status of image matching in photogrammetry", *Photogrammetric Record*, vol. 27, no. 137, pp. 36–57, 2012.

[GUI 08] GUIZAR-SICAIROS M., THURMAN S.T., FIENUP J.R., "Efficient subpixel image registration algorithms", *Optics Letters*, vol. 33, no. 2, pp. 156–158, 2008.

[HEI 12] HEID T., KÄÄB A., "Evaluation of existing image matching methods for deriving glacier surface displacements globally from optical satellite imagery", *Remote Sensing of Environment*, vol. 118, pp. 339–355, 2012.

[HIR 09] HIRSCHMULLER H., SCHARSTEIN D., "Evaluation of stereo matching costs on images with radiometric differences", *IEEE Transactions on Pattern Analysis and Machine Intelligence*, vol. 31, no. 9, pp. 1582–1599, 2009.

[HOF 58] HOFMANN W., *Bestimmung von Gletschergeschwindigkeiten aus Luftbildern*, Wichmann, 1958.

[HOL 12] HOLLINGSWORTH J., LEPRINCE S., AYOUB F. *et al.*, "Deformation during the 1975–1984 Krafla rifting crisis, NE Iceland, measured from historical optical imagery", *Journal of Geophysical Research: Solid Earth (1978–2012)*, vol. 117, no. B11, 2012.

[HOR 81] HORN B.K., SCHUNCK B.G., "Determining optical flow", *Technical Symposium East: International Society for Optics and Photonics*, pp. 319–331, 1981.

[IMM 14] IMMERZEEL W., KRAAIJENBRINK P., SHEA J. *et al.*, "High-resolution monitoring of Himalayan glacier dynamics using unmanned aerial vehicles", *Remote Sensing of Environment*, vol. 150, pp. 93–103, 2014.

[ING 07] INGLADA J., MURON V., PICHARD D. *et al.*, "Analysis of artifacts in subpixel remote sensing image registration", *IEEE Transactions on Geoscience and Remote Sensing*, vol. 45, no. 1, pp. 254–264, 2007.

[JEO 16] JEONG S., HOWAT I.M., AHN Y., "Improved multiple matching method for observing glacier motion with repeat image feature tracking", *IEEE Transactions on Geoscience and Remote Sensing*, available at: https://bpcrc.osu.edu/sites/bpcrc.osu.edu/files/gdg-publicationsimproved-multiple-matching-method.pdf, 2016.

[KAA 00] KÄÄB A., VOLLMER M., "Surface geometry, thickness changes and flow fields on creeping mountain permafrost: automatic extraction by digital image analysis", *Permafrost and Peroglacial Processes*, vol. 11, pp. 315–326, 2000.

[KAA 02] KÄÄB A., "Monitoring high-mountain terrain deformation from repeated air- and spaceborne optical data: examples using digital aerial imagery and ASTER data", *ISPRS Journal of Photogrammetry and Remote Sensing*, vol. 57, nos. 1–2, pp. 39–52, 2002.

[KAA 14] KÄÄB A., LEPRINCE S., "Motion detection using near-simultaneous satellite acquisitions", *Remote Sensing of Environment*, vol. 154, pp. 164–179, 2014.

[KAU 12] KAUFMANN V., "The evolution of rock glacier monitoring using terrestrial photogrammetry: the example of Äußeres Hoch-ebenkar rock glacier (Austria)", *Austrian Journal of Earth Sciences*, vol. 105, no. 2, pp. 63–77, 2012.

[KHA 07] KHAZENDAR A., RIGNOT E., LAROUR E., "Larsen B ice shelf rheology preceding its disintegration inferred by a control method", *Geophysical Research Letters*, vol. 34, no. 19, 2007.

[KLI 06] KLINGER Y., MICHEL R., KING G.C., "Evidence for an earthquake barrier model from Mw~ 7.8 Kokoxili (Tibet) earthquake slip-distribution", *Earth and Planetary Science Letters*, vol. 242, no. 3, pp. 354–364, 2006.

[LAC 15] LACROIX P., BERTHIER E., MAQUERHUA E.T., "Earthquake-driven acceleration of slow-moving landslides in the Colca valley, Peru, detected from Pléiades images", *Remote Sensing of Environment*, vol. 165, pp. 148–158, 2015.

[LEE 71] LEESE J.A., NOVAK C.S., CLARK B.B., "An automated technique for obtaining cloud motion from geosynchronous satellite data using cross correlation", *Journal of Applied Meteorology*, vol. 10, no. 1, pp. 118–132, 1971.

[LEP 07a] LEPRINCE S., BARBOT S., AYOUB F. *et al.*, "Automatic and precise orthorectification, coregistration, and subpixel correlation of satellite images, application to ground deformation measurements", *IEEE Transactions on Geoscience and Remote Sensing*, vol. 45, no. 6, pp. 1529–1558, 2007.

[LEP 07b] LEPRINCE S., AYOUB F., KLINGER Y. *et al.*, "Co-registration of optically sensed images and correlation (COSI-Corr): an operational methodology for ground deformation measurements", *IEEE International Geoscience and Remote Sensing Symposium*, IGARSS'07, Barcelona, Spain, pp. 2700–2702, 23–28 July 2007.

[LEP 14] LEPRINCE S., AYOUB F., LIN J. *et al.*, "COSI-Corr: a solution for 3D change detection using optical imagery", *Joint Agency Commercial Imagery Evaluation-Civil Commercial Imagery Evaluation Workshop*, Louisville, 2014.

[LEW 95] LEWIS J., "Fast normalized cross-correlation", *Vision Interface*, vol. 10, no. 1, pp. 120–123, 1995.

[LOW 04] LOWE D.G., "Distinctive image features from scale-invariant keypoints", *International Journal of Computer Vision*, vol. 60, no. 2, pp. 91–110, 2004.

[LUC 81] LUCAS B.D., KANADE T., "An iterative image registration technique with an application to stereo vision", *7th International Joint Conference on Artificial Intelligence (IJCAI)*, Vancouver, pp. 674–679, 24–28 August 1981.

[LUC 86] LUCCHITTA B., FERGUSON H., "Antarctica: measuring glacier velocity from satellite images", *Science*, vol. 234, no. 4780, pp. 1105–1108, 1986.

[LUC 14] LUCIEER A., JONG S.M.D., TURNER D., "Mapping landslide displacements using structure from motion (SfM) and image correlation of multi-temporal UAV photography", *Progress in Physical Geography*, vol. 38, no. 1, pp. 97–116, 2014.

[MAK 16] MAKSYMIUK O., MAYER C., STILLA U., "Velocity estimation of glaciers with physically-based spatial regularization – Experiments using satellite SAR intensity images", *Remote Sensing of Environment*, vol. 172, pp. 190–204, 2016.

[MCN 12] MCNABB R., HOCK R., O'NEEL S. *et al.*, "Using surface velocities to calculate ice thickness and bed topography: a case study at Columbia Glacier, Alaska, USA", *Journal of Glaciology*, vol. 58, no. 212, pp. 1151–1164, 2012.

[MES 15] MESSERLI A., GRINSTED A., "Image georectification and feature tracking toolbox: ImGRAFT", *Geoscientific Instrumentation, Methods and Data Systems*, vol. 4, no. 1, pp. 23–34, 2015.

[MIC 06] MICHEL R., AVOUAC J.P., "Coseismic surface deformation from air photos: the Kickapoo step over in the 1992 Landers rupture", *Journal of Geophysical Research: Solid Earth (1978–2012)*, vol. 111, no. B3, 2006.

[MUS 11] MUSTE M., HO H.C., KIM D., "Considerations on direct stream flow measurements using video imagery: outlook and research needs", *Journal of Hydro-environment Research*, vol. 5, no. 4, pp. 289–300, 2011.

[NEC 09] NECSOIU M., LEPRINCE S., HOOPER D.M. *et al.*, "Monitoring migration rates of an active subarctic dune field using optical imagery", *Remote Sensing of Environment*, vol. 113, no. 11, pp. 2441–2447, 2009.

[NIS 14] NISSEN E., MARUYAMA T., ARROWSMITH J.R. *et al.*, "Coseismic fault zone deformation revealed with differential LiDAR: examples from Japanese Mw~ 7 intraplate earthquakes", *Earth and Planetary Science Letters*, vol. 405, pp. 244–256, 2014.

[PET 82] PETERS W., RANSON W., "Digital imaging techniques in experimental stress analysis", *Optical Engineering*, vol. 21, no. 3, 213427, 1982.

[PIE 13] PIECZONKA T., BOLCH T., JUNFENG W. *et al.*, "Heterogeneous mass loss of glaciers in the Aksu-Tarim Catchment (Central Tien Shan) revealed by 1976 KH-9 Hexagon and 2009 SPOT-5 stereo imagery", *Remote Sensing of Environment*, vol. 130, pp. 233–244, 2013.

[PIE 06] PIERROT-DESEILLIGNY M., PAPARODITIS N., "A multiresolution and optimization-based image matching approach: an application to surface reconstruction from Spot 5- HRS stereo imagery", *Proceedings of ISPRS Workshop "Topographic Mapping from Space"*, Part 1/W41, Ankara, Turkey, vol. 36, 14–16 February 2006.

[PIN 16] PINEL V., RAUCOULES D., "The contribution of SAR data to volcanology and subsistence studies", in BAGHDHDI N., ZRIBI M. (eds), *Land Surface Remote Sensing*, ISTE Press Ltd, London and Elsevier Ltd, Oxford, 2016

[POU 14] POUPARDIN A., DE MICHELE M., RAUCOULES D. *et al.*, "Water depth inversion from satellite dataset", *IEEE International Geoscience and Remote Sensing Symposium*, IGARSS'14, Québec, Canada, pp. 2277–2280, 13–18 July 2014.

[RIC 46] RICHTER H., "Photogrammetrische Eisbeobachtung und Eismessung", *Polarforschung*, vol. 16, nos. 1–2, pp. 94–97, 1946.

[ROS 15a] ROSENAU R., SCHEINERT M., DIETRICH R., "A processing system to monitor Greenland outlet glacier velocity variations at decadal and seasonal time scales utilizing the Landsat imagery", *Remote Sensing of Environment*, vol. 169, pp. 1–19, 2015.

[ROS 15b] ROSU A.M., PIERROT-DESEILLIGNY M., DELORME A. *et al.*, "Measurement of ground displacement from optical satellite image correlation using the free open-source software MicMac", *ISPRS Journal of Photogrammetry and Remote Sensing*, vol. 100, pp. 48–59, 2015.

[SCA 92] SCAMBOS T.A., DUTKIEWICZ M.J., WILSON J.C. *et al.*, "Application of image cross-correlation to the measurement of glacier velocity using satellite image data", *Remote Sensing of Environment*, vol. 42, no. 3, pp. 177–186, 1992.

[SCH 08] SCHERLER D., LEPRINCE S., STRECKER M.R., "Glacier-surface velocities in alpine terrain from optical satellite imagery – Accuracy improvement and quality assessment", *Remote Sensing of Environment*, vol. 112, no. 10, pp. 3806–3819, 2008.

[SCH 11] SCHERLER D., BOOKHAGEN B., STRECKER M.R., "Spatially variable response of Himalayan glaciers to climate change affected by debris cover", *Nature Geoscience*, vol. 4, no. 3, pp. 156–159, 2011.

[STU 14a] STUMPF, A., MALET J.P., ALLEMAND P. *et al.*, "Surface reconstruction and landslide displacement measurements with Pléiades satellite images", *ISPRS Journal of Photogrammetry and Remote Sensing*, vol. 95, pp. 1–12, 2014.

[STU 14b] STUMPF A., CANCOUËT R., PIETE H., *et al.*, "Change detection and dynamics toolkit", IQmulus Deliverable D4.5.1, 2014.

[STU 15a] STUMPF A., DELACOURT C., MALET J.P., "Towards time-series processing of VHR satellite images for surface deformation detection and measurements", *GeoBigData 2015 - The International Archives of the Photogrammetry, Remote Sensing and Spatial Information Sciences*, La Grande Motte, France, 2015.

[STU 15b] STUMPF A., DELACOURT C., MALET J.P., "Towards time-series processing of VHR satellite images for surface deformation detection and measurements", *International Archives of the Photogrammetry, Remote Sensing and Spatial Information Sciences*, vol. XL–3/W3, pp. 595–599, 2015.

[STU 15c] STUMPF A., MALET J.P., ALLEMAND P. *et al.*, "Ground-based multi-view photogrammetry for the monitoring of landslide deformation and erosion", *Geomorphology*, vol. 231, pp. 130–145, 2015.

[TRA 11] TRAVELLETTI J., Multi-parameter and multi-resolution imaging for the observation and characterization of clayey flow-like landslides, PhD Thesis, University of Caen Basse-Normandie, 2011.

[TRA 12] TRAVELLETTI J., DELACOURT C., ALLEMAND P. *et al.*, "Correlation of multi-temporal ground-based optical images for landslide monitoring: application, potential and limitations", *ISPRS Journal of Photogrammetry and Remote Sensing*, vol. 70, pp. 39–55, 2012.

[TRA 14] TRAVELLETTI J., MALET J.P., DELACOURT C., "Image-based correlation of laser scanning point cloud time series for landslide monitoring", *International Journal of Applied Earth Observation and Geoinformation*, vol. 32, pp. 1–18, 2014.

[TUR 15] TURNER D., LUCIEER A., DE JONG S.M., "Time series analysis of landslide dynamics using an unmanned aerial vehicle (UAV)", *Remote Sensing*, vol. 7, no. 2, pp. 1736–1757, 2015.

[VAL 15] VALLAGE A., KLINGER, Y., GRANDIN R. *et al.*, "Inelastic surface deformation during the 2013 Mw 7.7 Balochistan, Pakistan, earthquake", *Geology*, vol. 43, no. 12, pp. 1079–1082, 2015.

[VAN 00] VAN PUYMBROECK N., MICHEL R., BINET R. *et al.*, "Measuring earthquakes from optical satellite images", *Applied Optics*, vol. 39, no. 20, pp. 3486–3494, 2000.

[VER 12] VERMEESCH P., LEPRINCE S., "A 45-year time series of dune mobility indicating constant windiness over the central Sahara", *Geophysical Research Letters*, vol. 39, no. 14, 2012.

[WAN 06] WANGENSTEEN B., GUÐMUNDSSON Á., EIKEN T. *et al.*, "Surface displacements and surface age estimates for creeping slope landforms in northern and eastern Iceland using digital photogrammetry", *Geomorphology*, vol. 80, no. 1, pp. 59–79, 2006.

[WEI 11] WEI S., FIELDING E., LEPRINCE S. *et al.*, "Superficial simplicity of the 2010 El Mayor-Cucapah earthquake of Baja California in Mexico", *Nature Geoscience*, vol. 4, no. 9, pp. 615–618, 2011.

[WOO 15] WOO J., STONE M., PRINCE J.L., "Multimodal registration via mutual information incorporating geometric and spatial context", *IEEE Transactions on Image Processing*, vol. 24, no. 2, pp. 757–769, 2015.

[ZIT 03] ZITOVÁ B., FLUSSER J., "Image registration methods: a survey", *Image and Vision Computing*, vol. 21, no. 11, pp. 977–1000, 2003.

6

Application of Optical Remote Sensing for Monitoring Environmental Impacts of Mining: From Exploitation to Postmining

6.1. The environmental and societal stakes of mining

6.1.1. *Environmental and societal impacts*

Mining has played a significant role in the development of many countries all over the world. The industry has been, and continues to be an important contributor to both national and regional economies and is critical to national defense. Mining, and the industries it supports, remains a fundament building block of modern societies.

There have been many benefits for mining but, in many countries these have come at a cost to the environment. Early mining operations have left a historical legacy of negative environmental impacts which have affected the common perception of mining. As countries have matured, there has been increasing acknowledgement that environmental protection is as fundamental to a healthy economy and society as is growth. The challenge is to simultaneously promote both economic growth and environmental protection.

Chapter written by Stéphane CHEVREL and Anne BOURGUIGNON.

Mining, active or abandoned, is one of man's more impactful activities on the environment. Mine waste is associated with a high concentration of toxic metals, low pH waters and soils, and an inadequate soil structure providing a low level of plant nutrients. The main contaminants are associated to different types of ores and their processing. Pollution caused by mine waste extends much further than the visible mounds of spoil due to dispersion by gravity, surface water and groundwater. Pollution is also caused by acid mine waters draining from the adits of abandoned mines and by mining residues. Such mining contamination can directly affect the health of the local communities. The impact on surrounding farmland is a concern, leading to livestock ingesting contaminated grass and soil.

Social impacts of mining are complex and controversial. Mining can generate wealth, while triggering significant disruptions. A mining project can generate employment, grow transport infrastructure, education facilities and increase the availability of goods and services in poor and remote areas; however, these benefits might be unequally distributed. Social tensions and conflict, sometimes riots, can rise from affected communities. The societal acceptability of a mining project is among the challenges the industry has to face. The concepts of "Social License to Operate" (SLO) and "responsible mining" reflect this need.

All phases of a mining project, from exploration to exploitation and closure, must be anticipated and monitored (Figure 6.1). The aim of this chapter is to demonstrate the contribution of remote sensing techniques.

6.1.2. *The source–pathway–receptor paradigm*

The "source–pathway–receptor" (SPR) paradigm is a commonly used model in environmental risk assessment and management [CUS 14].

It is based on the principle that the existence of risk implies the concomitant presence (Figure 6.2) of:

– a dangerous or hazardous source (S);

– a transfer mode to and from the transfer medium (Pathway : P);

– a target or receptor (R).

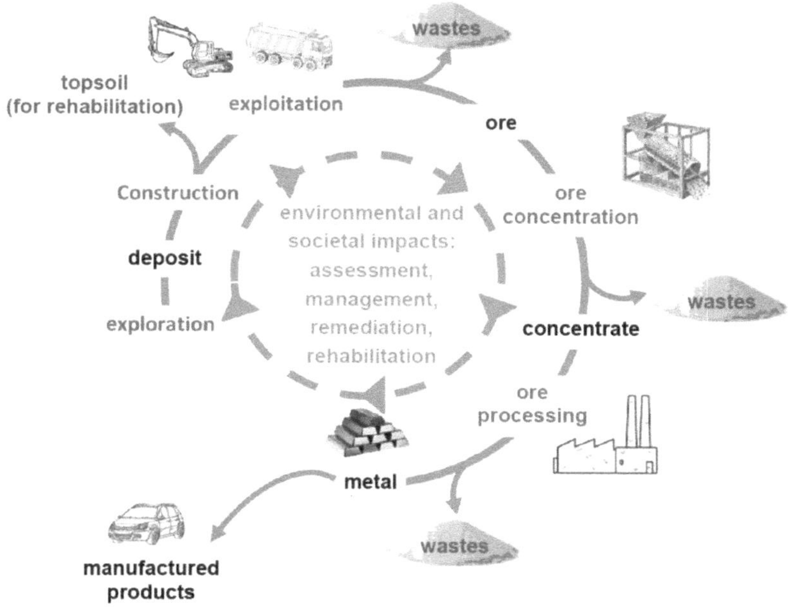

Figure 6.1. *Phases of production of a metal from a metallic ore (source: White Paper on Responsible Mining, French Ministry of Ecology, Sustainable Development and Energy, 2016). For a color version of this figure, see www.iste.co.uk/baghdadi/6.zip*

There must be a source of potentially harmful material in sufficient concentrations to pose a significant potential risk to people or the environment. There must also be a pathway linking the harmful material with the receptors at risk (e.g. humans, livestock, and important resources such as clean soil or groundwater). The presence of all three elements is also referred to as a pollutant linkage.

As soon as one of these factors is not present, the risk becomes irrelevant and a risk assessment for this milieu and its planned use is not necessary. This mode of thought provides the basis for various risk management options such as source removal, elimination of an exposure pathway by capping a site or changing the use of the site [CUS 14].

Figure 6.2. *The pollutant linkage*

6.2. The role of Earth Observation

6.2.1. *Introduction*

Earth Observation (EO) offers a unique opportunity and varieties of methods and tools to collect and process spatial information to monitor and assess each step of the mining cycle. It can contribute towards improving policies surrounding raw materials and better exploiting mineral resources from mineral supplying countries, as well as to demonstrate how to improve their capacity in setting up new mining sites.

EO includes:

– spaceborne and airborne imagery;

– ground and airborne geophysics;

– geochemistry;

– *in situ* measurements and monitoring networks;

– 3D modeling;

The combination of EO and geographic information systems (GIS) could be an appropriate response to decision maker needs since these tools can easily associate ongoing monitoring data, environmental database layers and regulatory requirements, information that can be accessed through the use of automatic and customized GIS queries.

In terms of risk assessment, EO can contribute to:

– the identification and mapping of pollution sources, e.g. tailings (resulting from ore processing, generally muddy), waste rocks (material displaced during exploitation without being processed), infrastructures, etc.;

– the identification and mapping of pollution dispersal pathways, e.g. natural and artificial drainage;

– the identification and mapping of at risk receptors;

– the identification, characterization and mapping of contaminated areas, e.g. contamination plumes, contaminated soils, contaminated surface waters, etc.

Moreover, EO can map not only their distribution but consider their spatial relationships.

6.2.2. *Study scales, from a regional approach to the mine site*

The applications of remote sensing at different scales focus on the processing of spectral information from optical sensors, whether for a regional scale inventory of mineralogical indicators of the presence of hazardous materials, or for the spectrometric characterization of minerals present in a mining site (Figure 6.3).

6.2.3. *Identification, characterization and mapping of pollutant linkage elements and their spatial distribution*

Table 6.1 lists the main sensors that can be used to identify and map the pollutant linkage elements and their spatial relationships. The choice of the sensor also depends on the scale of the study (Figure 6.3).

Very high resolution means sensors with a sub-meter ground resolution (e.g. Pleiades, SPOT 6/7, IKONOS, QuickBird, WorldView-2, etc.). Very high spectral resolutions sensors (hyperspectral sensors or imaging spectrometers) provide more than 100 narrow, contiguous spectral bands. The advantage of a LiDAR DTM (digital terrain model) over a DEM (digital elevation model) from stereo pairs lies in its ability to provide the altitude of the ground and not only the elevation of the first object encountered (canopy, building, etc.).

The choice of the sensor is very important as it will determine how well the land surface features are identified and mapped. The data price (from free Landsat or Sentinel data up to several thousand euros for very high resolution images and several tens of thousands of euros for an airborne acquisition) is also to be considered.

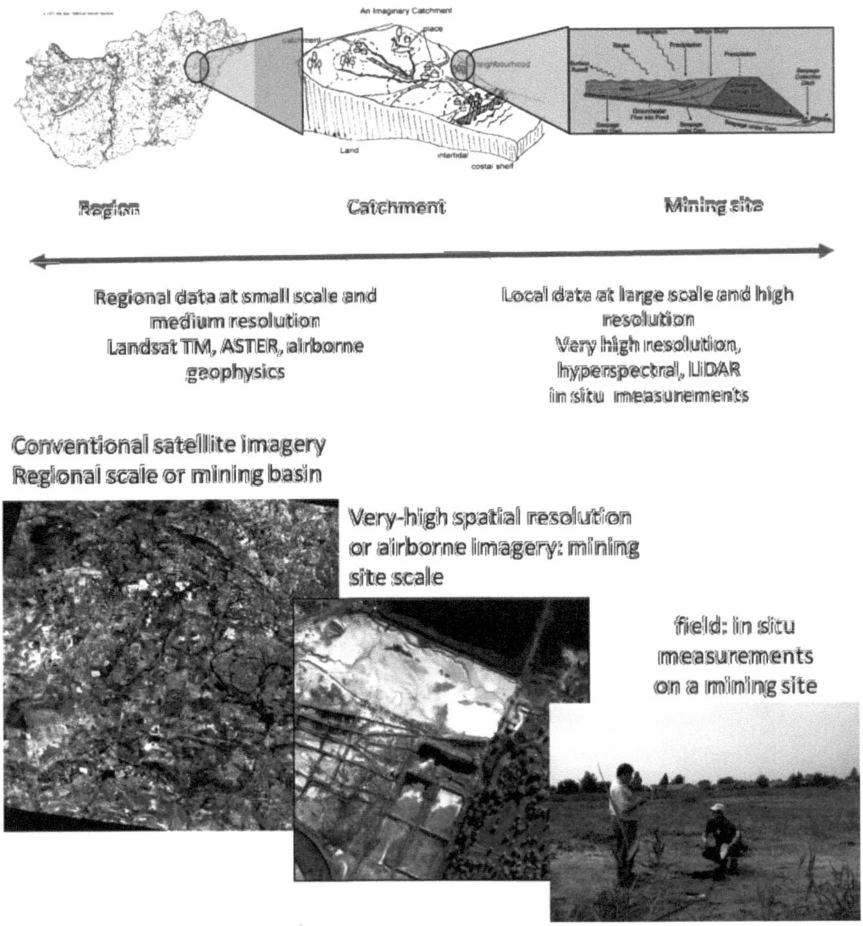

Figure 6.3. *Main data sources for integrated assessments of mining environments at different scales [CHE 07]*

Aim	Appropriate Sensors
Identify and map	
– Pollution sources	
– tailings dams and heaps	– Medium (> 10 m) resolution multispectral sensors: Sentinel 2, Landsat TM, ASTER – Very high resolution (metric to <1 m): Pleiades, IKONOS, Spot 6/7, WorldView 2 – LiDAR sensors
– waste rocks	– idem
– mining infrastructures	– Very high resolution (metric to <1 m): Pleiades, IKONOS, Spot 6/7, WorldView 2
– Pollution dispersal pathways	
– natural and artificial drainage	– Digital Elevation Models (DEMs) from Very High Resolution stereo pairs, Digital Terrain Models (DTMs) from LiDAR
– Receptors at risk	– Medium (> 10 m) resolution multispectral sensors: Sentinel 2, Landsat TM, ASTER – Very high resolution (metric to <1 m): Pleiades, IKONOS, Spot 6/7, WorldView 2 – LiDAR sensors
Identify, characterise and map contaminations	
– Contamination plumes	– Hyperspectral sensors, airborne geophysics
– Contaminated soils	– Hyperspectral sensors, airborne geophysics
– Contaminated waters	– Hyperspectral sensors, ASTER, WorldView-2
Spatialisation	
– Spatial distribution	– All sensors
– Spatial Relationships	– Very high resolution

Table 6.1. *Selection of sensors in a mining environmental management perspective*

6.3. Application examples

6.3.1. *Visual identification of mining infrastructures from very high spatial resolution imagery*

Very high spatial resolution optical images (1 m or less) are valuable in identifying infrastructures on active or abandoned mine sites and in monitoring their temporal evolution and their impact on their immediate environment.

6.3.1.1. *East Rand, Witwatersrand, South Africa*

Figure 6.4 (IKONOS image acquired in 2000 over the East Rand, Witwatersrand goldfield) shows a pumping station that pumps tailings from a tailings dam. Tailings leaks are collected in a discharge ditch toward surface waters and contaminate them.

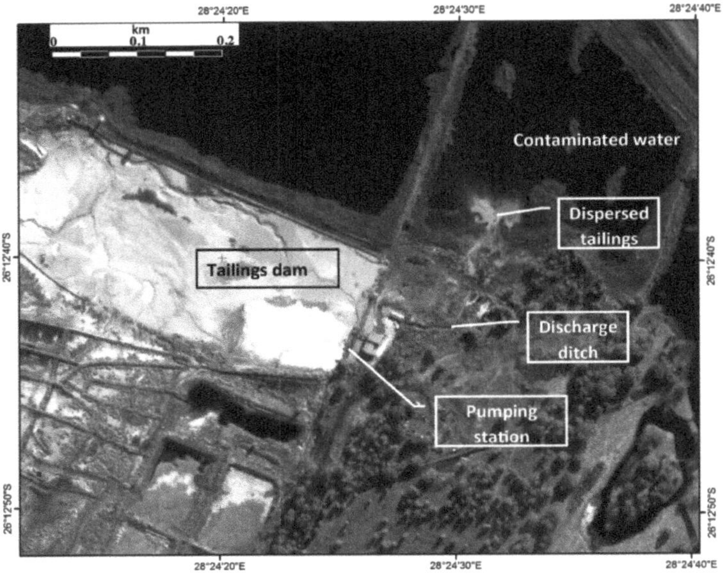

Figure 6.4. *IKONOS image showing surface water contamination by a discharge ditch (Witwatersrand, South Africa). For a color version of this figure, see www.iste.co.uk/baghdadi/6.zip*

The same IKONOS image (Figure 6.5) shows tailings dispersed by leaks from a tailings transport pipeline that contaminates surface water trough runoff.

6.3.1.2. *Monitoring glacier advance in response to mining-related debris loading – Kumtor mine, Kirghizstan*

Located at an altitude of more than 4,000 m, the Kumtor gold mine excavates a 4.5 km^2 area of rock and glaciers and deposits rock and ice on the surrounding glaciers, over an area of 9.3 km^2. In Figure 6.6, the parts of the Davidov glacier and Lysii glacier covered with material from the mine clearly appear in dark gray.

Figure 6.5. *IKONOS Image showing surface water contamination by tailings leakage from a tailings transportation pipeline (Witwatersrand, South Africa). For a color version of this figure, see www.iste.co.uk/baghdadi/6.zip*

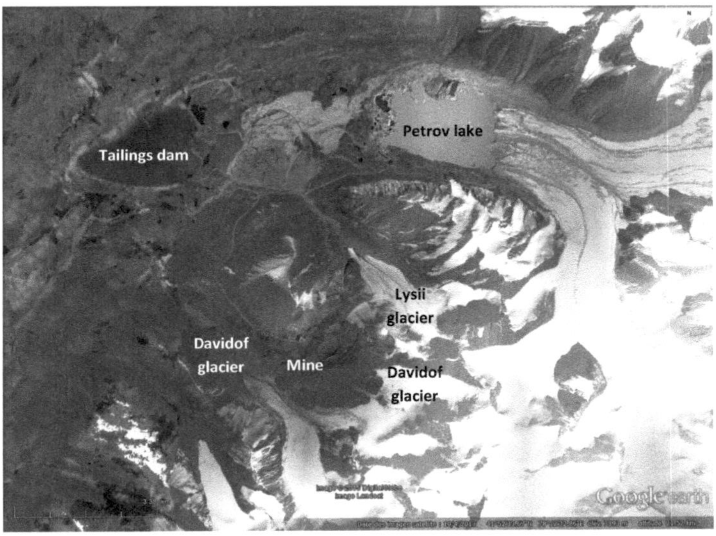

Figure 6.6. *Google Earth Image of Kumtor mine and its surroundings. For a color version of this figure, see www.iste.co.uk/baghdadi/6.zip*

Jamieson *et al.* [JAM 15] used a series of very high resolution images (IKONOS, Quickbird, WorldWiew) to measure the accelerating advance of two glaciers in response to the load induced by the deposition of mining residues on these glaciers between 2002 and 2014. The classification of land use from this series of images has helped highlight a rapid advance of Davidov and Lysii glaciers, due to the overload of tailings and sterile deposits on or besides the glaciers. Figure 6.7 shows that the Davidov glacier, under the pressure from anthropogenic deposits, moved sideways in a narrower space. The authors suggest a traveling speed of about 3.8 m per year.

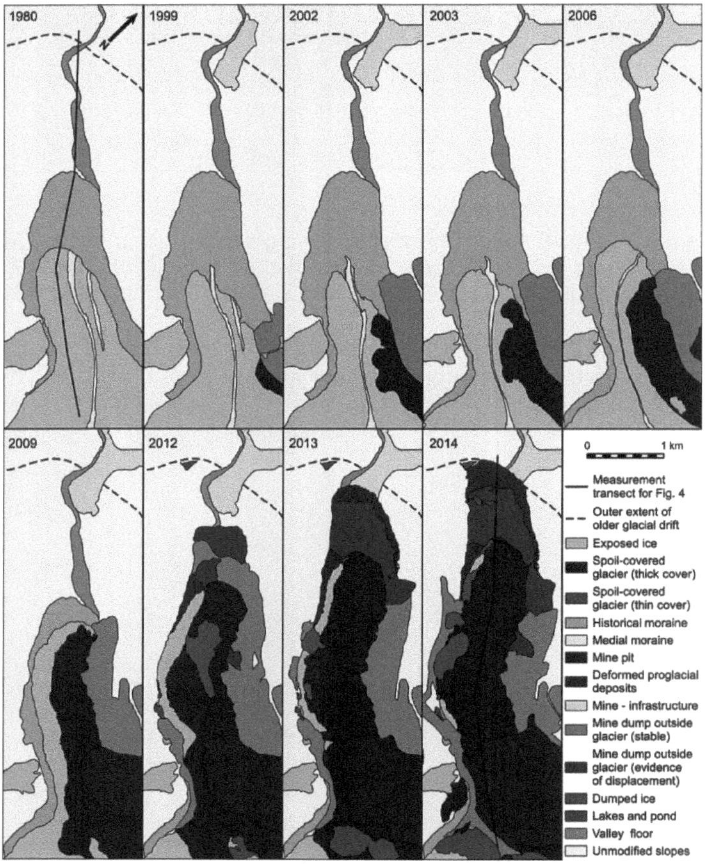

Figure 6.7. *Time series of classifications from very high resolution satellite imagery, Davidov glacier [JAM 15]. For a color version of this figure, see www.iste.co.uk/baghdadi/6.zip*

6.3.2. *Mineral characterization and contamination mapping from very high spectral resolution imagery, Mpumalanga coal field, South Africa*

Using a field spectroradiometer enables us to measure the reflectance spectra of mineral associations of interest over a spectral range covering 2,150 spectral bands from 450 to 2500 nm, with a spectral resolution of less than 5 nm (ASD spectroradiometer FieldSpec 3). It is then possible to generate a spectral library of reference field spectra dedicated to the study area. From this, we can identify the spectral characteristics of each mineral or each group of minerals recognized in the field. These reference spectra are then used to classify hyperspectral images available over the study area.

This preliminary field approach is used to map the different minerals present on a study site from hyperspectral images. It has been implemented to map the acid mine drainage (AMD) minerals on the Mpumalanga coal field in South Africa, where coal is rich in pyrite.

Sulfide minerals (pyrite being the most common) are naturally subject to oxidation by contact with air and water. This slow phenomenon leads to the production of sulfuric acid. When mining operations disrupt the chemical balance of deep metallic sulfide deposits by brutally putting them under oxidizing conditions, this amplified reaction creates acidic runoff called AMD. Such effluents contain sometimes toxic concentrations of metals and are one of the major environmental problems of mining.

The main chemical reactions that generate AMD are the oxidation of iron and sulfur from pyrite (Figure 6.8(a)). The oxidation produces sulfuric acid and sulfates and is responsible, depending the acidity of the water, for solubilization, transport and redeposition of the iron and toxic metals such as arsenic, lead and heavy metals, mercury and cadmium. It is not possible to spectrally distinguish between these metals, it is however possible to recognize, based on their spectral characteristics, the different secondary minerals resulting from oxidation processes and weathering of pyrite and whose stability depends on water pH (Figure 6.8(b)). It is therefore essential to be able to map these minerals to indirectly assess the associated environmental risk.

With the exception of the pyrite itself, minerals from the oxidation suite of iron-bearing sulfides are identifiable by their spectral characteristics [SWA 00]. In Figure 6.9, we can thus identify spectral absorption features characteristic of the following materials:

– Jarosite (olive green spectrum): 0.437 and 2.267 µm;

– Goethite +kaolinite (orange spectrum): 2.170 and 2.206 µm + absorption characteristics of goethite in the visible–near infrared;

– Hematite + gypsum (cyan spectrum): 1.450 and 1.490 µm + characteristic absorptions of hematite in the visible–near infrared.

The eMalahleni area of in the Mpumalanga coalfield was covered, on behalf of the Council for Geoscience (South African Geological Survey), by hyperspectral images acquired with the ProspecTIR VS3 sensor from SpecTIR, a company operated by Southern Mapping. This sensor acquires images in 360 spectral bands covering the range 390–2,450 nm, with a nominal spectral resolution of 5 nm. At a flight altitude of about 2,500 m above the ground, the sensor resolution allows pixels of 1 m.

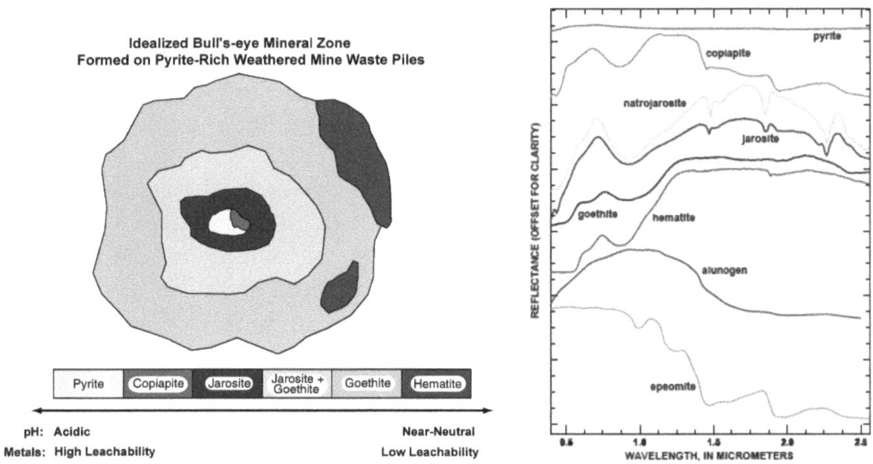

Figure 6.8. *a) AMD minerals formed on pyrite-rich mining waste piles, and b) corresponding reflectance spectra (source: (USGS) United States Geological Survey). For a color version of this figure, see www.iste.co.uk/baghdadi/6.zip*

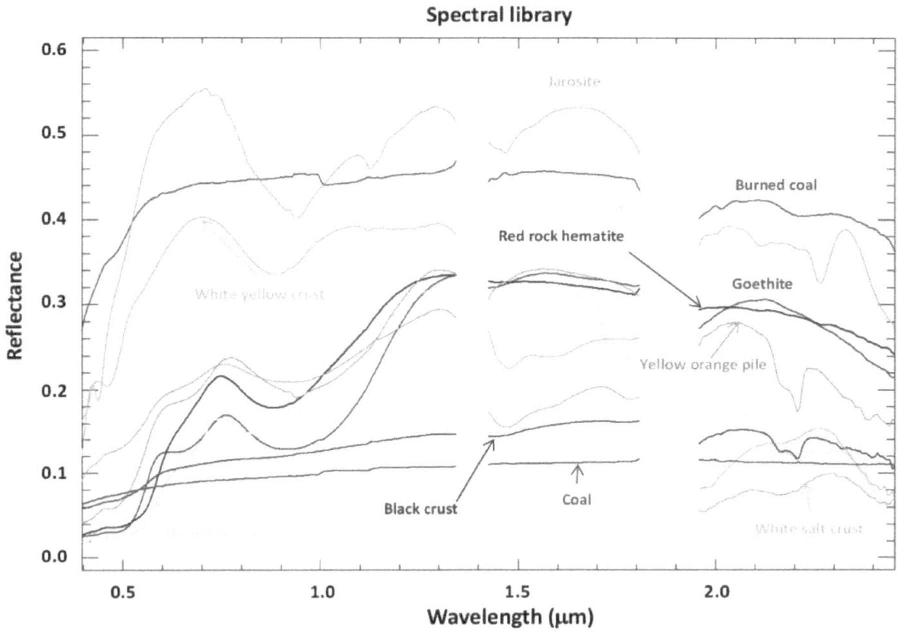

Figure 6.9. *Field reflectance spectra of various mineral associations related to mining activity in the Mpumalanga coalfield (South Africa). For a color version of this figure, see www.iste.co.uk/baghdadi/6.zip*

The Spectral Angle Mapper (SAM) technique [KRU 93] is a method of automatic classification that compares image spectra to known spectra. It determines the similarity between an image spectrum (representing an unknown material) and a reference spectrum (a known material) calculating the spectral angle to each other, treating them as vectors in a n-dimensional spectral space, where n is the number of spectral bands or channels (Figure 6.10). A spectrum is represented as a vector where enlightenment is characterized by the vector length, SAM hence is insensitive to light and the effects of albedo, meaning that all possible illuminations (including areas in shadow) are treated equally.

The resulting classified image represents, for each pixel, the best match to a reference spectrum, corresponding to a given material. The smaller angles represent the best match with the reference spectrum while pixels outside a threshold angle are not classified.

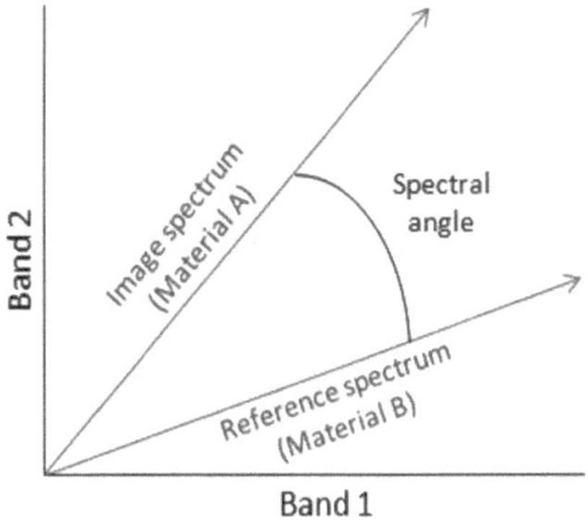

Figure 6.10. *Principle of the Spectral Angle Mapper*

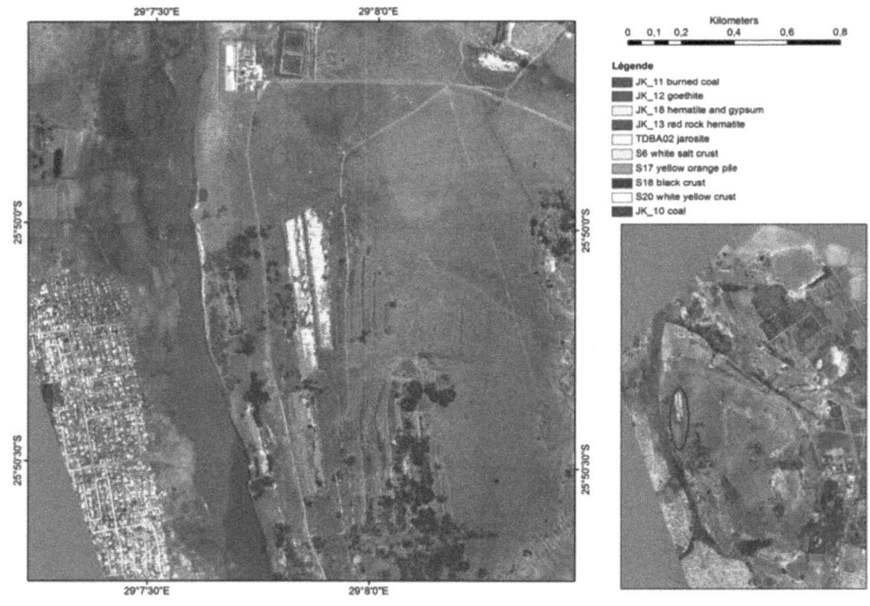

Figure 6.11. *Classification of the hyperspectral image acquired in November 2013 (Mpumalanga coal field, South Africa) from field spectra of Figure 6.9. For a color version of this figure, see www.iste.co.uk/baghdadi/6.zip*

The image in Figure 6.11 represents the SAM classification of a hyperspectral image of the region of eMalahleni obtained from the previous spectral library of Figure 6.9. Classification highlights (bottom right) are as follows: in purple coal (mining residues); in orange anthropised goethite-rich areas; in blue hematite-rich areas and in yellow two jarosite-rich areas. The latter correspond (zoom in red surrounded zone) to retention ponds that collect runoff where jarosite resulting from the upstream leaching of contaminated areas is concentrated, avoiding their transfer to the surface network.

6.4. Multisensor integrated approaches

6.4.1. *Fusion of optical imagery and airborne geophysics*

The gold and uranium mining in the Witwatersrand gold field in South Africa, in the region of Johannesburg, has left a legacy of over 40,000 ha of tailings impoundments (lagoon of ore processing residues) with significant radioactivity and that are sources of AMD responsible for the contamination of surface waters and groundwaters.

The *Bureau de Recherches Géologiques et Minières*[1] (BRGM) and its South African counterpart have developed a simple but robust methodology [CHE 00, COE 02] to highlight these contaminations and their release into the environment on the western part (West Rand) of the gold field, based on the processing and integration of the following:

– SPOT XS image (20 m resolution) and Landsat TM image (30 m resolution)

– SPOT DEM obtained from a panchromatic steropair (10 m resolution);

– High-resolution airborne gamma spectrometry (250 m line spacing).

A land-use classification has been obtained from the combination of the Spot XS (Figure 6.12(a)) and Landsat TM (Figure 6.12(b)) images. This classification (Figure 6.12(c)) allows us to map the different types of vegetation (in green), savanna areas (in light and dark gray), urban areas (in red) and townships (in light brown), the wetlands associated with the river system (in blue) and most importantly the mining areas and tailings ponds (in yellow and orange), considered source of contamination.

1 French Geological Survey.

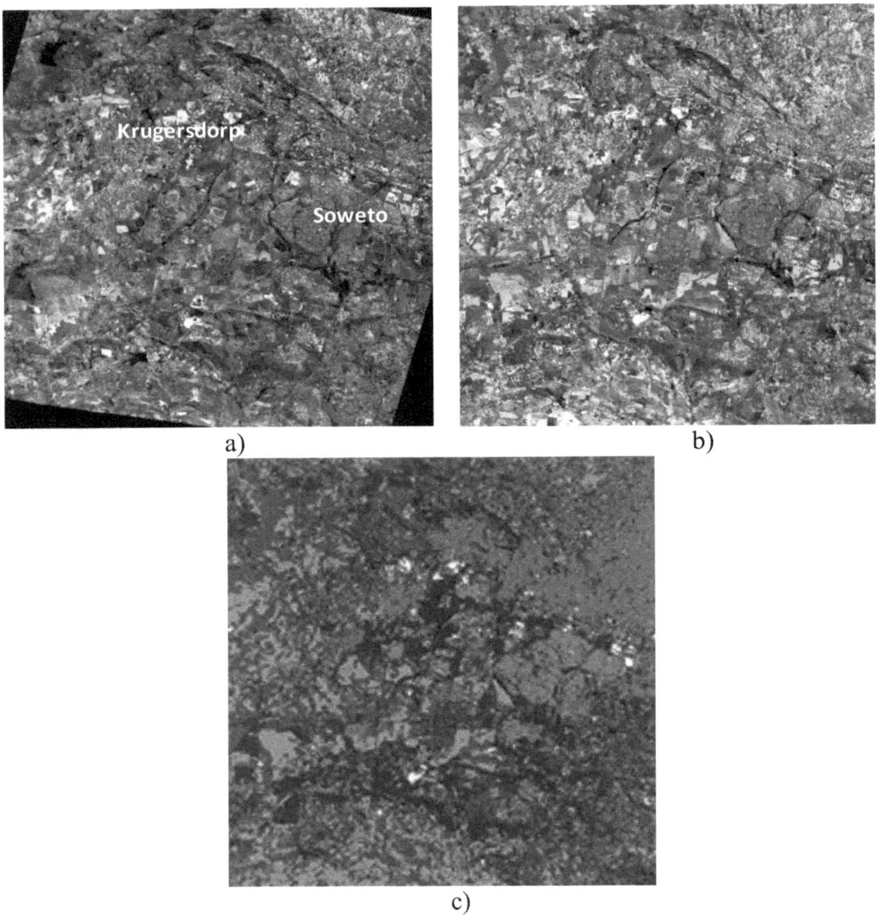

Figure 6.12. *Spot image a), Landsat TM image b) and resulting land-use classification c), West Rand, South Africa, center coordinates 26° 14′33" S and 27° 44′55" E. For a color version of this figure, see www.iste.co.uk/baghdadi/6.zip*

Radioactivity acts as a tracker for pollution, and gamma spectrometry allows us to identify areas of high concentration in radionuclides (in red, Figure 6.13(a), total count) related to mining activities. The ternary image of potassium, uranium and thorium concentrations (Figure 6.13(b)) helps differentiate between lithological units and the discrimination between natural radioactivity (blue and violet in Figure 6.13(b)) and the anthropogenic (uranium in green, Figure 6.13(b)) corresponding to waste rocks and tailings resulting from underground mining.

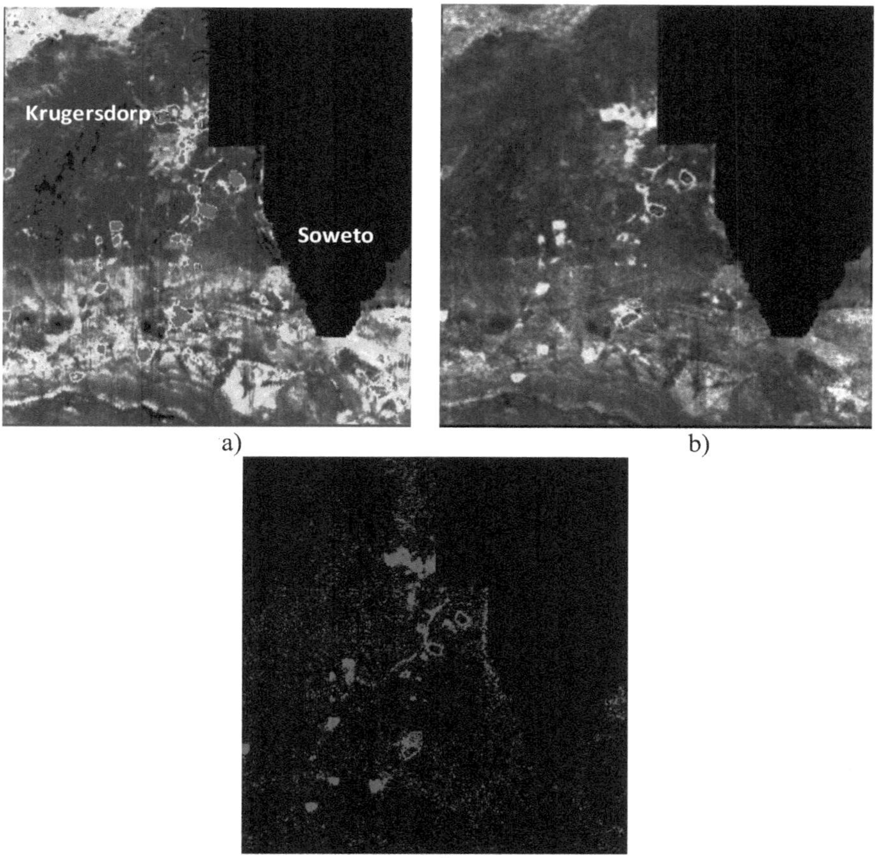

c)

Figure 6.13. *Airborne gamma spectrometry: a) total count, b) ternary K-U-Th image, c) ratio U/Th > 1, West Rand, South Africa. For a color version of this figure, see www.iste.co.uk/baghdadi/6.zip*

The U/Th > 1 ratio (Figure 6.13(c)) indicates uranium-rich thorium-poor areas, typical of gold and uranium mining residues brought to the surface (while the U/Th ratio remains between 0.1 and 0.5 in natural background) and therefore provides an accurate map of the contaminated areas (around the tailings dams and immediately downstream river system). Overlaid on the SPOT image (Figure 6.14), this ratio clearly shows radioactive contamination around the dams and of the downstream river system (wetlands).

Figure 6.14. *Ratio U/Th > 1 overlaid on SPOT image. For a color version of this figure, see www.iste.co.uk/baghdadi/6.zip*

Applying a modified flow accumulation algorithm [FAI 91] to the SPOT MNE (Figure 6.15(a)) makes it possible to model the runoff from the contamination sources identified by classification of SPOT and Landsat images (Figure 6.15(b)). Compared with the image of U/Th > 1 ratio (Figure 6.15(c)), it highlights the role of the drainage system in collecting the pollution.

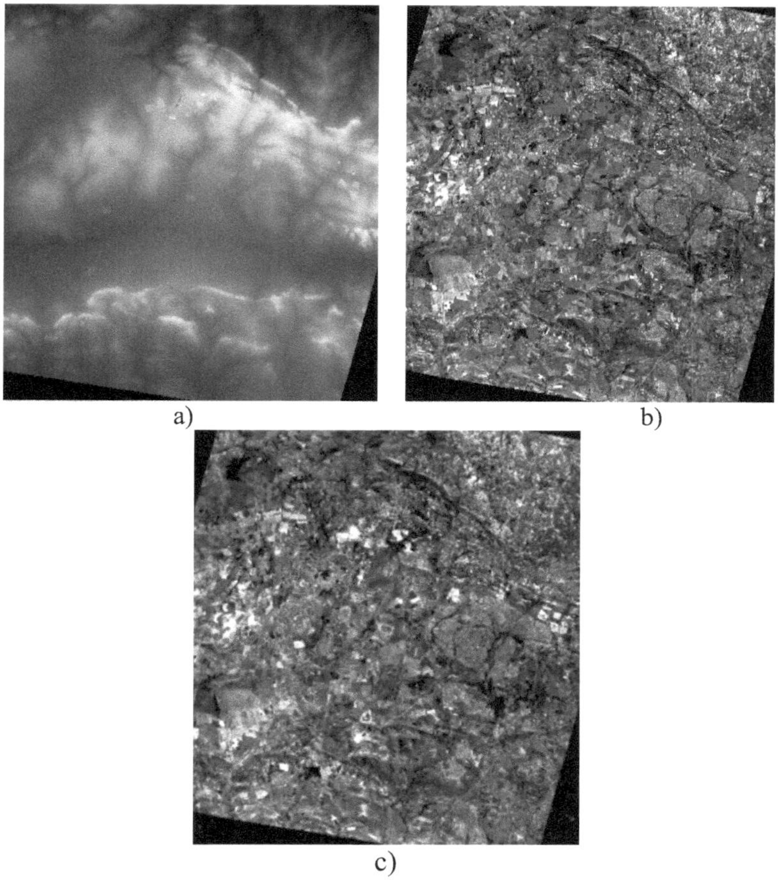

Figure 6.15. *Flow accumulation algorithm applied to SPOT DEM (West Rand, South Africa). a) SPOT DEM, b) modeling surface flow from contamination sources identified from SPOT and Landsat images, c) highlighting the role of the drainage system as collector of pollution by comparison with the U/Th > 1 image. For a color version of this figure, see www.iste.co.uk/baghdadi/6.zip*

The superposition of such a drainage pattern over the entire study area highlights sections of the regional water system (in orange, Figure 6.16(a)) that are likely to be affected by the identified contamination sources (in green). Gamma spectrometry confirms this model, with elevated radiometry levels of anthropogenic origin (U/Th > 1) in the river system (in red, Figure 6.16(b)) in the northeast (south of Krugersdorp) and, to a lesser extent, in the southwest (Carletonville).

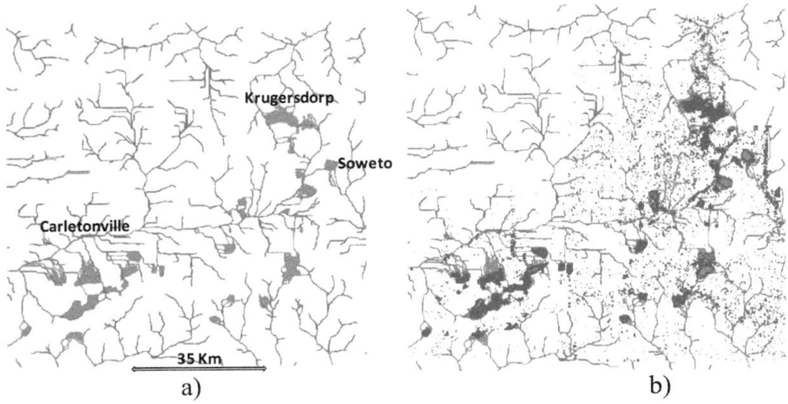

Figure 6.16. *a) Hydrographic network potentially contaminated by tailings dams, b) U/Th > 1 ratio, West Rand, South Africa (center coordinates: 29° 19' S, 27° 33' E). For a color version of this figure, see www.iste.co.uk/baghdadi/6.zip*

6.4.2. *Modeling potential pollution transfer from a tailings dam*

The Makmal mine in central Kyrgyzstan extracts the gold contained in skarn formations (calcareous-silicate rock resulting from the processing of carbonates in contact with a magma intrusion). The ore is processed by cyanidation in a plant located on an alluvial terrace over hanging the city of Kazarman (15,000 inhabitants) and the tailings are stored in a lagoon occupying a valley enclosed by a dam. The question was whether possible seepages of water containing cyanide were likely to contaminate the city via their transfer by the drainage system.

A hierarchized drainage system was computed from a very high resolution DEM (1 m), obtained from a pair of WorldView-2 panchromatic stereoscopic shots (gray in Figure 6.17). This network was then numbered in upstream/downstream drains to allow subsequent GIS queries (Figure 6.17(b), ascending drains in red and descending in blue). From this we can identify (Figure 6.17(a)), for each point of the drainage network, the downstream drain at this point (blue) and all upstream drains (red).

Applying this technique to the regional water system enables us to model the downstream flow of a possible seepage of contaminated water at the base of the dam.

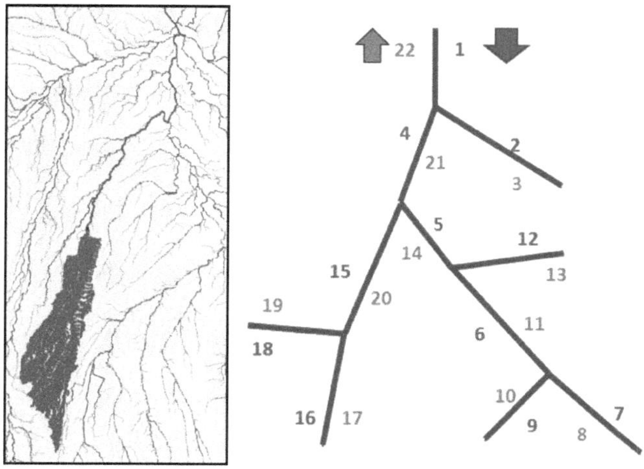

a) b)

Figure 6.17. *a) Identification of downstream drain (in blue) and upstream drains (in red) for a given point of the drainage system; b) organization of the drainage system in upstream and downstream drains. For a color version of this figure, see www.iste.co.uk/baghdadi/6.zip*

Figure 6.18 shows in red the downstream drain of the dam that would likely carry contaminated water. It highlights that despite the higher elevation of the tailings pond, a leak of contaminated water would not affect the city of Kazarman as the water flows to the northwest and not to the northeast toward the city.

6.4.3. *Modeling mud flow propagation after a tailings dam failure*

Kazarman inhabitants worried about the consequences of the dam failure and if the city would be affected by a resulting mud flow. Downstream sections perpendicular to the flow (Figure 6.19), characterized by their geometry (width, depth, symmetry, etc.) are computed from the same DEM as previously.

We can then establish a model of the extent of a water or mud flow, at any point on the surface network, of a given thickness (Figures 6.19 and 6.20). From this we are able to theorize the consequences of a 5 m mud flow downstream of the tailings dam in the event of a dam failure. Despite this conservative assumption (mud flow would get thinner flowing downstream), the Kazarman city would not be affected by the flow that would stretch in the alluvial plain to the northwest (in orange, Figure 6.21).

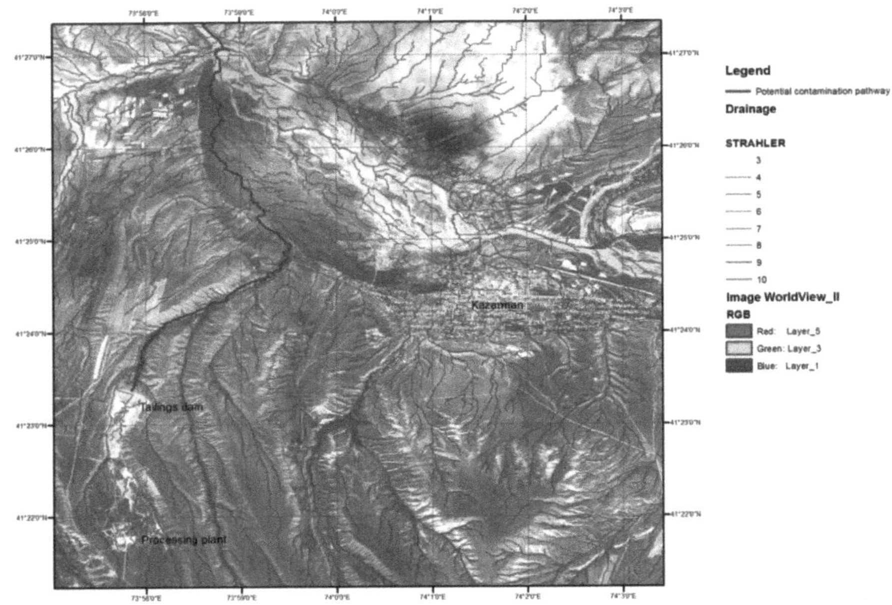

Figure 6.18. *Modeling a possible contamination of the drainage network from a leakage at the bottom of the tailings dam (Makmal mine, central Kyrgyzstan). For a color version of this figure, see www.iste.co.uk/baghdadi/6.zip*

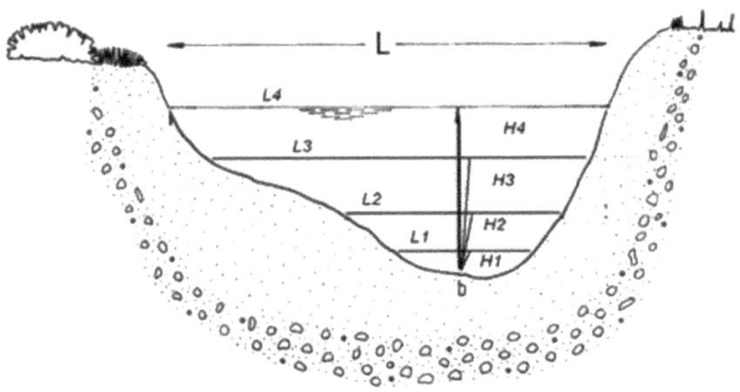

Figure 6.19. *Vertical slice of overland flow*

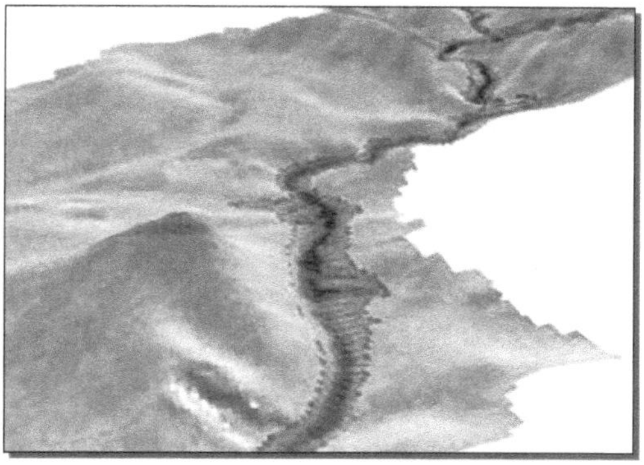

Figure 6.20. *Modeling a water or mud flow for a given thickness. For a color version of this figure, see www.iste.co.uk/baghdadi/6.zip*

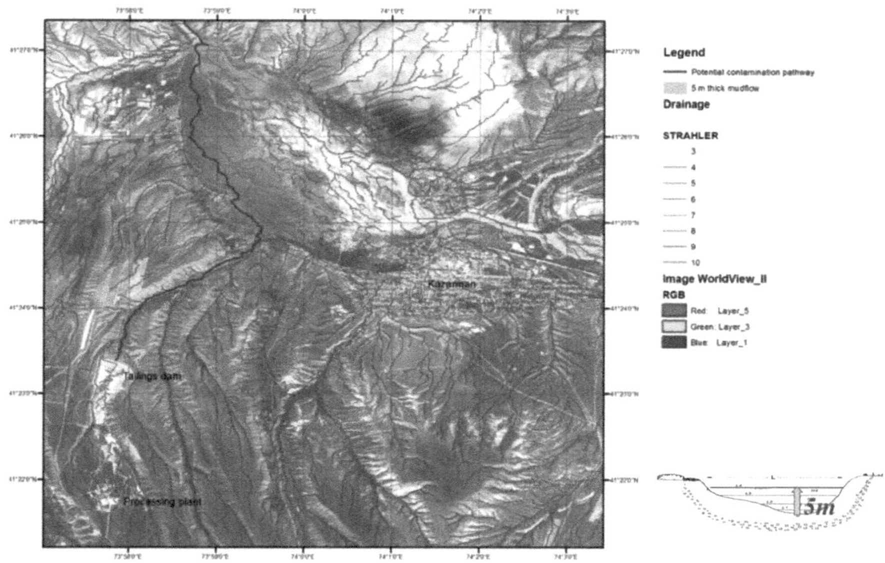

Figure 6.21. *Modeling maximum probable extension of a 5-m-thick mud flow (Makmal mine, central Kyrgyzstan). For a color version of this figure, see www.iste.co.uk/baghdadi/6.zip*

6.5. Exploitation of an airborne LiDAR DTM

6.5.1. *Subsidence*

The advantage of a LiDAR DTM over a DEM from pairs of stereoscopic images lies in its ability to provide the altitude of the ground and not only the elevation of the first object encountered (canopy, building, etc.).

A LiDAR DTM provides topographical information on small differences in ground elevation and is thus well suited to mapping mining-related surface disturbances, where the topography has been substantially modified, such as subsidence, mining operations or volume assessment of waste heaps. In addition, LiDAR DTM allows the calculation of precise drainage systems undisturbed by vegetation and buildings.

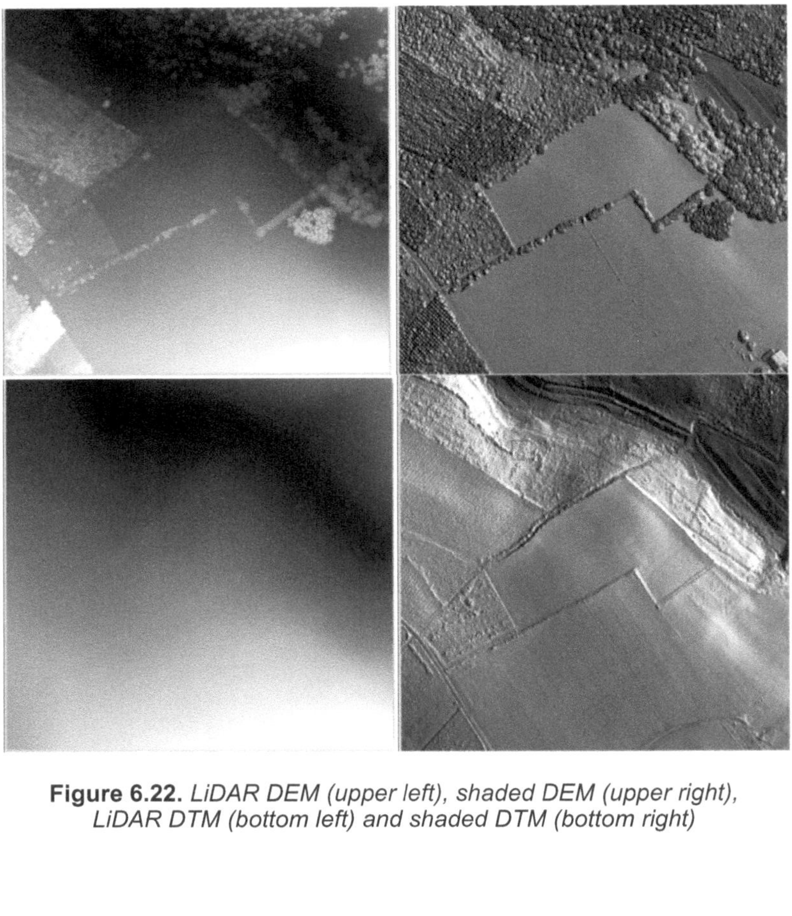

Figure 6.22. *LiDAR DEM (upper left), shaded DEM (upper right), LiDAR DTM (bottom left) and shaded DTM (bottom right)*

Underground mining of coal by board and pillar technique in the coalfields of Mpumalanga (South Africa) results in surface collapses of metric magnitude over several meters in length. Poorly visible on optical images, even very high resolution ortho-mosaics (Figure 6.23, (top) there), too high in magnitude to be addressed by radar interferometry, these collapses, however, are clearly visible and measurable on the very high resolution LiDAR DTM (Figure 6.23, (bottom) there). The north-south profile (Figure 6.24) shows the magnitude of the collapses and highlights the related surface cracks.

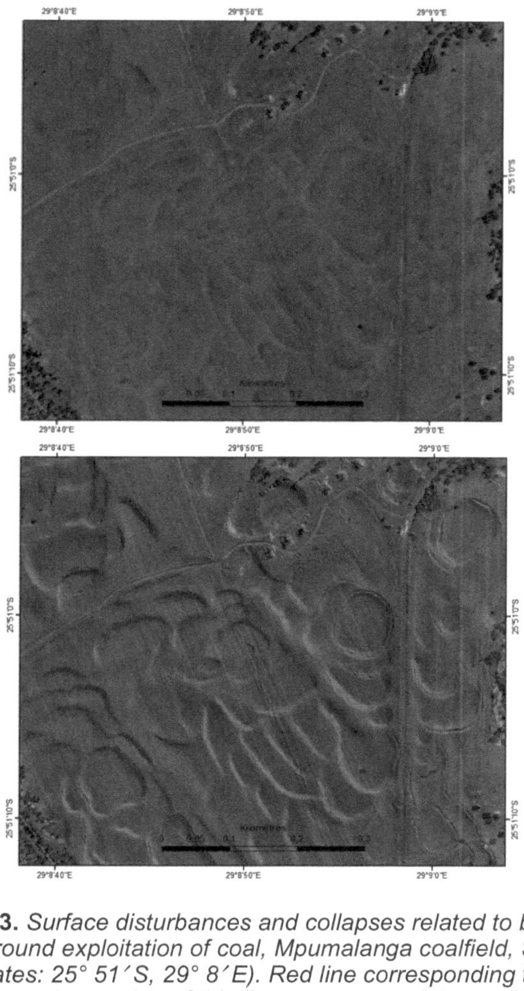

Figure 6.23. *Surface disturbances and collapses related to board and pillar underground exploitation of coal, Mpumalanga coalfield, South Africa (center coordinates: 25° 51′ S, 29° 8′ E). Red line corresponding to the profile of Figure 6.24. For a color version of this figure, see www.iste.co.uk/baghdadi/6.zip*

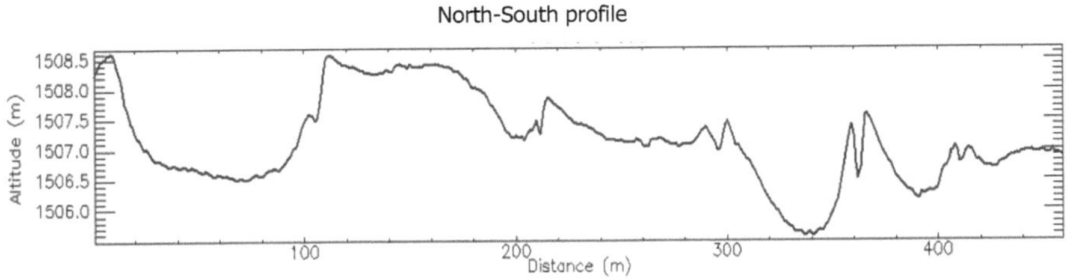

a)

b)

Figure 6.24. *a) North–south topography (in red in Figure 6.23) showing surface cracks, b) photography of a subsidence-related surface fissure. For a color version of this figure, see www.iste.co.uk/baghdadi/6.zip*

6.5.2. Assessment of the potential for groundwater contamination by surface waters

The extension and depth of closed watershed areas ("endorheic areas", i.e. areas without surface water outflow) are computed from LiDAR DTM. The top part of Figure 6.25 shows shaded terrain and we see terrain elevation and depth of endorheic areas below. The general altitude decreases from east to southwest, in alignment with regional topography. Endorheic areas correspond either to artificially constructed zones (waste piles and tailings, retention basins, etc.) or to localized shallow collapse areas, or deeper depressions of larger scale (e.g. large red zone in the northwest corner).

The profile in Figure 6.26 corresponds to the red line in Figure 6.25. It shows the altitude of the ground (black line), the extension of the endorheic zones (in blue) and their depth (in red). This type of model allows the drainage system calculated from the same DTM to be overlayed in order to highlight the areas of potential ingress of contaminated surface water to groundwater.

Figure 6.27(a) shows a convergence of surface drains toward the center of an endorheic area (orange and red), revealing an area where a contaminated surface runoff could affect groundwater. Figure 6.27(b), however, shows a man-made trench system to avoid such a flow to the endorheic area (in red) on the west.

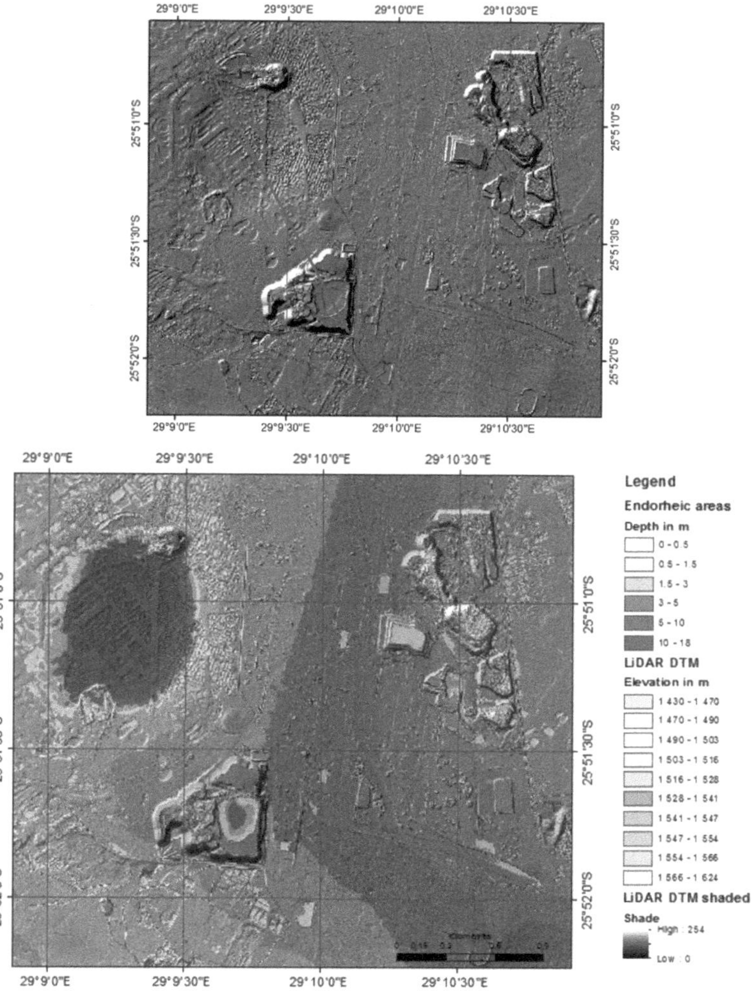

Figure 6.25. *Shaded LiDAR DTM (top) and extension and depth of endorheic areas, Mpumalanga coalfield. For a color version of this figure, see www.iste.co.uk/baghdadi/6.zip*

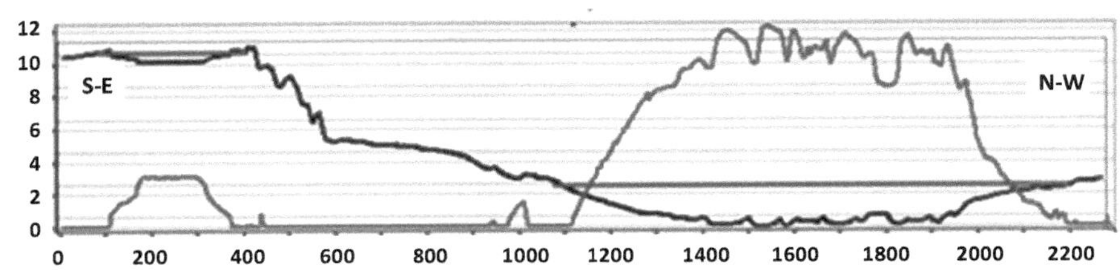

Figure 6.26. *Elevation profile (red line in Figure 6.25) and endorheic areas (in meters). Ground elevation (black), extension (in blue) and depth (in red). For a color version of this figure, see www.iste.co.uk/baghdadi/6.zip*

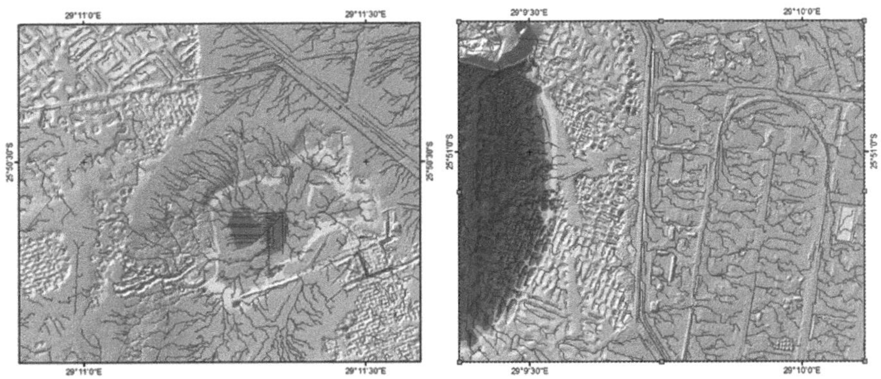

Figure 6.27. *Surface water ingress potential. a) Convergence of surface drains toward the center of an endorheic area (orange and red), b) man-made trench system to avoid flow to the endorheic area (in red) in the west. For a color version of this figure, see www.iste.co.uk/baghdadi/6.zip*

6.6. Conclusions and perspectives

Twenty years of methodological developments showed the operational potential of remote sensing and EO to mining activities and their surroundings. This potential is largely due to improved spatial resolution (high resolution) and spectral resolution (imaging spectroscopy) and GIS developments that allow integration with other measuring techniques (*in situ* measurements, geophysical ground and airborne, LiDAR, etc.).

Due to their obvious applications for societal issues, the current trend is for the launch of space sensors with very high spatial resolution with powerful revisit and stereoscopic capabilities; very high spectral resolution, however, is currently limited to expensive airborne acquisitions. Despite its demonstrated ability, this hampers the use of imaging spectroscopy in the monitoring of contamination from mining.

However, the development of drones and the miniaturization of sensors open interesting prospects in this field for the 5–10 coming years.

6.7. Key points

The developments of remote-sensing sensors with high spatial resolutions and high spectral resolutions opens new perspectives in the development of tools and methods that deal, directly or indirectly, with the essential parameters in the monitoring and evaluation of the environmental and societal impacts of mining throughout the mining cycle, from exploration to closure and decommissioning of the sites.

The integration of satellite data, airborne remote sensing and *in situ* measurements with other relevant EO data allows mapping of these parameters and the development of integrated, objective and enforceable products, and can thus enhance the interaction between the mining industry and relevant stakeholders, in view of a better social acceptability of mining.

6.8. Acknowledgments

This work was carried out through the contribution of Anne Bourguignon, Vincent Mardhel, Francis Cottard, Olivier Rouzeau (BRGM, French Geological Survey) and Henk Coetzee (Council For Geoscience, South Africa).

6.9. Bibliography

[CHE 00] CHEVREL S., COETZEE H., "A new tool for assessing the environmental impact of mining activities. Application to the surface and groundwater sensitivity analysis of the West Rand area, Gauteng, South Africa", *Proceedings of the 28th International Symposium on Remote Sensing of Environment*, Cape Town, South Africa, Paper RN 5.12 March 2000.

[CHE 07] CHEVREL S., SOMMER S., "The use of remote sensing as a tool to enforce EU regulation on mining waste", *Proceedings of the 22nd International Conference on Solid Waste Management*, Philadelphia, PA March 2007.

[CHE 09] CHEVREL S., COETZEE H., BOURGUIGNON A. *et al.*, "Very high spatial and spectral resolution remote sensing in mining-related environmental assessment of the Witwatersrand Gold Field", *IEEE International Symposium on Geoscience and Remote Sensing*, Cape Town, South Africa, 2009

[COE 02] COETZEE H., CHEVREL S., COTTARD F., "Inter-disciplinary studies of the impact of gold and uranium mining in the Witwatersrand Goldfield", in MERKEL B.J. *et al.* (eds), *Uranium in the Aquatic Environment*, available at: http://link.springer.com/book/10.1007/978-3-642-55668-5, 2002.

[CUS 14] CUSHMAN-ROISIN B., *Environmental Fluid Mechanics*, John Wiley & Sons, 2014.

[FAI 91] FAIRFIELD J., LEYMARIE P., "Drainage networks from grid digital elevation models", *Water Resources Research*, vol. 27, no. 5, pp. 709–717, 1991.

[JAM 15] JAMIESON S.S.R., EWERTOWSKI M.W., EVANS D.J.A., "Rapid advance of two mountain glaciers in response to mine-related debris loading", *Journal of Geophysical Research: Earth Surface*, vol. 120, pp. 1418–1435, 2015.

[KRU 93] KRUSE F.A., LEFKOFF A.B., BOARDMAN., J.B. *et al.*, "The spectral image processing system (SIPS) – interactive visualization and analysis of imaging spectrometer data", *Remote Sensing of Environment*, vol. 44, pp. 145–163, 1993.

[SWA 00] SWAYZE G.A., SMITH K.S., CLARK R.N. *et al.*, "Using imaging spectroscopy to map acidic mine waste", *Environmental Science and Technology*, vol. 34, pp. 47–54, 2000.

7

The Contribution of SAR Data to Volcanology and Subsidence Studies

7.1. Introduction

Measuring ground surface displacement is essential for studying and understanding many phenomena related to both natural hazards (volcanism, seismicity and landslides) and anthropogenic hazards (subsidence due to mining operations or fluid extraction/injection). The significance of these phenomena in terms of risks – to both people and infrastructures – is considerable; the resulting surface disturbances must therefore be surveyed in order to prevent future incidents (in the case of anthropogenic hazards, surveillance may even be required by the regulations in force).

Before the advent of satellite SAR (synthetic aperture radar) interferometric techniques (which began to develop in earnest after the launch of the European ERS-1 mission in 1991), soil movement was measured using ground-based geodetic measurements – except when stereophotogrammetry was used. These measurements are generally on a point basis and require instruments to be set up (mainly distance meters and GPS – later Global Navigation Satellite System (GNSS) – receivers, but also inclinometers and extensometers). SAR data make it possible to supplement or replace ground-based measurements that have been difficult to obtain (or even impossible at reasonable cost, depending on the accessibility of the survey site) with space-based information. Although SAR sensors have a

Chapter written by Virginie PINEL and Daniel RAUCOULES.

much lower temporal resolution than continuous ground-based measurement, SAR data provide far better spatial coverage than other techniques. Because of its similarity to geodetics, use of SAR data to measure displacement is a specific discipline of remote sensing. Furthermore, its ability to provide information regardless of the degree of sunlight or cloud cover gives it an advantage over optical and thermal imaging.

This chapter focuses on the displacement measuring techniques that use satellite SAR data (specifically correlation of SAR images and interferometric techniques) and on two specific applications: volcanism and anthropogenic subsidence. Particular attention has been given to the benefits and limitations of these tools for these applications.

7.2. Principles of using satellite SAR data to measure ground displacement

Soil displacement can be estimated using a minimum of two SAR images acquired in very similar conditions (same mode of acquisition, same angle of incidence and similar orbits). Two techniques, using amplitude and phase, respectively, are described here. The displacements are obtained in the acquisition geometry of the SAR images (Figure 7.1) and can then be converted into a displacement field in the terrain geometry (in geographical coordinates). Sub-pixel correlation of the amplitude images provides a displacement in the direction of both the satellite's trajectory (azimuth) and the line of sight (distance) to a precision of around 1/10th of a pixel (around 1 m or even 1 dm for high-resolution images). Differential interferometry (DInSAR), which uses the phase of the radar signal, provides a displacement in the direction of the line of sight with centimetric precision.

7.2.1. *Measuring ground displacement through sub-pixel correlation of SAR images (offset tracking)*

7.2.1.1. *Principle*

The aim of this technique is to estimate, through local correlation, the disparities (offsets in lines and columns) – which can be interpreted as displacements – between homologous points in two images acquired on different dates that have undergone an initial rough overall registration (translation and rotation). Measurement of the disparities is based on correlation calculations. Within small windows corresponding to each of the images (thumbnails), the relative offset of the data can be estimated. This is done by evaluating the offset, which makes it possible to maximize the correlation between these two thumbnails (whose size is usually a power of 2 for algorithmic reasons). The correlation is often estimated in the frequency domain (although methods that work in the spatial domain can also be successful) through Fourier transform (FT) of the contents of the two windows. The displacement extraction in the frequency domain is detailed in Appendix 1 (section 7.6).

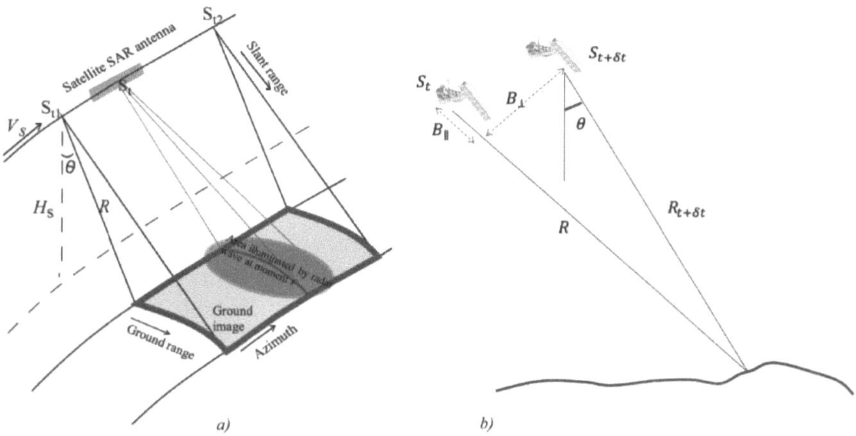

Figure 7.1. *a) Acquisition geometry of a radar image. S_{t1} and S_{t2} represent the positions of the satellite's antenna at the beginning (t_1) and at the end (t_2, typically a few seconds later) of image acquisition, respectively. Vs corresponds to the speed at which the satellite is moving, Hs to the satellite's altitude, θ to the viewing angle and R to the distance between the satellite and the ground, which increases with the position in "range" in the image. The ground image directions are the projection onto the ground of the direction of the line of sight (range) and of the satellite's trajectory (azimuth); b) interferometric geometry. S_t and $S_{t+\delta t}$ are the satellite's positions on two different dates (t and t + δt) in the plane perpendicular to the orbit. $B_\perp$ and $B_\|$ are the perpendicular and parallel components of the orbital spacing, respectively. It should be noted that the orbital spacings in this figure are exaggerated as $R \sim Rt \sim R_{t+\delta t}$*

In favorable conditions, this type of measurement is precise between 1/10th and 1/20th of a pixel at best. Although until recently radar tools had low resolution (azimuth around 4 m or more and distance around 15 m for the most commonly used sensors), radar has an advantage over optical imaging for this type of application in that radar sensors can acquire data regardless of the meteorological conditions, ascending as well as descending, thus providing two points of view and four displacement components. Sensors of this type (ERS, Envisat/ASAR, ALOS/PALSAR, RADARSAT, Sentinel-1, etc.) are therefore useful for mapping metric displacements (landslides and major earthquakes). Furthermore, it is important to note that the geometrical properties (notably targeted/satellite remote sampling as depicted in Figure 7.1(a)) of radar images differ greatly from that of images produced through optical imaging. A horizontal component of the displacement (parallel to the orbit) and a line-of-sight component are therefore measured, rather than planimetric displacements such as those produced by correlation on optical imaging. Because of its "line-of-sight" component, SAR is therefore more sensitive to vertical displacement than optical imaging, which is usually available on satellite platforms.

With the advent of high-resolution sensors (such as COSMO-SkyMed and TerraSAR-X), which achieve resolutions better than 1 m, the possibility of radar image correlation techniques became very promising, nearing (without yet reaching) the precision performance of conventional differential interferometry. Furthermore, the use of significant datasets may even make it possible to contemplate multi-temporal processing, which could improve the precision of deformation speed estimations by combining many displacement estimations.

Finally, the combination of ascending and descending acquisition configurations makes it possible to consider extracting the three displacement components (each configuration being able to provide only two components) [DEM 10a, DEM 10b] in certain conditions.

7.2.1.2. *Constraints and limitations*

Following are some of the constraints:

– Pixel size: correlation algorithms perform differently. Nonetheless, pixel size is a determining factor, because correlation processing aims to match similar objects on two images and estimate the positional offsets. The typical precision is around 1/10th of a pixel, depending on the properties of the images. In this respect, the technique targets decimetric to decametric surface movements according to the type of sensor used.

– Correlation window: using extended correlation windows makes it possible to better identify homologous points on the two images. Nonetheless, this presupposes that the majority of pixels within the correlation window move coherently among themselves. It is therefore desirable that the window remains smaller than the soil surface whose displacement is being measured.

– Topography: as SAR sensors are designed to produce (among other things) interferometric series, they make it possible to produce series with low orbital spacings between acquisitions (typically well below 500 m for satellite/targeted distances 1,000 km away). The contribution of topography (in other words, offsetting due to the "stereoscopic" effect caused by orbital spacings between acquisitions) is therefore usually negligible. Nonetheless, when the aim is to provide evidence of low displacement in areas with significant reliefs, this phenomenon must be taken into account, either through a digital surface model (DSM, a digital representation of the topography), which simulates the relief effect on displacement measurements, or through interpretation, to prevent movement from being falsely detected. This phenomenon relates to offsets in the line of sight, but to a lesser extent also offsets in azimuth if there is little difference between the properties of the acquisitions on the two dates.

7.2.2. *Differential radar interferometry (DInSAR)*

7.2.2.1. *Principle*

Displacement is measured by comparing the phases of two radar images acquired on different dates and coregistered [HAN 01, MAS 08]. For the procedure to work, the two images must be overlaid to the nearest pixel fraction (typically a precision of around 1/10th of a pixel is required). The difference between the phases of the two images (interferogram) then reveals the movements of the soil surface. The interferogram is an image that can be geocoded and can therefore provide a synoptic representation of the displacement.

Therefore, for a given pixel, the phase difference $\delta\Phi$ [FER 16a, FER 16b] between two acquisitions may be expressed as:

$$\delta\Phi = \frac{4\pi\,\delta l}{\lambda}\,[2\pi] \qquad [7.1]$$

where λ is the wavelength of the signal emitted and δl the path difference from sensor to soil surface (for the measured pixel) between the two acquisitions. The term δl incorporates various contributing factors, such as ground displacement, but also geometric effects caused by orbital spacings and topography, as well as atmospheric components, since wave propagation time is determined by physical atmospheric properties that may change between two successive acquisitions.

In short, the phase difference, $\delta\Phi_{1-2}$, between two points (corresponding to subscripts 1 and 2) representing the same ground target seen on the two successive images may be represented as the sum of many contributing factors [RAU 07]:

$$\delta\Phi_{1-2} = \delta\Phi_{displac.\ 1-2} + \delta\Phi_{topo.\ 1-2} + \delta\Phi_{atm.\ 1-2} + \delta\Phi_{flat\ ground\ 1-2} + n\ [2\pi] \quad [7.2]$$

where:

– $\delta\Phi_{displac.\ 1-2} = \frac{4\pi\ \delta r}{\lambda}\ [2\pi]$ is the displacement component and δr the displacement in the satellite's line of sight of pixel 2 in relation to pixel 1;

– $\delta\Phi_{topo.\ 1-2} = \frac{4\ \pi}{\lambda\ R \sin i} B_{\perp} h_{1-2}\ [2\pi]\ = 2\pi \frac{h_{1-2}}{E_a}\ [2\pi]$ is the component showing the impact of the relief. h_{1-2} is the altitude of the target point. i is the incidence angle (in relation to the local vertical, a value close to the viewing angle). $B_{\perp}$ (perpendicular component of orbital spacing) is the projection of the spacing between the two orbits on the normal in the direction of the line of sight (Figure 7.1(b)). E_a, which is inversely proportional to $B_{\perp}$, is called the "altitude of ambiguity" and corresponds to the variation in the altitude associated with a phase variation of 2π rad. In terms of size, for ERS-1/2 satellites, as well as the I2 mode (incidence of around 23°) of the Envisat/ASAR sensor, $E_a \approx \frac{9{,}000}{B_{\perp}}$ m;

– $\delta\Phi_{flat\_ground\ 1-2}$ is the component associated with satellite geometry/supposed Earth surface without relief. It can be estimated using precise knowledge of the orbits corresponding to the two images used, and is generally removed from the interferometric phase;

– $\delta\Phi_{atm.\ 1-2}$ and n are the atmospheric component and the noise associated with measuring, respectively.

It should be noted that the movement component and the topographic component are considered to be decorrelated in the previous representation. This means, rather counter-intuitively that the ground surface and its variation are measured independently. This explains why the DInSAR satellite technique makes it possible to produce DSMs with a metric to decametric altitude precision, whereas displacements may be measured to a centimetric precision or even better. Because the aim is to measure ground surface displacements rather than to produce DSMs, $\delta\Phi_{topo.\ 1-2}$ must be estimated and eliminated. The most common solution is to use preexisting DSMs to model $\Phi_{topo.\ 1-2}$, when the perpendicular component of the orbital spacing is known ($B_{\perp}$). Interferometric pairs with low $B_{\perp}$ have lower phase sensitivity with regard to topography. Therefore, weak $B_{\perp}$ are preferable, particularly if the DSM used for correction is not very precise: altitude uncertainty must, as far as possible, be lower than the altitude of ambiguity.

Producing a simulation of the phase's topographic component from a DSM and orbital information requires a transition from radar geometry (described in Figure 7.1) to cartographic geometry [FER 16a]. The most important thing is that the DSM be transposed in the sensor's geometric reference (azimuth and distance).

The use of interferograms stripped of their topographic component in this way is known as differential interferometry and specifically targets the cartography of ground surface displacements. These displacements appear in the form of "fringes" representing the sensor's movement in the line of sight modulo half the wavelength.

7.2.2.2. *Unwrapping of fringes*

Although interpretation of a simple interferogram is often possible, depending on the properties of the displacements, fringe unwrapping is often required for a quantitative characterization of the displacement being studied. Indeed, the displacement measurements provided by the interferograms are obtained modulo half a wavelength. The multi-temporal approaches described here require, directly or indirectly, fringe unwrapping.

For a measured $\Phi \in]-\pi, \pi]$ value, the φ value verifying $\Phi = \varphi\ [2\pi]$, which is the best representation of the displacement, must be found. Without this last condition (which has no general mathematical form), we would be left with a very ill-posed problem with unlimited possible solutions (in the form $\varphi + 2k\pi$, where φ is a solution and k any integer). The condition generally used constrains the phase difference between two adjacent pixels to a value lower than π radians (that is $|\varphi_i - \varphi_{i+1}| < \pi$, i and $i + 1$ representing two adjacent pixels). In other words, the smallest possible difference is selected, ensuring that the surface described by the function $\varphi(x,y)$ (phase to be estimated, depending on the position) is as regular as possible (Figure 7.2). On this basis, "local" techniques [GOL 88] entail making step-to-step corrections to the interferogram (adding or removing 2π), while "global" techniques [GHI 94] seek to minimize one functional characterizing the regularity of the function $\varphi(x,y)$ to be estimated.

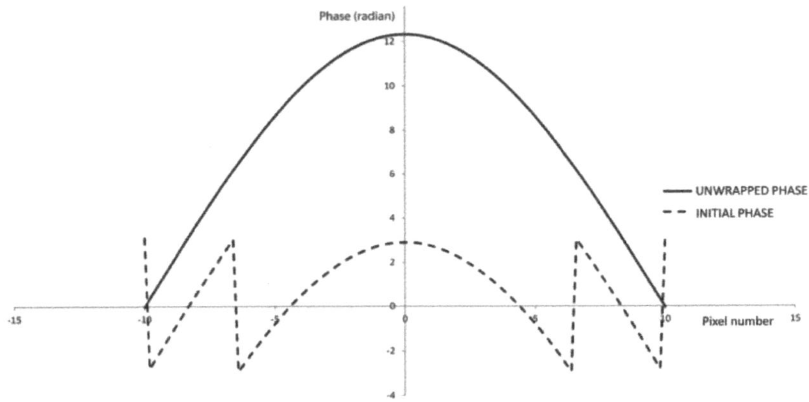

Figure 7.2. *Fringes are unwrapped by passing from one known phase modulo 2 π to an unwrapped phase that does not present a "phase jump" (in other words, that is continuous) and represent the displacement as best as possible*

Inconsistencies (called "residual errors" by Goldstein *et al.* [GOL 88]) may appear. Indeed, the fact that a looped route may alter the initial point value is characteristic of the unwrapping errors that may spread across the whole area of an image. This may be a result of noise on the phase, or of the fact that the regularity condition is false locally (due to topography or movement characteristics). Various existing algorithms offer solutions to reduce the spread of errors. Furthermore, it should be noted that advance knowledge of the function $\varphi(x,y)$ may facilitate unwrapping. For example, this may be the case if deformation modeling has produced a simulation of the displacement component. It is therefore wise to unwrap the difference between interferogram and simulation if this produces fewer fringes [RAU 10b].

Finally, the subject of unwrapping makes it easy to introduce one of the limitations of DInSAR. The acceptable gap between two adjacent pixels is lower than π rad (half a fringe). Beyond this, the result becomes ambiguous (deformations that produce gaps between adjacent pixels from π or $-\pi$ cannot be distinguished). This means that a deformation gradient greater than a quarter of a wavelength per pixel cannot be observed through interferometry. For example, for a C-band sensor with 15 m resolution, the (maximum) resulting detectability threshold is almost 1 cm per 10 m. This constraint on the maximum deformation gradient that can be measured is specific to the radar interferometric technique and limits certain applications characterized by overly rapid or localized movements (Figure 7.3).

7.2.3. *Performance and limitations of SAR data in measuring displacement*

For the first 15 years during which SAR data were used to measure ground displacement – in other words, until 2010 – most studies were carried out using data acquired by European Space Agency satellites, the ERS-1 and ERS-2 satellites and then the Envisat satellite [PIN 14]. These now decommissioned C-band satellites (wavelength 5.66 cm) acquired images corresponding to a ground area of 10,000 km^2, with a resolution of 4 m azimuth and 20 m distance, and had a sampling period of 35 days. Figure 7.3 gives an idea of the displacements that could potentially be measured by this type of sensor (the white area). Extremely localized displacements covering less than 10 m, such as some collapsed cavities (Figure 7.3(c)), cannot be measured. This is also true of very extensive displacements (covering more than 100 km) or those characterized by steep gradients (greater than 1 cm in 10 m, for example). Towards the end of the first decade of the 21st century, data were obtained from the Japanese ALOS satellite – which produced a massive amount of L-band data (wavelength 23.5 cm) – which work better than shorter wavelengths (as discussed later on) in highly vegetated areas, and very high spatial resolution (HR) X-band satellites (wavelength 3.1 cm), TerraSAR-X and COSMO-SkyMed, were developed, which improved temporal resolution by a factor of 2 or 3 and spatial resolution by a factor of almost 10. The arrival of high resolution (HR) sensors made it possible to measure displacements that were more localized in space or transitory in time. Of course, the presence of topography can locally alter the capacities of SAR images, particularly steep slopes that create shaded areas or, on the contrary, layover areas [FER 16a] that are not correctly imaged by the satellite.

Figure 7.3 summarizes the technical limitations of interferometry on the basis of the sensors' spatial and temporal resolutions, but disregards an important aspect, namely that using interferometry to measure displacement is possible only if the phase remains coherent (coherence is a statistical indicator that locally characterizes the stability of the phase of an interferogram, see Appendix 2, section 7.7) between two acquisitions – that is if the soil backscattering properties do not change [FER 16b]. The presence of vegetation may cause a rapid loss of coherence between successive acquisitions [FER 16b], making it more difficult for InSAR to measure displacement. This limitation can be reduced by using a longer wavelength (e.g. using L-band rather than C-band data when the existing archive – generally less well supplied with available sensors – allows it), which enables the wave to penetrate vegetation more effectively and reach – within certain limitations – elements underneath the vegetation that are more stable in time. Longer wavelengths (e.g. P-band: ~ 70 cm) could prove more successful in the presence of vegetation (although these long wavelengths are less sensitive, which renders this sort of measurement unhelpful for slow movements), but no satellite sensor of this type currently exists. Tests using airborne data have made it possible to prove that P-band data are significantly better than L-band data in terms of coherence ([LON 11] – TropiSAR campaign). It should be noted that these tests have provided evidence of a temporal decorrelation that is considerably higher in HH than in VV polarization, suggesting that VV polarization might be preferable. This type of phenomenon is not observed in the most commonly used wavelengths (X, C, L).

The final limitation of differential interferometry is because of the variations in the atmosphere's physical properties (temperature, humidity and pressure) between two radar acquisitions, which leads to variation in the propagation speed of the electromagnetic waves. This phenomenon of atmospheric refraction, which affects the phase, must not be confused with the very slight atmospheric attenuation of the amplitude of satellite radar images (one of the notable advantages of radar imaging in the presence of cloud cover). This effect produces a phase difference that is caused by the atmosphere and that can be wrongly interpreted as a ground displacement up to several dozen centimeters [ZEB 97, HAN 01]. The phase known as "atmospheric" can be divided into a turbulent component and a stratified component. The turbulent component falls under the dynamics of the atmosphere. It is not easy to estimate, but can usually be considered random on each SAR image acquisition date. It can therefore be greatly reduced, even eliminated, by using sets of interferograms. The stratified component is caused by variation in pressure, temperature and water vapor content in the troposphere between radar acquisitions. It leads to atmospheric fringes that are first and foremost related to topography. In the absence of additional information, this effect can be reduced empirically using the correlation between phase and altitude observed in the images in areas without deformation [REM 03]. This technique is based on the assumption that there are no lateral variations in stratification in the image. Furthermore, it can cause partial elimination of the displacement signal, particularly on volcanoes, where ground displacement caused by a magma chamber is usually strongly correlated to topography. If there is a sufficiently dense GPS network in the studied area, this can be used to estimate the stratified atmospheric and turbulent delay [WEB 02, LI 06]. Meteorological data acquired locally or from global models can also be taken into account in estimating the tropospheric delay [DEL 98, BEA 00, DOI 09].

Finally, the contribution of the ionosphere to this atmospheric effect must also be considered. The ionosphere is the partially ionized upper layer of the atmosphere. It is a dispersive medium for the envisaged frequency bands, which means that it consequently has the greatest effect on the longest wavelengths (the resulting ionospheric extension increases with the wavelength of the signal). L-band interferograms are therefore more likely to be affected by this phenomenon than C-band or X-band data (an example of this type of phenomenon is given in [RAU 10a]).

Some of the limitations caused by decorrelation or the atmospheric effects listed previously can be reduced by using not only a single interferogram obtained from two images, but also information extracted from a temporal series of images.

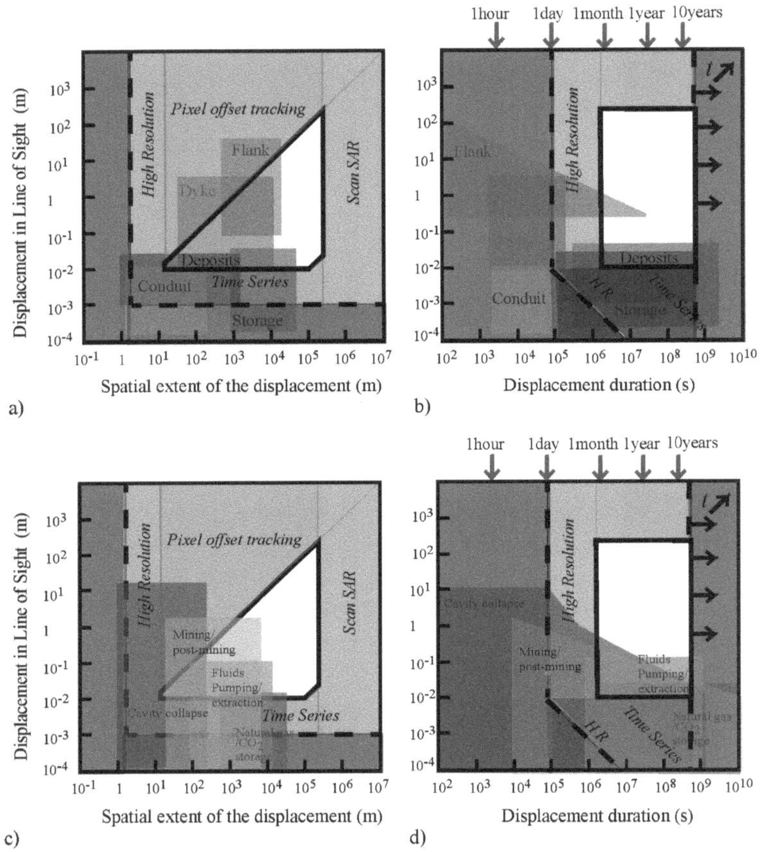

Figure 7.3. *Field of application of SAR data for measuring deformation in volcanology (a and b) and in the study of subsidence (c and d). The classic technical limitations of the sensors (spatial and temporal resolution, period covered by the archive) are compared to deformation characteristics in volcanic regions and in areas of subsidence. The ability of high spatial and temporal resolution (HR) sensors to overcome these technical limitations, the methods developed for processing time series (see section 7.2.4) and the increase in the period covered by the available archive with time (t ↗) are highlighted. Furthermore, InSAR's additional pixel offset tracking technique (see section 7.2.1) makes it possible to expand the range of application of SAR data in the event of steep deformation gradients. The white area bordered by solid black lines represents the area of validity in InSAR with classic Envisat data, and the shaded area bordered by dashed lines shows the limits of SAR data, including pixel offset tracking, HR data and the use of time series. (a and c) Amplitude of deformation according to spatial coverage (deformation gradients). (b and d) Amplitude of deformation according to duration (deformation rate). For a color version of this figure, see "http://www.iste.co.uk/baghdadi/6.zip*

7.2.4. *Multi-temporal processing*

7.2.4.1. *Techniques for post-processing interferograms*

Multi-temporal processing has two purposes: first, to use the large number of SAR acquisitions to measure deformation speed more precisely, compensating in part for the biases discussed in the previous section, and second, to highlight and measure variations in deformation speed over relatively long periods [HOO 12].

The simplest types of multi-temporal processing consist of averaging interferograms:

– for a sudden movement (on a precise date), it is best to combine interferograms produced from pairs of images acquired one before and one after the event. If the interferograms are independent of one another (no common image), their deformation components will be alike, while their atmospheric components will be different. A simple average will reproduce the deformation component, while the contribution of the atmospheric component to uncertainty in the deformation estimation will be reduced as the inverse of the square root of the number of averaged interferograms;

– for movements that are fairly regular in time, the aim is to estimate the average displacement speed (*V*) from a set of interferograms covering variable durations over a long or short period. Therefore, a given pixel within the interferograms is sought, the value *V*, which best verifies the relation:

$$\Phi_i = \frac{4\pi}{\lambda} V \Delta t_i \qquad [7.3]$$

where *i* is the interferogram index, Φ_i and Δt_i are the interferometric phase and the duration associated with interferogram *I*, respectively, and λ is the wavelength of the sensor.

An estimation of *V* (displacement speed) can be obtained as a solution for an inversion problem through least squares formulation as follows:

$$\hat{V} = \frac{\lambda\, \boldsymbol{\Delta t}^t \boldsymbol{\Phi}}{4\pi(\boldsymbol{\Delta t}^t\, \boldsymbol{\Delta t})} = \frac{\lambda \sum \Delta t_i \Phi_i}{4\pi \sum {\Delta t_i}^2} \qquad [7.4]$$

where Φ and Δt are the column vectors made up of Φ_i and Δt_i.

This estimation corresponds to a phase average weighted by the square of the duration. Ideally (in a least squares sense), the noise on the phase would be independent of the duration Δt_i. This is not the case here, notably due to the phenomenon of temporal decorrelation. Nonetheless, if interferograms presenting a low coherence are excluded from the set, this estimation can provide reliable information.

The methods described previously do not always provide information on the evolution of the displacement. An initial method for detecting possible increase or reduction in displacement speed is to create pseudo-interferograms for each of the dates in relation to the first date of the series from a more significant set of (previously created) interferograms. There is therefore a very simply formulated overdetermined linear inverse problem that aims to reproduce the temporal evolution of the phase (which can be interpreted as displacement), which is solved through least squares [USA 99, BER 02, LEM 05]. The advantage is that this produces a simplified dataset containing the useful information extracted from the interferograms. When the set of pseudo-interferograms has been created, temporal filtering can be applied to it. The result is therefore less affected by noise and atmospheric effects than the initial interferograms. More simply, average speeds can be extracted through linear regression for several dates within a given interval. It is useful to note that the direct stacking of interferograms (totals, time-weighted averages, etc.) can nonetheless be useful, making it possible to have control over the interferograms that actually affect the estimation of average speeds.

The simplest formulation for the estimation of these pseudo-interferograms is:

$$Y = AX \qquad [7.5]$$

where Y is a column vector containing, for a given pixel of the SAR image, the phases of the M interferograms produced and selected as reliable, X is the vector corresponding to the reconstructed phases for the N acquisition dates. A is a matrix containing the values –1, 0 or +1, reflecting the fact that the interferograms of Y must be element differences of the vector X.

A simple estimation of X can be obtained from:

$$X = (A^t A)^{-1} A^t Y \qquad [7.6]$$

Expression [7.6] assumes that A^tA is reversible. This assumes that every image in set X can be linked to another by a sum (absolute value) of interferograms of the set Y, or that there are no groups of images isolated from the others due to the absence of interferograms among them. Furthermore, more sophisticated estimation methods can use the average coherence of the various interferograms to weight their contribution to X.

Different variants therefore exist to address the obvious limitations of the approach. It is required to estimate X (e.g. use of singular value decomposition approaches rather than least squares), but also to choose the interferograms to invert well so that they are only slightly affected by any execution errors or major atmospheric effects that could greatly influence the result. On this second point, although a visual selection is feasible for sets of several dozen interferograms, an initial automatic selection should be carried out for larger sets. Finally, any gaps in the set of interferograms should be filled.

One of these variants is the SBAS (small baselines) [BER 02] method, which involves selecting (automatically) the interferometric pairs associated with low orbital and temporal spacing (spatial and temporal baselines), taking into account the deformation speed characteristics to be measured. First, spatial and temporal decorrelations increase with the baselines (also, according to land use, only fairly short spatial baselines are used, as decorrelation is less significant in urban areas), and second, as we have seen, extracting the topographic component requires a DSM whose uncertainty is much lower than the altitude of ambiguity (typically $E_a/3$). Therefore, unless a particularly precise DSM is available, it is unlikely that interferograms with an altitude of ambiguity lower than 30 m can be used.

7.2.4.2. *Persistent scatterer interferometry (PSI)*

This approach is different because it has been developed to deal with some of the limitations mentioned previously. The technique was first used at the end of the 1990s [FER 99]. The analysis focuses on particular points of the image, the PSs (persistent scatterers), which are major individual backscatterers compared to the rest of its pixels. Pixels containing a PS are identified primarily by the fact that they have points whose backscattering coefficients are higher than their surroundings, but which are also temporally stable. For a pixel, one criterion that enables the selection is a threshold on the link between the standard deviation (temporal) of the amplitude and its average (also temporal). Other criteria (including retrospectively examining the evolution of the estimated displacement component) can be used to eliminate possible false detections of PS. It should be noted that the various thresholds are to some extent arbitrary and the user's experience affects how they are fixed so as to obtain a network of points that are reliable but also dense enough that they adequately represent the displacement field.

The advantage of working on such points is as follows. Since they are temporally very stable in terms of radiometry, they are also assumed to be stable in terms of phase. They are therefore affected only minimally by temporal decorrelation.

Because of its point nature (in other words, it is much smaller than the pixel containing it), the PS is not very sensitive to spatial decorrelation. Indeed, its structure is perceived no differently if the viewing angle is slightly changed (within the limits of the slight variations of angle permitted by a series of interferometric images).

To summarize, an ideal PS would be an object that backscatters strongly (much more than the other scatterers within its pixel – if not, it will be masked), but that is punctual (therefore not subject to spatial decorrelation) and that does not evolve over time (therefore not subject to temporal decorrelation), with the exception of its movement, which must be less than a quarter of a wavelength compared to the nearest other PSs (otherwise the result of the measurement will be ambiguous). A actual PS must resemble this type of object in the domain defined by the acquisition geometry (viewing angles, maximum perpendicular baselines of the image sets) and the duration of the imaged period. Indeed, the "lifespan" of a PS (i.e. the duration during which the identified scatterer has not undergone any change that would alter its nature as a PS) is supposed to be longer than the duration covered by the dataset. However, it should be noted that using limited-lifespan PSs for part of the surveyed period (or partial PSs) can be an advantage [BAS 04].

The phase of one PS in relation to another chosen as a reference for the study can be expressed in the following form at a time in relation to the reference date (which may be the first date of the dataset):

$$\Phi\,(\Delta t) = \frac{4\pi}{\lambda} V.\Delta t + \frac{2\pi}{Ea}\Delta h + \Phi_{residual}(\Delta t) + \Phi_{plane}\ [2\pi] \qquad [7.7]$$

where V is the average speed of the monitored PS in relation to the reference point. Δh is the gap between the real altitude difference between the PS and the reference and the altitude difference provided by the initial DSM. $\Phi_{residual}(\Delta t)$ is the sum of the atmospheric phase difference between the corresponding acquisition Δt and the reference date and a possible irregular movement component. Φ_{plane} corresponds to an uncertainty that may be caused by both imprecision in the orbital coordinates and the difficulty of identifying atmospheric phenomena, even spatial wavelength deformations much greater than the image size. These two phenomena are reflected in spatial trends on the phase similar to the planes. Therefore, to eliminate this irregular movement component, an average plane estimated on the map of $\Phi\ (\Delta t)$ is removed (except in specific cases).

The first step is therefore linear regression of Φ in relation to the variables Δt and $B_{\perp}$ (inversely proportional to E_a), to extract an initial estimation of V and Δh and to separate the first three components of the phase. It should be noted that the simultaneous estimation of V and Δh is necessary because one of the advantages of the PSI technique is that it can accentuate the largest values of $B_{\perp}$. High values of $B_{\perp}$ result in the fact that the interferometric phase is very sensitive to topography. In other words, an error in the estimation of Δh leads to an error in estimation of V greater with large $B_{\perp}$. Also, as much as possible, the altitude of the point must be known in order to correct the topographic component: therefore, V cannot be precisely estimated without the estimation of Δh.

The remainder is therefore made up of the atmosphere as well as the nonlinear displacement component:

$$\Phi_{residual}(\Delta t) = \Phi_{atm}(\Delta t) + \Phi_{nonlin}(\Delta t) \qquad [7.8]$$

The two terms in equation [7.8] are separated according to the following criteria:

– the atmosphere is supposed to be temporally decorrelated (the state of the atmosphere changes significantly in a few hours – the case of correlated phenomena on a relief is not considered here), but its fluctuations show a certain spatial extension;

– the movement – although nonlinear – is supposed to have a certain temporal continuity. Thus, by applying temporal high-pass filters and spatial low-pass filters to $\Phi_{residual}(\Delta t)$, $\Phi_{atm}(\Delta t)$ can be estimated to extract a nonlinear deformation component that will be added to the linear deformation component.

It should be noted that the linear regression in expression [7.7] is not straightforward, because of the ambiguity of 2π on the phase. Indeed, due to this difficulty, which is somewhat similar to the unwrapping problem, for a number of values that are not infinite, type $\frac{4\pi}{\lambda}V.\Delta t \quad [2\pi]$ linear regressions in relation to Δt do not guarantee a unique solution for V. Also, when the atmospheric component has been estimated, it can be used to correct the data and perform several iterations in order to better estimate all the components of the phase. Furthermore, *a priori* knowledge of the deformation speed to be evaluated and of the magnitude of the altitude difference (in other words, the uncertainty of the initial DSM) can be useful for directing the deformation estimation and avoiding outliers.

7.2.4.3. *Interferometry using diffuse scatterers*

This technique can be seen as an extension of the PS technique for non-punctual objects (in other words, small groups of pixels) whose properties evolve in a similar way over time. An insufficiently dense network of PS points is supplemented with diffuse scatterers (DSs).

The first step is to assemble pixels in DS so that they have similar properties. The method chosen by Ferretti *et al.* [FER 11] to assemble these points consists of applying a Kolmogorov–Smirnov test (which facilitates comparison of the distribution functions of two samples) to the temporal evolution of the amplitude of two adjacent pixels. The samples associated with two temporal series of amplitudes are compared. If the test is positive, the two pixels in the same DS are then matched. The DS grows step-by-step in this way until the size limit set by the user (usually a few dozen pixels) is reached or there are no more candidates that have sufficiently similar starting points to be included in the DS.

For each DS constructed in this way, all possible interferograms from the dataset are calculated. One phase value for each acquisition date is then reconstructed and assigned to each DS on the basis of the interferometric phase values assigned to the pixels contained in these DSs. The coherence values (see Appendix 2, section 7.7) of the various interferograms estimated on the DSs themselves are used to weight the contributions of each interferogram to the estimation of the phases of the DSs.

When the phases on the various acquisition dates have been reconstructed for all the DSs in the image, they can be used to estimate temporal series through PSI processing: although DSs are not PSs, deformation estimations can be applied to them, albeit with lower reliability than for PSs. The reconstructed phases can also be directly used for convential differential interferometry (calculation of some interferograms) in order to perform a sort of smart multi-view processing: because the pixels are not assembled in rectangular groups independent of their properties but according to their similarity, the result will be more coherent.

7.3. The contribution of InSAR to the understanding and monitoring of volcanoes

7.3.1. *Types of deformation observed through InSAR on volcanoes*

There are many potential sources of deformation on volcanoes [DZU 07]. Most of them are directly linked to magmatic activity, but some are related to tectonics or landslides.

As far as deformation of magmatic origin is concerned, the main cause identified is inflation or deflation of the volcanic area caused by a magma chamber, in other words a magma storage area located in the Earth's crust, at depth or within the volcanic edifice. Inflation is due to an increase in pressure. This increase can be caused by an influx of magma into the storage zone or by the crystallization of the stored magma. Deflation, by contrast, corresponds to a decrease in pressure caused by magma leaving the storage zone, to potentially feed an eruption, or by physical changes of the magma stored in the crust (cooling or release of gas). The influx of magma at depth can occur over several months or several dozen years. The amplitude of the deformation caused (from a few millimeters to a few dozen centimeters) depends on the variation of pressure. Deformation can affect an area of around 1 km to several dozen kilometers, depending on the depth of the storage zone (labeled *storage* in Figure 7.3). When it is observed, this deformation, caused by the accumulation of magma at depth, is a long-term precursor of volcanic activity.

When overpressurization in a storage zone becomes sufficiently large, it causes the surrounding rock to rupture and magma to move. The magma propagates toward the surface, along planar structures. These magmatic intrusions are called *dykes* when they cross-cut the geological layers of the crust (e.g. in a vertical path) and *sills* when they propagate along the interfaces between geological layers (usually horizontally). A magma intrusion propagates over a period of a few hours to a few days, causing a ground displacement of a few centimeters to several meters over an area of around 100 m to a dozen kilometers (labeled *dyke* in Figure 7.3). The propagation phase is therefore generally shorter than the storage phase and the resulting geophysical signals (surface deformation, seismicity) are short-term precursors of eruptive activity.

In the case of andesite volcanoes, near the surface, the magma flows in an open conduit, causing transient deformation over a period of a few hours to a few days, of low amplitude (less than a few centimeters) and across an area of around 100 m at most (labeled *conduit* in Figure 7.3). This type of signal is difficult to measure with InSAR.

A volcanic structure grows through the emplacement of magmatic intrusions at depth and the accumulation of eruption products at the surface. The destabilization of this structure, due to tectonic or gravitational forces, can lead to landslides, which begin suddenly (in the space of a few minutes) but can sometimes last for several months. Landslides are characterized by dramatic movements (from a dozen centimeters to a dozen meters), affecting areas between 1 km and several kilometers (labeled *flank* in Figure 7.3). On a smaller spatial scale (around 1 m to 1 km), the emplacement of volcanic deposits (lava flows, explosive deposits) causes downward subsidence of a few millimeters to a few centimeters over a period of several months to several years (labeled *deposits* in Figure 7.3). This displacement is caused by the compaction of deposits and subsidence of the surrounding crust due to the applied load.

Finally, on most volcanoes, the interaction between the meteoric waters and the magmatic system leads to the development of a hydrothermal system that is regularly disrupted by the injection of hot fluids. This system produces surface displacements of variable duration and spatial expanse and of centimetric to pluri-decimetric amplitude.

When it comes to volcanic risk management, it is important to be able to distinguish deformations induced by the magmatic system from other types of deformation. It is also essential to measure the accumulation of magma at depth and to know when it will rise toward the surface. It is also useful to know how much magma is available and where the magma will reach the surface. By making it possible to measure surface displacement, InSAR can improve risk management. Indeed, this surface displacement can be interpreted in terms of pressure sources at depth, with the help of mechanical models. It provides information on the existing volume of magma, its geometry and its depth. This information is obtained with a degree of uncertainty, due to noise in the available data, imprecision of the models used in terms of the rheological properties considered and non-uniqueness of the inversion solutions. The numerous studies that have been performed using this remote sensing technique since the 1990s have made it possible to observe deformations on more than 160 volcanoes all over the world [BIG 14], including all types of deformation (Figure 7.4). Using the archive of SAR data has made it possible to perform statistical studies, thereby improving our understanding of how volcanoes work. It has been shown that although there is a strong link between eruption and deformation, not all volcanic eruptions produce a deformation signal that can be measured by InSAR [BIG 14, PIN 14]. This is the case, in particular, for transient and small-scale displacements caused by andesite magma flow in a near-surface conduit, which are not usually detected through InSAR but can be measured by near-field tiltmeters. The InSAR technique has been shown to be particularly effective for monitoring deformation caused by eruptive deposits or temporal monitoring of magma accumulation at depth. This effectiveness has been strengthened by the development and use of processing methods for SAR time series (see section 7.2.4). The relatively poor temporal resolution of InSAR limits its use for monitoring magma movement somewhat, but the advent of high-resolution data has strengthened these capacities (see section 7.2.3).

Finally, over the last few years, InSAR has made it possible to "image" exceptional volcanic events, such as the Piton de la Fournaise flank displacement linked to the large-volume eruption of April 2007 [FRO 15], as well as the rifting events in Afar [WRI 06, GRA 09] and Iceland [SIG 15]. In the event of rifting, sub-pixel correlation of SAR images provides supplementary near-field information (near the opening area), where the displacement gradients are significant and exceed InSAR's resolution limit [GRA 09].

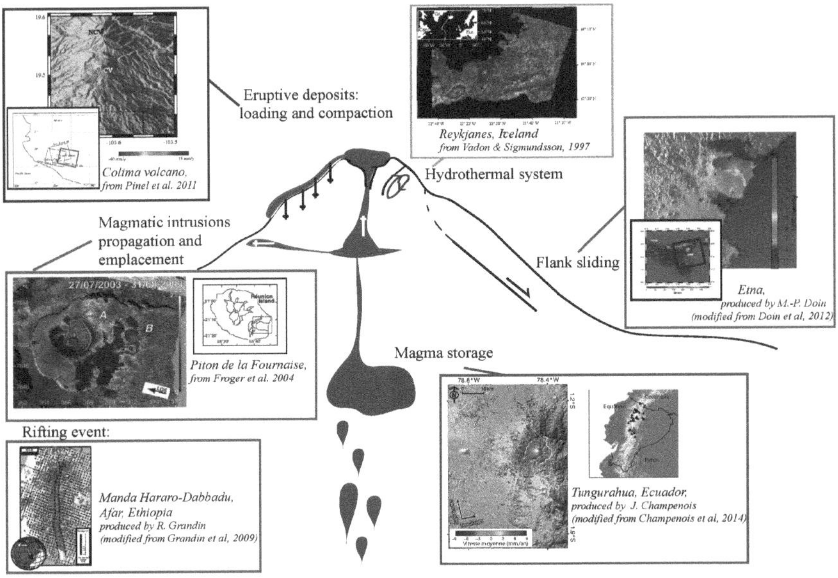

Figure 7.4. *Various types of deformation related to volcanic activity, observed through InSAR (figure modified from Pinel et al. [PIN 14] using different studies [VAD 97, FRO 04, GRA 09, PIN 11, DOI 12, CHA 14]). All examples of displacement fields shown in this figure were obtained through DInSAR. The displacement field of the Manda Hararo–Dabbahu rift also results from information provided through sub-pixel correlation of SAR as well as optical Systeme probatoire pour l'observation de la Terre (SPOT) images. For a color version of this figure, see "http://www.iste.co.uk/baghdadi/6.zip*

7.3.2. *Quantification of magma storage at depth and link to eruptive activity*

The advantage of InSAR lies in its capacity to detect centimetric displacement zones in volcanic arcs, making it possible to obtain information on volcanoes that have hardly been instrumented or that have not been instrumented at all. It has therefore been possible to observe magma storage zones, for example in the Andes [PRI 02, FOU 10]. Magma chambers have also been discovered near to volcanoes that are well instrumented but whose geodetic networks are focused on the summit of the structure, while the storage zone is a few kilometers away. This was the case, for example, at South Sister in Oregan, USA [WIC 02]. This observation through InSAR made it possible to install ground-based sensor networks across the terrain, in order to improve surveying.

InSAR also made it possible to show that most of the magma chambers detected were located in the surface zone of the Earth's crust at a depth of less than 10 km [BIG 14, PIN 14] and that this storage depth seemed to be influenced by the tectonic context, storage being deeper in areas of thick, old crust characterized by a compressive stress field [CHA 14b]. Furthermore, the large spatial area covered by InSAR made it possible to observe deep sources of deformation located around 20 km deep into the Andes [PRI 02, SPA 08, HEN 13] and in Iceland [DEZ 04, OFE 11].

As explained previously, statistical studies have also shown that on volcanic arcs like that of the Andes, although deformations associated with eruptive activity are often observed, some eruptions do not lead to deformations that InSAR can measure. Conversely, InSAR has also observed cases of deformation over several dozen years without any eruptive activity at the surface [HEN 13]. This last scenario poses the problem of volcanic risk associated with the accumulation of large volumes of magma at depth [SPA 08]. Even in classic cases where surface deformation is linked to eruptive activity, comparing the volume of magma emitted at the surface with the InSAR estimation of the volumes accumulated at depth is essential for understanding the evolution of eruptive activity. Therefore, at Tungurahua, an Ecuadorian volcano active since 1999, Champenois *et al.* [CHA 14a] observed inflation of the volcano at a constant rate during the period 2003–2009 (Figure 7.5), by studying a temporal series of SAR images acquired by the Envisat satellite. Inversion of the data using a spherical pressure source model in an elastic crust that takes into account the topography of the area (use of a DSM) makes it possible to obtain model parameters that explain relatively well the displacement field measured at the surface. Inversion makes it possible to attribute the origin of this signal to an accumulation of magma at 11.5 km below sea level. This accumulation occurs at a rate of 7 million m^3/year. Over the period in question, InSAR estimated that the total volume of magma stored at depth was 46.5 million m^3, a minimal value that does not take into account the compressibility of magma. Over the same period, the volume emitted at the surface was estimated to be 30 million m^3. This means that the volume of magma emitted during that period was more than compensated for by the supply of magma at depth. Despite uncertainties in the estimation of the volumes emitted and stored at depth, InSAR suggests that, judging by the rate of magma production at depth, eruptive activity could intensify in the future.

In the case of Tungurahua, InSAR revealed a single storage zone with a spherical geometry, but it sometimes reveals a complex storage system beneath the volcanic structure. In Kilauea, Hawaii, a study of SAR data has identified many storage zones of varied geometries [BAK 12].

In short, InSAR has made it possible to observe magma storage zones at depth, but also to characterize magma feeding systems in terms of geometry, volume and number of storage zones.

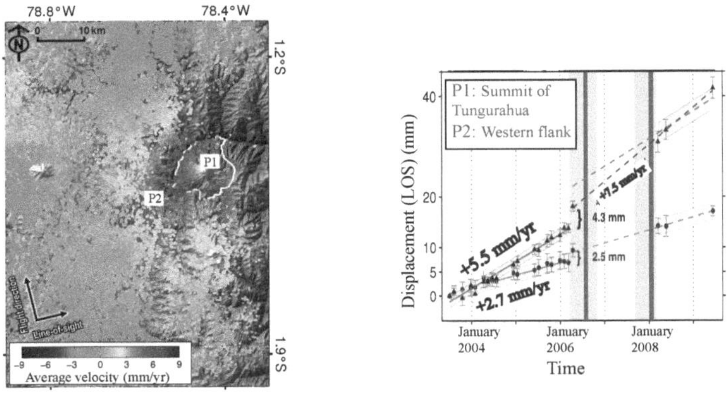

a) b)

Figure 7.5. *Observation of magma accumulation at depth by InSAR: Tungurahua, Ecuador, over the period 2003–2009: a) average velocity measured by the persistent scatterers technique around Tungurahua; inflation is observed; b) temporal evolution of the displacement for two points (P1 near the summit of the volcano and P2 on the west side around 20 km away from the summit). For a color version of this figure, see "http://www.iste.co.uk/baghdadi/6.zip*

7.3.3. *Temporal monitoring of the magma's path toward the surface*

Propagation of magma toward the surface generally occurs over a few hours or a few days. The relatively low temporal resolution of InSAR (at best a few days with high-resolution data) does not make it the ideal tool for monitoring the progression of magma toward the surface, and this monitoring usually occurs through measurement of the deformation by ground-based sensors (GNSS, extensometers and/or tiltmeters) or through studying seismicity. However, the progression of magmatic intrusions has been observed when the magmatic progression has been slow [PED 06] or when a SAR acquisition has taken place a few hours before an eruption [BAG 13]. Furthermore, the advent of sensors and sensor constellations (e.g. COSMO-SkyMed) with better temporal resolution and the use of several satellites together now greatly improve the InSAR's performances when monitoring the temporal evolution of the geometry of a magmatic intrusion and its volume [LUN 13].

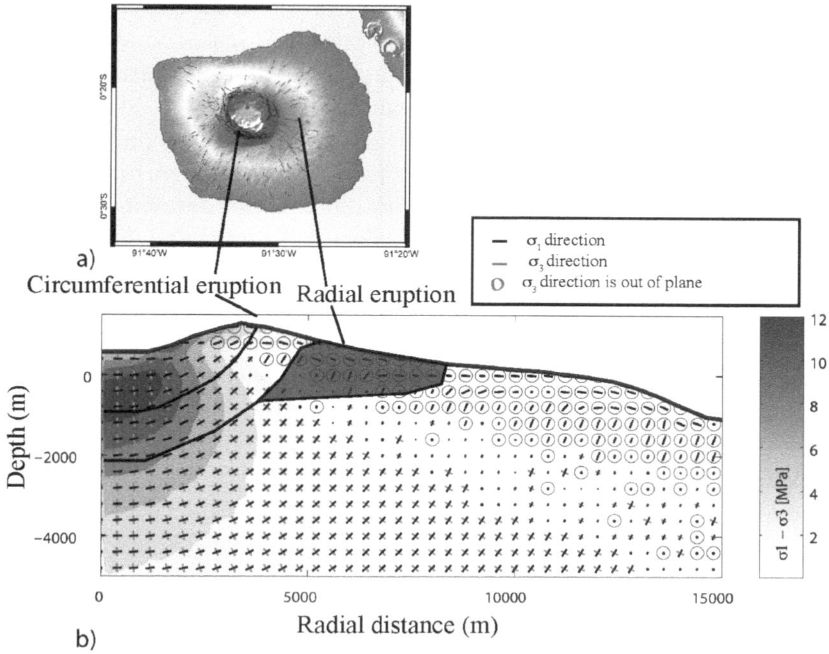

Figure 7.6. *Magma plumbing system of Fernandina volcano (Galapagos); a) distribution of eruptive fissures on the surface; b) radial cross-section of the volcanic structure showing the stress field caused by the caldera formation – that is the collapse of the crater. The geometry of the various magmatic intrusions estimated through inversion of the InSAR data is also represented in black. The geometry of the intrusions is therefore explained by a stress field dominated by the formation of the caldera; in other words, by the collapse of the summit crater [COR 15]*

The progression or just the final geometry of a magmatic intrusion is "imaged" by InSAR. The essential advantage of this technique is the spatial density of the measurements. Because it produces a surface displacement map, InSAR is the ideal technique for characterizing the geometry of the magmatic intrusions through inversion of the surface displacement. In particular, on the island of Fernandina, Galapagos, this technique has made it possible to show how circumferential eruptions (whose vents are localized near the edge of the summital crater), or radial eruptions (whose vents are localized on the flanks of the volcano), were always produced from *sills* (horizontal planar intrusions), which either curve up vertically or turn 90° to propagate laterally in a vertical plane in the form of *dykes* [BAG 13]. Figure 7.6(b) represents the geometry of the intrusions observed through inversion of InSAR data and the stress field caused by formation of the caldera. Intrusions that lead to circumferential eruptions begin at the shallowest depth and propagate in a plane perpendicular to the cross-section plane in Figure 7.6(b), whereas intrusions that lead to radial eruptions propagate first in a plane perpendicular to the cross-section plane before turning 90° (when they are in the area where σ_3 is out of plane) to enter the vertical plane represented in Figure 7.6(b). All intrusions open onto a plane perpendicular to σ_3, as theory predicts that they will when emplacement is controlled by the stress field. This geometry of magmatic intrusions observed through InSAR is explained by the stress field in the volcanic structure caused by the caldera formation (collapse of the summit crater) [COR 15] (Figure 7.6). The knowledge that InSAR provides about the geometry of magmatic intrusions is essential because it provides information regarding the stress field within the edifice when they are emplaced [HOO 11, COR 15]. Knowledge of this geometry also makes it possible to measure changes in the local stress field caused by the influx of magma and their influence on the progression of future intrusions [AME 07, GRA 10].

7.4. Contribution of InSAR to the characterization of anthropogenic subsidence

Since the end of the 1990s, differential SAR interferometry has been used to observe and monitor man-made vertical movements (subsidence and sometimes soil heave; see [AME 99] on subsidence caused by pumping in Las Vegas, and [CAR 00] on deformation due to mining on the site of Gardanne in France). This usually involves providing information that makes it possible to characterize the surface deformation and potentially to better constrain the phenomenon in the subsoil (e.g. using digital modeling of the deformation). It should be noted that for movements that have had observable consequences (such as damage to buildings), ground-based instrumentation (e.g. installation of a leveling network) is often installed only after these observations have been made. In such cases, because of images recorded before the phenomenon occurred, InSAR can be the only source of information on the first phase of the deformation [COL 05].

Furthermore, InSAR can be part of the surveillance of specific sites/phenomena. However, even with the recent advent of sensors with repeatability of better than a week, it is still difficult to imagine using the tool as a warning system, because of the delay caused by insufficient temporal sampling and acquisition and processing times. However, interferometric observations provide information that can be combined with ground-based measurements (usually punctual), providing better temporal sampling, but not making it possible to understand the deformation geometry as a whole. Also, satellite InSAR can provide additional information on top of that provided by a ground-based leveling system, and can make it possible to optimize the layout of network stations, depending on the geometry of the zone affected by the deformation.

In the framework of risk analysis, the purpose of measuring ground surface deformation is to better identify (in addition to punctual ground-based observations and measurements) the extent of the area that may be affected and potentially to evaluate the dangers it faces in the future by understanding its evolution in the past (acceleration/slowing). InSAR presents an advantage for many applications, which are detailed in the following and shown in Figure 7.7.

7.4.1. *Deformation due to mining and post-mining*

This mainly concerns deformation connected to present or past activity in underground mines. It should be noted that in French territory, although using DInSAR to map deformation resulting from underground mine operations (for the period covered by the satellite SAR tools, this mainly concerns coal mining) presented a great advantage before the last coal mines were closed in 2004, this type of technique is now applied instead of the post-mining risk – that is possible deformations that remain after the mines have been closed.

In modern long wall mining operation, where subsidence is deliberately caused by collapsing the cavity (roof materials filling up the empty space behind the mining face), this involves first correlating the deformations observed on the surface with the advance of the mining face and its extension, and then identifying anomalies in relation to the expected movements.

Deformations resulting from mining operations cover an area the size of an exploited panel and its area of influence (often at least 250–500 m wide and up to a few kilometers long). Deformation occurs in three stages [GUE 07]:

– an initial subsidence phase corresponding to 10–15% of the total deformation, lasting less than a month;

– a main subsidence phase lasting several months and corresponding to around 80% of the total deformation (this stage may correspond to deformation speeds of a few decimeters per year to several meters per year);

– residual subsidence corresponding to 5–10% and occurring over several years.

The first two phases correspond to the period when the point monitored on the surface is within the area of influence of the mining work in progress, while the final phase corresponds to the remaining deformation when the monitored point is outside the area of influence of the work in progress (even, in some cases, after operations have ceased). The deformation is therefore due to the gradual compacting of the backfilling materials. Depending on the environments of the monitored sites (the maximum deformation amplitude must not be too significant), InSAR can map deformation during site activity [CAR 00, ENG 11]. It is sometimes also possible to monitor the residual deformation [RAU 09].

For room-and-pillar mining, the effects of deformation are more limited when work is actually in progress. In this type of mining, useable rock is deliberately left in place to form pillars, which limits surface deformations (but mining is less productive because some coal remains in place). However, the rise of the piezometric level after the mine closes and the pumping required for ceasing the work can lead to long-term consequences in terms of surface deformation. The flooding of the mine can gradually degrade the pillars and alter the mechanical properties of any backfilling materials. Therefore, surface phenomena can occur with variable amplitudes (cumulative deformations from a fraction of a centimeter to several meters) decades after operations have ceased, and can be observable through radar interferometry [COL 05, GUE 09].

7.4.2. *Pumping*

Extraction of fluids can cause subsidence (and, vice versa, injection can cause inflation). Because of the expansion of urban areas that began in the 20th Century, and the increasing associated extraction of water resources due to the growing needs of populations, agriculture and industry, subsidence phenomena of great amplitude are fairly frequent. In these cases, the drop in the piezometric level due to pumping leads to more or less significant subsidence according to the properties of the aquifer and the materials located between the aquifer and the surface. In terms of the application of InSAR to urban subsidence related to the population's water requirements, a good example is subsidence in Mexico City (Mexico [STR 99, YAN 12]), with deformation that may reach 40 cm/year across an area of around 20 km. Incidences of subsidence in Jakarta (Indonesia [CHA 13]) and Manila (Philippines [RAU 13b]) are lesser but still significant (>10 cm/year). In terms of pumping linked to industry, in the town of Prato (Italy [RAU 03a]), InSAR has revealed movements of around 10 cm/year. Finally, subsidence phenomena are frequently linked to the use of groundwater for agricultural requirements. In Lorca (Spain [DEM 13]) and the San Joaquin valley (California [BAC 13]), InSAR has revealed substantial deformation (more than 1 dm/year).

In all cases, measuring surface displacement makes it possible (or could make it possible) for operators to better anticipate the possible consequences of their work and to take any possible corrective measures (modification of pumping rates, change of extraction site, etc.) and for local authorities to better identify and manage these consequences after the event (even to identify the cause of the damage observed on the surface).

Extraction of other types of fluid (hydrocarbons, brine, etc.) for industrial purposes should also be mentioned. Van der Kooij [VAN 97] has mapped displacements on the Belridge Oil Field (California), observing deformations of around 30 to 40 cm/year over an area of a few kilometers. Similarly, on the Vauvert salt mine (Gard, France), mined through brining (water is injected into drilled holes to dissolve the salt layer and brine is extracted at other points), interferometric measurements have made it possible to map deformations of up to 2 cm/year over an area of 8 km [RAU 03b].

Subsidence phenomena related to pumping are very variable in scale (less than 1 km to several dozen kilometers) and amplitude (a few millimeters per year to around 50 cm/year for the most significant cases) according to the properties of the subsoil, the depth at which the liquid is located and the extraction flow, and occur in areas with a variety of different land uses (urban, vegetation). The choice of conventional interferometry or PSI depends greatly on the properties of the deformation and of the site studied.

7.4.3. *Gas storage*

This concerns inflation related to the injection of gas at significant depths (hundreds of meters or kilometers). This may be natural gas or CO_2 (in the case of geological repositories aiming to reduce emissions into the atmosphere). On the one hand, the resulting deformations are low (a fraction of a millimeter per year to a few centimeters per year), but on the other, stores of natural gas present cyclical (often seasonal) deformations. The combination of these two aspects can render InSAR rather unsuccessful, and this is compounded by the fact that this type of installation is usually located on sites with non-urban soil occupations (where temporal decorrelation phenomena are therefore more sensitive). However, in some situations, it can be used. The case of the CO_2 injection demonstrator at In Salah (Algeria) is particularly significant. A series of CO_2 injection tests were performed on this natural gas mining site between 2004 and 2011 (globalccsinstitute.com/projects/salah_co2_storage). Among the surface observations made, InSAR (PSI and classic interferometry) made it possible to map the resulting deformation [VAS 10, ONU 09]. Over the period monitored, soil heave rates of 5 to 7 mm/year over an area of 4 to 5 km were observed. This deformation can be linked to the activities of the various injection pits over the given period. Furthermore, the asymmetry of the surface deformation geometry has been interpreted by these authors by the presence of faults.

The success of the technique in this type of application should nonetheless be put into perspective due to the specific properties of the site. Desert areas are conducive to interferometry, and the properties of the subsoil (notably the permeability of the reservoir) produce, for the same injection rate, soil heave rates much higher than those of a site with different properties in relation to the injection of CO_2 [ROH 12]. Also, in the case of an injection site such as those currently envisaged for Europe, the deformation – if it is significant enough to be observed through InSAR – would be especially observable during the first years of the injection program (the maximum speed diminishing gradually as the deformed zone expands) and the zone to be surveyed could reach a radius of 20 km or more around the injection point. Furthermore, vegetated land covers of the envisaged sites often make precise measurement of such slight deformations difficult (due to temporal decorrelation).

7.4.4. *Localized subsidence (collapse of cavities, dissolution, underground work, etc.)*

This concerns phenomena of varying speeds, often limited in time (a few months to a few years) and of small spatial coverage (less than a few hundred meters). Other than the mine-related effects described earlier, the collapse of quarries and mines, gypsum and salt dissolution [RAU 13a] and underground work (e.g. tunnel digging) should be mentioned. From the point of view of the InSAR technique, this type of deformation is problematic because it can be associated with significant deformation gradients or changes in the surface leading to loss of coherence (see section 7.2.3). Collapses, which cause sinkholes on the surface, can generally not be measured using this technique, although sensors with the highest temporal and spatial resolution (e.g. TerraSAR-X and COSMO-SkyMed) now make it possible to observe phenomena that are both more rapid and more localized than was possible with Envisat/ASAR and ALOS sensors. However, in cases where deformation is moderate and far larger than the pixel size of the SAR image, observations are possible.

Underground work causes surface deformations, either directly, because of the excavation of cavities, or due to the lowering of the piezometric level, which is required for work to be performed (pumping). This type of deformation has recently been monitored with PSI on digging work for railway tunnels in the town of Bologna (Italy) by the TRE (Tele-Rilevamento Europa: http://treuropa.com/construction-engineering/tunneling) team, which observed subsidence and soil heave of around 1 cm/year. Similar phenomena (subsidence and soil heave) had been observed near Saint-Lazare station (Paris) during work to extend the metro [FRU 00, LEM 02].

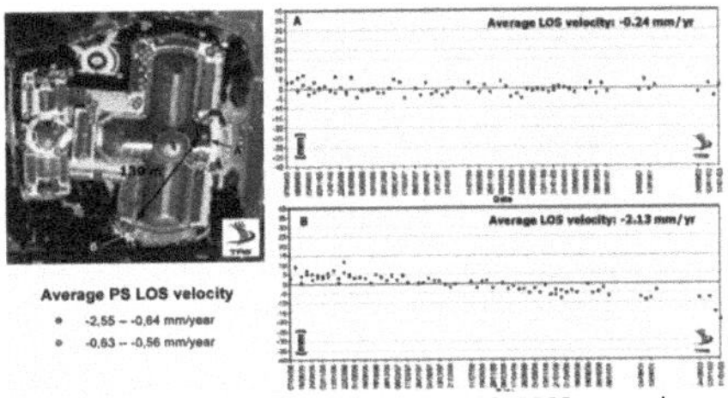

Deformation on the Grand Palais between 1995–2003, seen through PSI interferometry ([RAU 07] – processing TRE). The south wing subsides at a rate of 2 mm/yr, while the north wing is stable throughout the monitored period. Time Series of the deformation in the Line of Sight of two significant PSs (A and B) are shown.

Vertical movement (soil heave) related to work on the Paris metro in July 1998 ([RAU 07]; after [LEM 02]).

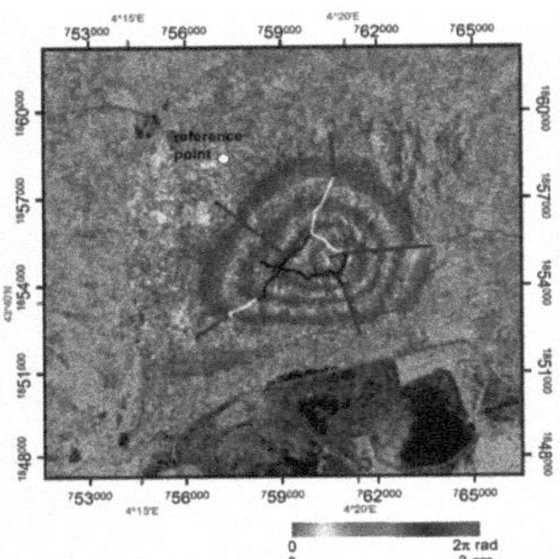

Subsidence related to salt mining in Vauvert (after [RAU 03b]) over the period 1993–1999).

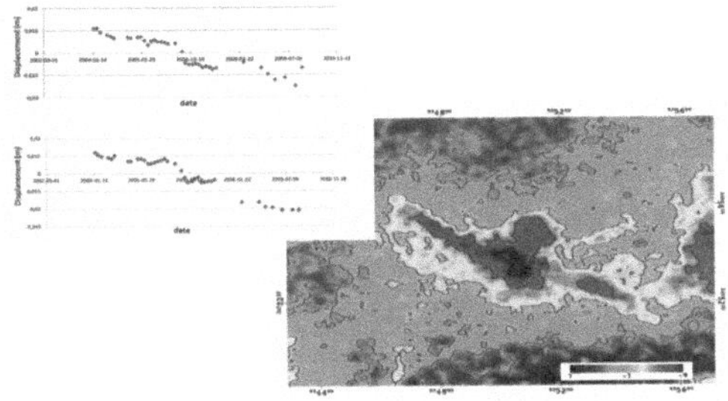

Subsidence related to dissolution of salt in the municipality of Hilsprich (after [RAU 13a]). Stack of PALSAR interferograms and ASAR Time Series on some PS points on the edge of a subsidence basin.

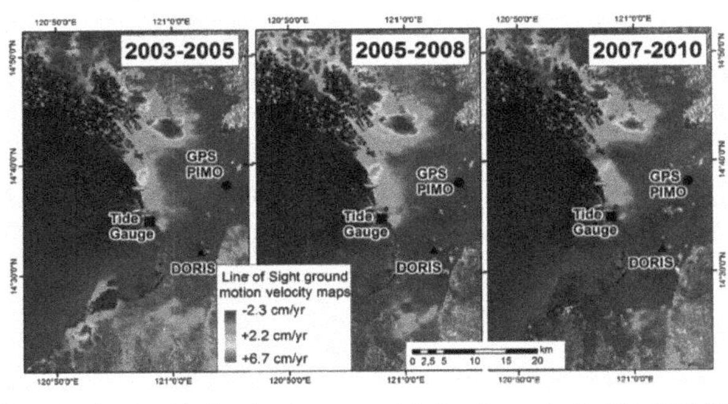

Deformations in urban Manila related to over-exploitation of groundwater (after [RAU 13b]).

Figure 7.7. *Some examples of subsidence phenomena with man-made or natural origins, monitored by DInSAR. The deformations presented are caused by the extraction of fluids (water, brine), the presence of cavities (mines) and the phenomenon of salt dissolution. Satellite InSAR (DInSAR and PSI) has made it possible to map the deformations and to monitor changes in their evolution (variable deformation speeds)*

7.5. Outlook

We currently have an archive of 20 years of SAR data, which can be used to understand the history of deformation over almost all the world. This archive, which has built up over time, makes it possible to observe low-amplitude displacements (of around a few millimeters per year) and their temporal variations. This growing archive represents a large volume of data that require improved storage capabilities and processing chains.

Until the present day, ground displacements were measured using SAR data acquired a long time before the event. But the improvement in the temporal resolution of SAR data brought about by the launch of the satellite constellations (COSMO-SkyMed, Tandem-X, Sentinel-1) opens up a fresh perspective on monitoring the deformation of the Earth's surface within a relatively short timeframe.

In the field of volcanology, the main challenge for the years to come is to successfully integrate, within the volcano observatories in charge of volcano monitoring, the information provided by remote sensing, and in particular by SAR imaging, with the additional field data. In the event of a crisis, field data can be temporarily but drastically limited by access difficulties and the destruction of instruments, which makes SAR data extremely important. But for optimal effectiveness in times of crisis, SAR data must be integrated into the hazard estimation procedures from the start. Furthermore, the availability of SAR data within a short timeframe (a few hours) will also require the development of data assimilation methods based on physical models to provide effective prediction tools.

For measuring anthropogenic deformation, other than the previously mentioned contribution of better temporal resolution – whether to ensure better reactivity in terms of monitoring the phenomenon, or simply to obtain more precise measurements because of a higher number of acquisitions – the potential of recent and future high spatial resolution sensors (better than 1 m) is high. Indeed, since the type of phenomena studied sometimes covers only small areas (a few hundred meters or less), high resolution ensures better characterization of phenomena for which multi-metric resolution data remain ill-adapted.

7.6. Appendix 1

7.6.1. *Estimation of offsets through correlation in the frequency domain*

The product of the Fourier transform (FT) of the content of one of the windows by the conjugate of the FT of the contents of the other makes it possible to directly obtain the correlation after inverse FT [LEP 07]. Indeed, to estimate the offset between two thumbnails i_1 and i_2 extracted from two images that are similar but locally offset around point (x,y) of the first image, the thumbnail i_2 can be represented as $i_2(x,y) \approx i_1(x-\Delta_x, y-\Delta_y)$, where Δ_x and Δ_y are the offsets (in pixel unity) to be determined.

The thumbnails will be represented in the frequency domain (after FT) as:

$$I_2(\omega_x,\omega_y) = TF(i_2) = \iint_{-\infty}^{+\infty} i_2(x,y)e^{-i(\omega_x x+\omega_y y)}\,dx\,dy \approx \iint_{-\infty}^{+\infty} i_1(x-\Delta_x,y-\Delta_y)\,e^{-i(\omega_x x+\omega_y y)}\,dx\,dy \qquad [7.9]$$

hence by changing a variable $x \rightarrow x-\Delta_x$ and $y \rightarrow y-\Delta_y$.

$$I_2(\omega_x,\omega_y) \approx I_1(\omega_x,\omega_y)e^{-i(\omega_x\Delta_x+\omega_y\Delta_y)} \qquad [7.10]$$

In other words, $f(\omega_x,\omega_y) = \omega_x\Delta_x + \omega_y\Delta_y$ is the phase of the complex function $I_1(\omega_x,\omega_y)\,I_2^*(\omega_x,\omega_y)$.

Thus, the derivatives (in relation to ω_x and ω_y) of the phase of $I_1(\omega_x,\omega_y)\,I_2^*(\omega_x,\omega_y)$ make it possible to rise to (Δ_x,Δ_y) – $\Delta_x = \frac{df}{dx}$ and $\Delta_y = \frac{df}{dy}$.

This is the basis of many calculation methods in the frequency domain.

7.7. Appendix 2

7.7.1. *Interferometric coherence*

The two images (acquired on two different dates) can be represented by the complex functions $S_1 = A_1 e^{i\Phi_1}$ and $S_2 = A_2 e^{i\Phi_2}$.

The complex coherence $\Gamma_{1,2}$ of the interferogram $\Phi_2 - \Phi_1$ at point (x,y) of the image will be defined as:

$$\Gamma_{1,2} = \frac{\langle S_1 . S_2^* \rangle}{\sqrt{\langle A_1^2 \rangle \langle A_2^2 \rangle}} \quad [7.11]$$

$\langle . \rangle$ being the average estimated in a small vicinity (generally a rectangular window with the size of a few pixels) of (x,y).

The module of $\rho_{1,2}$ of this complex value will simply be called coherence. It will be considered to be an indicator of the stability of the phase between the two acquisitions and therefore of the quality of the interferogram produced (with values between 0 and 1; 1 corresponding to the ideal case where the two images are identical).

7.8. Key points

Differential radar interferometry is a technique for mapping surface deformation from archived SAR data, suitable primarily for measuring movements of several millimeters to a few dozen centimeters that cover an area greater than one hundred meters. The theoretical precision is around one 20th of the radar wavelength, but in practice this is reduced by the noise of measuring and by atmospheric effects. Offset tracking techniques can be used on the same data to measure displacements of larger amplitude. The main advantage of SAR data is its good spatial coverage, which severely constrains the geometry of the deformation sources.

In the field of volcanology, SAR data offers a unique opportunity for global monitoring of volcanoes, making it possible to better understand the geometry and behavior of their magma feeding systems, but also enabling early detection of magma storage below the surface that could potentially cause an eruption.

The advantage of SAR data in the study of subsidence phenomena is its capacity to provide a precise measurement and a view of the entirety of the deformation, in addition to or in the absence of ground-based instrumentation (or even to provide mapping prior to any ground-based observation).

7.9. Bibliography

[AME 99] AMELUNG F., GALLOWAY D.L., BELL J.W. *et al.*, "Sensing the ups and downs of Las Vegas: InSAR reveals structural control of land subsidence and aquifer-system deformation", *Geology*, vol. 27, no. 6, pp. 483–486, 1999.

[AME 07] AMELUNG F., YUN S.H., WALTER T.R. *et al.*, "Stress control of deep rift intrusion at Mauna Loa volcano, Hawaii", *Science*, vol. 316, pp. 1026–1030, 2007.

[BAC 13] BACQUES G., Etude de variations spatio-temporelles de glissements asismiques le long de failles majeures par interférométrie Radar satellitaire, PhD Thesis, University of Grenoble, 2013.

[BAG 13] BAGNARDI M., AMELUNG F., POLAND M.P., "A new model for the growth of basaltic shields based on deformation of Fernandina volcano, Galapagos Islands", *Earth and Planetary Science Letters*, vol. 377–378, pp. 358–366, 2013.

[BAK 12] BAKER S., AMELUNG F., "Top-down inflation and deflation at the summit of Kilauea Volcano, Hawaii observed with InSAR", *Journal of Geophysical Research: Solid Earth*, available at http://dx.doi.org/10.1029/2011JB009123, vol. 117, 2012.

[BAS 04] BASILICO M., FERRETTI A., NOVALI F. *et al.*, "Advances in permanent scatterers analysis: semi and temporary PS", *Proceedings of European Conference on Synthetic Aperture Radar*, Ulm, Germany, pp. 349–350, 25–27 May 2004.

[BEA 00] BEAUDUCEL F., BRIOLE P., FROGER J.L., "Volcano wide fringes in ERS synthetic aperture radar interferograms of Etna (1992-1999): deformation or tropospheric effect?", *Journal of Geophysical Research*, vol. 105, no. B7, pp. 16391–16402, 2000.

[BER 02] BERARDINO P., FORNARO G., LANARI R. *et al.*, "A new algorithm for surface deformation monitoring based on small baseline differential SAR interferograms", *IEEE Transactions on Geoscience and Remote Sensing*, vol. 40, pp. 2375–2383, 2002.

[BIG 14] BIGGS J., EBMEIER S.K., ASPINALL W.P. *et al.*, "Global link between deformation and volcanic eruption quantified by satellite imagery", *Nature*, vol. 5, 2014.

[CAR 00] CARNEC C., DELACOURT C., "Three years of mining subsidence monitored by SAR interferometry, near Gardanne, France", *Journal of Applied Geophysics*, 43, p. 43–54, 2000.

[CHA 13] CHAUSSARD E., AMELUNG F., ABIDIN H. *et al.*, "Sinking cities in Indonesia: ALOS PALSAR detects rapid subsidence due to groundwater and gas extraction", *Remote Sensing of Environment*, vol. 128, pp. 150–161, 2013.

[CHA 14a] CHAMPENOIS J., PINEL V., BAIZE S. *et al.*, "Large scale inflation of Tungurahua volcano (Ecuador) revealed by persistent scatterers SAR interferometry", *Geophysical Research Letters*, vol. 41, pp. 5821–5828, 2014.

[CHA 14b] CHAUSSARD E., AMELUNG F., "Regional controls on magma ascent and storage in volcanic arcs", *Geochemistry, Geophysics, Geosystems*, vol. 15, pp. 1407–1418, 2014.

[COL 05] COLESANTI C., LE MOUELIC S., BENNANI M. *et al.*, "Detection of mining related ground instabilities using the permanent scatterers technique – a case study in the East of France", *International Journal of Remote Sensing*, 26, no. 1, pp. 201–208, 2005.

[COR 15] CORBI F., RIVALTA E., PINEL V. *et al.*, "How caldera collapse shapes the shallow emplacement and transfer of magma in active volcanoes", *Earth and Planetary Science Letters*, vol. 431, pp. 287–293, 2015.

[DEL 98] DELACOURT C., BRIOLE P., ACHACHE J., "Tropospheric corrections of SAR interferograms with strong topography: application to Etna", *Geophysical Research Letters*, vol. 25, pp. 2849–2852, 1998.

[DEM 10a] DE MICHELE M., RAUCOULES D., LASSERRE C. *et al.*, "The Mw 7.9, 12 May 2008 Sichuan earthquake rupture measured by sub-pixel correlation of ALOS PALSAR amplitude images", *Earth Planets and Space*, vol. 62, pp. 875–879, 2010.

[DEM 10b] DE MICHELE M., RAUCOULES D., DE SIGOYER J. *et al.*, "Three-dimensional surface slip of the 12 May 2008 Sichuan earthquake (China) from synthetic aperture radar", *Geophysical Journal International*, vol. 183, pp. 1097–1103, 2010.

[DEM 13] DE MICHELE M., BRIOLE P., RAUCOULES D. *et al.*, "Revisiting the shallow Mw 5.1 Lorca earthquake (southeastern Spain) using C-band InSAR and elastic dislocation modeling", *Remote Sensing Letters*, vol. 4, pp. 863–872, 2013.

[DEZ 04] DE ZEEUW-VAN DALFSEN E., PEDERSEN R., SIGMUNDSSON F., "Satellite radar interferometry 1993–1999 suggests deep accumulation of magma near the crust–mantle boundary at the Krafla volcanic system, Iceland", *Geophysical Research Letters*, available at http://dx.doi.org/10.1029/2004GL020059, vol. 31, 2004.

[DOI 09] DOIN M., LASSERRE C., PELTZER G. *et al.*, "Corrections of stratified tropospheric delays in SAR interferometry: validation with global atmospheric models", *Journal of Applied Geophysics*, vol. 69, no. 1, pp. 35–50, 2009.

[DOI 12] DOIN M.-P., LODGE F., GUILLASO S. *et al.*, "Presentation of the small baseline NSBAS processing chain on a case example: the Etna deformation monitoring from 2003 to 2010 using Envisat data", *Fringe Workshop Proceedings*, ESA Special Publication SP-697, Frascati, Italy, 2012.

[DZU 07] DZURISIN D., *Volcano Deformation: Geodetic Monitoring Techniques*, Springer-Praxis Books, 2007.

[ENG 11] ENGELBRECHT J., INGGS M. R., MAKUSHA G., "Detection and monitoring of surface subsidence associated with mining activities in the Witbank coalfields, South Africa, using differential radar interferometry", *South African Journal of Geology*, vol. 114, no. 1, p. 77, 2011.

[FER 99] FERRETTI A., PRATI C., ROCCA F., "Non-uniform motion monitoring using the permanent scatterers technique", *Proceedings of FRINGE'99*, Liège, Belgium, pp. 1–6, 10–12 November 1999.

[FER 11] FERRETTI A., FUMAGALLI A., NOVALI F. *et al.*, "A new algorithm for processing interferometric data stacks: Squeesar", *IEEE Transactions on Geoscience and Remote Sensing*, vol. 49, no. 9, pp. 3460–3470, 2011.

[FER 16a] FERRO-FAMIL L., POTTIER E., "Synthetic aperture radar imaging", in BAGHDADI N., ZRIBI M. (eds), *Microwave Remote Sensing of Land Surfaces*, ISTE Press, London and Elsevier, Oxford, 2016.

[FER 16b] FERRO-FAMIL L., POTTIER E., "Coherent diversity mode SAR imaging: SAR polarimetry, interferometry and tomography", in BAGHDADI N., ZRIBI M. (eds), *Microwave Remote Sensing of Land Surfaces*, ISTE Press Ltd, London and Elsevier Ltd, Oxford, 2016.

[FOU 10] FOURNIER T.J., PRITCHARD M.E., RIDDICK S.N., "Duration, magnitude, and frequency of subaerial volcano deformation events: new results from Latin America using InSAR and a global synthesis", *Geochemistry, Geophysics, Geosystems*, vol. 11, 2010.

[FRO 04] FROGER J.L., FUKUSHIMA Y., BRIOLE P., *et al.*, "The deformation field of the August 2003 eruption at Piton de la Fournaise, Reunion Island, mapped by ASAR interferometry", *Geophysical Research Letters*, vol. 31, 2004.

[FRO 15] FROGER J.L., FAMIN V., CAYOL V., *et al.*, "Time-dependent displacements during and after the April 2007 eruption of Piton de la Fournaise, revealed by interferometric data", *Journal of Volcanolgy and Geothermal Research*, vol. 296, pp. 55–68, 2015.

[FRU 00] FRUNEAU B., SARTI F., "Detection of ground subsidence on the city of Paris using radar interferometry: isolation, of deformation from atmospheric artifacts using correlation", *Geophysical Research Letters*, vol. 27, no. 24, pp. 3981–3984, 2000.

[GHI 94] GHIGLIA D.C. and ROMERO L.A., "Robust two-dimensional weighted and unweighted phase unwrapping that uses fast transforms and iterative methods", *Journal of Optical Society of America A*, vol. 11, pp. 107–117, 1994.

[GOL 88] GOLDSTEIN H., ZEBKEN A., WERNER C.L., "Satellite radar interferometry: two-dimensional phase unwrapping", *Radio Science*, vol. 23, pp. 713–720, 1988.

[GRA 09] GRANDIN R., SOCQUET A., BINET R. *et al.*, "September 2005 Manda Hararo–Dabbahu rifting event, Afar (Ethiopia): constraints provided by geodetic data", *Journal of Geophysical Research*, vol. 114, 2009.

[GRA 10] GRANDIN R., SOCQUET A., JACQUES E. *et al.*, "Sequence of rifting in Afar, Manda-Hararo rift, Ethiopia, 2005–2009: time space evolution and interactions between dikes from interferometric synthetic aperture radar and static stress change modeling", *Journal of Geophysical Research*, vol. 115, 2010.

[GUE 07] GUÉGUEN Y., Etude des mouvements de surface en environnement minier à partir d'interférométrie radar et identification des origines de la déformation, PhD Thesis, University Paris Est, Marne la Vallée, 2007.

[GUE 09] GUÉGUEN Y., DEFFONTAINES B., FRUNEAU B. *et al.*, "Monitoring residual mining subsidence of Nord/Pas-de-Calais coal basin from differential and persistent scatterer interferometry (Northern France)", *Journal of Applied Geophysics*, vol. 69, pp. 24–34, 2009.

[HAN 01] HANSSEN, R.F., *Radar Interferometry – Data Interpretation and Error Analysis*, Kluwer Academic Publishers, vol. 2, p. 328, 2001.

[HEN 13] HENDERSON S. T., PRITCHARD M. E. "Decadal volcanic deformation in the Central Andes Volcanic Zone revealed by InSAR time series", *Geochemistry Geophysics Geosystems*, vol. 14, pp. 1358–1374, 2013.

[HOO 11] HOOPER A., OFEIGSSON B., SIGMUNDSSON F. *et al.*, "Increased crustal capture of magma at volcanoes with retreating ice cap", *Nature Geosciences*, vol. 4, pp. 783–786, 2011.

[HOO 12] HOOPER A., BEKAERT D., SPAANS K., *et al.*, "Recent advances in SAR interferometry time series analysis formeasuring crustal deformation", *Tectonophysics*, vol. 514, pp. 1–13, 2012.

[LEM 02] LE MOUÉLIC S., RAUCOULES D., CARNEC C., *et al.*, "A ground uplift in the city of Paris (France) detected by satellite radar interferometry", *Geophysical Research Letters*, vol. 29, no. 17, 2002.

[LEM 05] LE MOUÉLIC S., RAUCOULES D., CARNEC C. *et al.*, "A least-squares adjustment of multi-temporal InSAR data – application to the ground deformation of Paris", *Photogrammetric Engineering and Remote Sensing*, vol. 71, pp. 197–207, 2005.

[LEP 07] LEPRINCE S., BARBOT S., AYOUB F. *et al.*, "Automatic and precise orthorectification, coregistration, and subpixel correlation of satellite images: application to ground deformation measurements", *IEEE Transactions on Geoscience and Remote Sensing*, vol. 45, pp. 1529–1558, 2007.

[LI 06] LI Z., FIELDING E.J., CROSS P. *et al.*, "Interferometric synthetic aperture radar atmospheric correction: GPS topography-dependent turbulence model", *Journal of Geophysical Research*, vol. 111, p. B02404, 2006.

[LON 11] LONGA I., Télédétection radar en bande P potentiel pour l'interférométrie radar différentielle, MSc thesis, University of Toulouse III, 2011.

[LUN 13] LUNDGREN P., POLAND M., MIKLIUS A., *et al.*, "Evolution of dike opening during the March 2011 Kamoamoa fissure eruption, Kilauea volcano, Hawaiii", *Journal of Geophysical Research: Solid Earth*, vol. 118, p. 897–914, 2013.

[MAS 08] MASSONNET D., SOUYRIS J.C., *Imaging with Synthetic Aperture Radar*, EPFL-CRC Press, 2008.

[OFE 11] OFEIGSSON B.G., SIGMUNDSSON F., HOOPER A. *et al.*, "InSAR time series analysis at Hekla volcano, Iceland: inflation periods and crustal deformation associated with the 2000 eruption", *Journal of Geophysical Research: Solid Earth*, vol. 116, p. 2156–2202, 2011.

[ONU 09] ONUMA T., OHKAWA S., "Detection of surface deformation related with CO_2 injection by DInSAR at In Salah, Algeria", *Energy Procedia*, vol. 1, pp. 2177–2184, 2009.

[PED 06] PEDERSEN R., SIGMUNDSSON F., "Temporal development of the 1999 intrusive episode in the Eyjafjallajökull volcano, Iceland, derived from InSAR images", *Bulletin of Volcanology*, vol. 68, pp. 377–393, 2006.

[PIN 11] PINEL V., HOOPER A., DE LA CRUZ-REYNA S. *et al.*, "The challenging retrieval of the displacement field from InSAR data for andesitic stratovolcanoes: case study of Popocatepetl and Colima Volcano, Mexico", *Journal of Volcanology and Geothermal Research*, vol. 200, pp. 49–61, 2011.

[PIN 14] PINEL V., POLAND M., HOOPER A., "Volcanology: lessons learned from synthetic aperture radar imagery", *Journal of Volcanology and Geothermal Research*, vol. 289, pp. 81–113, 2014.

[PRI 02] PRITCHARD M.E., SIMONS M., "A satellite geodetic survey of large-scale deformation of volcanic centres in the central Andes", *Nature*, vol. 418, pp. 167–171, 2002.

[RAU 03a] RAUCOULES D., LE MOUÉLIC S., CARNEC C. *et al.*, "Urban subsidence in the city of Prato (Italy) monitored by radar satellite radar interferometry", *International Journal of Remote Sensing*, vol. 24, pp. 891–898, 2003.

[RAU 03b] RAUCOULES D., MAISONS C., CARNEC C. *et al.*, "Monitoring of slow ground deformation by ERS radar interferometry – The case of Vauvert (France)", *Remote Sensing of Environment*, vol. 88, pp. 468–480, 2003.

[RAU 07] RAUCOULES D., COLESANTI C., CARNEC C., "Use of SAR interferometry for detecting and assessing ground subsidence", *Comptes Rendus Geosciences*, vol. 339, pp. 289–302, 2007.

[RAU 09] RAUCOULES D., BOURGINE B., DE MICHELE M. *et al.*, "Validation and intercomparison of persistent scatterers interferometry: PSIC4 project results", *Journal of Applied Geophysics*, vol. 68, pp. 335–347, 2009.

[RAU 10a] RAUCOULES D., DE MICHELE M., "Assessing ionospheric influence on L-band SAR data: implications on co-seismic deformation measurements of the 2008 Sichuan earthquake", *IEEE Geosciences and Remote Sensing Letters*, vol. 7, no. 2, pp. 286–290, 2010.

[RAU 10b] RAUCOULES D., RISTORI B., DE MICHELE M. *et al.*, "Surface displacement of the Mw 7 Machaze earthquake (Mozambique): complementary use of multiband InSAR and radar amplitude image correlation with elastic modelling", *Remote Sensing of Environment*, vol. 114, no. 10, pp. 2211–2218, 2010.

[RAU 13a] RAUCOULES D., CARTANNAZ C., MATTHIEU F. *et al.*, "Combined use of space-borne SAR interferometric techniques and ground based measurements on a 0.3 km^2 subsidence phenomenon", *Remote Sensing of Environment*, vol. 139, pp. 331–339, 2013.

[RAU 13b] RAUCOULES D., LE COZANNET G., WOPPELMANN G. *et al.*, "High nonlinear urban ground motion in Manila (Philippines) from 1993 to 2010 observed by DInSAR: implications for sea-level measurement", *Remote Sensing of Environment*, vol. 139, pp. 386–397, 2013.

[REM 03] REMY D., BONVALOT S., BRIOLE P. *et al.*, "Accurate measurements of tropospheric effects in volcanic areas from SAR interferometry data: application to Sakurajima volcano (Japan)", *Earth and Planetary Science Letters*, vol. 213, pp. 299–310, 2003.

[ROH 12] ROHMER J., RAUCOULES D., "On the applicability of persistent scatterers interferometry (PSI) analysis for long term CO2 storage monitoring", *Engineering Geology*, vol. 147/148, pp. 137–148, 2012.

[SIG 15] SIGMUNDSSON F., HOOPER A., HREINSDOTTIR S. *et al.*, "Segmented lateral dyke growth in a rifting event at Bardarbunga volcanic system, Iceland", *Nature*, vol. 517, no. 7533, pp. 191–195, 2015.

[SPA 08] SPARKS S.R.J., FOLKES C.B., HUMPHREYS M.C.S. *et al.*, "Uturuncu volcano, Bolivia: volcanic unrest due to mid-crustal magma intrusion", *Asian Journal of Geoinformatics*, vol. 308, pp. 727–769, 2008.

[STR 99] STROZZI T., WEGMÜLLER U., "Land subsidence in Mexico City mapped by ERS differential SAR interferometry", *Proceedings International Geoscience and Remote Sensing Symposium, IGARSS'99*, pp. 1940–1942, 1999.

[USA 99] USAI S., DELGAUDIO C., BORGSTROM S. *et al.*, "Monitoring terrain deformations at Phlegrean Fields with SAR interferometry", *Proceedings of FRINGE'99*, ESA SP-478, Liège, Belgium, 10–12 November 1999.

[VAD 97] VADON H., SIGMUNDSSON F., "Crustal deformation from 1992 to 1995 at the Mid-Atlantic Ridge, Southwest Iceland, mapped by satellite radar interferometry", *Science*, pp. 194–197, 1997.

[VAN 97] VAN DER KOOIJ M., "Land subsidence measurements at the Belridge Oil Fields from ERS InSAR data", *Proceedings of 3rd ERS Symposium*, Florence, Italy, 1997.

[VAS 10] VASCO D.W., RUCCI A., FERRETTI A. *et al.*, "Satellite-based measurements of surface deformation reveal fluid flow associated with the geological storage of carbon dioxide", *Geophysical Research Letters*, vol. 37, p. L03303, 2010.

[WEB 02] WEBLEY P.W., BINGLEY R.M., DOBSON A.H. *et al.*, "Atmospheric water vapour correction to InSAR surface motion measurements on mountains: results from a dense GPS network on Mount Etna", *Physics and Chemistry of the Earth*, vol. 27, pp. 363–370, 2002.

[WIC 02] WICKS C.W., DZURISIN D., INGEBRITSEN S. *et al.*, "Magmatic activity beneath the quiescent Three Sisters volcanic center, central Oregon Cascade Range, USA", *Geophysical Research Letters*, vol. 29, pp. 26-1–26-4, 2002.

[WRI 06] WRIGHT T., EBINGER C., BIGGS J. *et al.*, "Magma maintained rift segmentation at continental rupture in the 2005 Afar dyking episode", *Nature*, vol. 442, pp. 291–294, 2006.

[YAN 12] YAN Y., DOIN M.-P., LOPEZ-QUIROZ P. *et al.*, "Mexico City subsidence measured by InSAR time series: joint analysis using PS and SBAS approaches", *IEEE Journal of Selected Topics in Applied Earth Observations and Remote Sensing*, vol. 5, pp. 1312–1326, 2012.

[ZEB 97] ZEBKER H.A., ROSEN P.A., HENSLEY S., "Atmospheric effects in interferometric synthetic aperture radar surface deformation and topographic maps", *Journal of Geophysical Research*, vol. 102, pp. 7547–7563, 1997.

8

Applications of Remote Sensing to Locust Management

8.1. Introduction

Locusts are grasshoppers that belong to the Acrididae family, of the Orthoptera order. They constitute one of the biggest threats to global food security. Their swarms can attack hundreds of species of plants and destroy most major crop varieties. Locusts are known to consume the equivalent of their own body weight in green vegetation daily [STE 88]. Besides the enormous distribution ranges, their voracity and polyphagy make locusts extremely dangerous pests. Outbreaks have occurred on every continent except Antarctica and affect 10% of the world's population [SIM 08]. Locusts have been the enemies of crops since the dawn of agriculture. They are mentioned in sacred books such as the Torah, the Koran and the Bible. Locust swarms have brought devastation and famine to entire nations. For example, in 1958 in Ethiopia, locusts destroyed 167,000 tonnes of grain, enough to feed 1 million people for a year [STE 88]. Locust plagues continue to be devastating in the 21st Century. In 2003–2005, 8 million people in more than 20 countries suffered from ravages caused by swarms of desert locusts, *Schistocerca gregaria*, with an estimated 80–100% of crops lost in afflicted regions, mostly in sub-Saharan Africa [BRA 06].

Locusts are characterized by their density-dependent behavioral, physiological and phenotypic polymorphism. When their population density

---

Chapter written by Alexandre LATCHININSKY, Cyril PIOU, Alex FRANC and Valérie SOTI.

is low, locusts exist in the so-called “solitarious phase”. Solitarious locusts have camouflage coloration, exhibit mostly cryptic behavior and produce very few social interactions [UVA 77]. When population density is high, locusts transform into the “gregarious phase”, with nymphs often colored in contrasting black and yellow, orange or red. Gregarious nymphs live in dense groups called “hopper bands” and exhibit a coordinated marching behavior. Adults of the gregarious phase often form huge swarms and produce long-distance migration flights [PEN 09]. Locust swarms may cover up to 800 km^2 and contain from several thousand to 40 billion individuals, making them the largest congregations of terrestrial animals on Earth [SIM 08]. They can cover distances of up to several thousand kilometers. For example, in 1988, desert locust swarms crossed the Atlantic Ocean from West Africa to the Caribbean Islands and northern shores of South America, covering more than 5,000 km in 6–10 days [RIT 89]. Average distances flown daily vary from 10 km for the Moroccan locust *Dociostaurus maroccanus* to 200 km for desert and migratory locusts, *Locusta migratoria* [UVA 77].

The most marked difference between the gregarious and solitarious locust phases is their behavior: while solitarious adults avoid each other, except to mate, gregarious adults migrate, feed, mate and lay eggs in groups. Other differences include morphology, food selection, nutritional physiology, metabolic rates, reproductive physiology, neurophysiology, endocrinology, pheromone production, longevity and molecular biology [PEN 09]. There are around a dozen locust species worldwide [LAT 10a], some with huge distribution ranges. For example, the invasion range of the desert locust covers more than 20% of the Earth’s dryland in more than 60 countries of the world [STE 88], while the distribution range of the migratory locust is even larger.

Locusts are also known for their prolific fecundity, high reproduction rates and enormous population densities. While temperate locusts have only one generation per year, in tropical and subtropical regions locusts develop continuously and may produce up to four annual generations. Locust lifecycle consists of three successive stages: egg, nymph and adults. Females lay eggs in the soil in batches called egg-pods. Depending on the species, each egg-pod contains a couple of dozen to more than 120 eggs. A female typically deposits one to four egg-pods throughout her lifetime. During egg-laying, gregarious females gather in groups of several thousand per square meter [UVA 77]. Record egg-pod densities were reported for the Moroccan locust and the Italian locust *Calliptamus italicus*, reaching 8,000 and even 10,000 egg-pods per square meter, respectively [LAT 13]. When the eggs have hatched, record first instar hopper densities have been known to exceed 30,000 individuals per square meter, in the case of the desert locust [UVA 77]. All of these traits place locusts among the most dangerous agricultural pests in the world.

The economic significance of locusts is not limited to direct damage to crops and pastures. During locust plagues, national and international plant protection organizations have to make tremendous and concerted efforts to control them. Currently, locust control strategies are still principally curative and rely on large-scale application of insecticides to locust-infested areas. In 2003–2005, to curtail a desert locust outbreak, 13 million ha of land in 26 countries were treated with broad-spectrum neurotoxic insecticides. The cost of the campaign, including food aid to the affected populations, amounted to half a billion US dollars [BEL 05, BRA 06].

Despite recent progress in research in this area, our understanding of locust population dynamics is not sufficient, and our ability to predict outbreaks is still inadequate. Since the 1970s, serious efforts have been made toward implementing a preventive desert locust management strategy [MAG 08], although some specialists have criticized the approach as inefficient, due to the difficulty of identifying targets to be treated [SYM 09]. Preventive control is advocated by the FAO (Food and Agriculture Organization of the United Nations) as a key strategy in locust management, regardless of species [FAO 15]. One of the goals of preventive control is to identify gregarization sites (locust aggregation zones) as early as possible, even if they are spread across vast areas. Targeted surveys of locust habitats is the key element of a preventive control strategy. Traditionally, locust monitoring is performed by ground teams of scouts. They collect environmental data on the static (elevation, topography, soil type, vegetation type), and dynamic (soil moisture, temperature, rainfall and rainfall, vegetation cover and growth) ecosystem parameters and report them to the national and, in some cases, international locust control units. Vegetation, which is the essential component of locust habitat, requires particular attention. But the immense scale of the potential locust habitats requires a much more significant monitoring organization than for most other pests dependent on crops. Millions of hectares of potentially locust-infested areas must be surveyed in a very narrow window of time (several weeks) dictated by meteorological conditions favorable to the locust lifecycle. For example, in 2014, more than 30 million ha of habitats of three locust species – the Moroccan, Italian and Migratory locusts – were surveyed in Central Asia [FAO 15]. Furthermore, during outbreaks, survey areas increase dramatically. For example, 34 million ha of Italian locust habitats had to be surveyed in 2000 in Kazakhstan alone [LAT 00]. Obviously, it is very difficult to obtain fast and reliable information on spatio-temporal locust distribution through ground-based monitoring. Remote sensing, and specifically satellite imagery, appears to be an adequate tool for addressing the challenge of scale and time in locust monitoring.

8.2. Potential of remote sensing in locust monitoring

The idea of using satellite images in locust monitoring is not new. It became possible as soon as satellite data became publicly available for non-military purposes. Pedgley [PED 74] was the first to apply satellite images to desert locust habitat monitoring, as early as in 1974. Subsequently, satellite data were primarily used for habitat monitoring of three locust species: the desert locust, the Australian plague locust *Chortoicetes terminifera* and the migratory locust. The methodological development for other locust species remains very limited, however. Detailed explanations of remote sensing and its history in locust control are available in Latchininsky and Sivanpillai [LAT 10b] and Latchininsky [LAT 13].

Researchers and managers use satellite images for different purposes. In particular, satellite data can be used for the detection and monitoring of vegetation, the estimation of soil moisture and the measurement of certain meteorological parameters, such as rainfall distribution. For example, satellite images are particularly effective in detecting zones of green vegetation that appear after sporadic rains and that have the potential to be colonized by locusts in the deserts and semi-deserts of Africa (desert locust) and Australia (Australian plague locust). With regard to the migratory locust, satellite images have been used, for instance, to map spatio-temporal distribution of its preferred vegetation, the common reed *Phragmites australis*, in Central Asia. They have also been used for post-outbreak damage assessment on vegetation in China [LAT 13].

In terms of operational use, the example of the desert locust shows that remote sensing has become important in early detection of habitat conditions conducive to locust outbreaks, making it a key element in locust preventive strategy. Preventive management relies on early detection of desert locust populations before they can expand into full-scale plagues [MAG 08]. To achieve this, survey teams from the affected countries visit potential breeding areas to assess the environmental conditions and the state of locust populations. However, in the case of the desert locust, the potential breeding habitats cover enormous areas that are often difficult to access. To direct ground survey teams precisely to the areas with highest probability of conditions favorable to desert locust breeding, satellite imagery is now regularly used to characterize the evolution of vegetation cover and soil moisture, which are the two main parameters. The Desert Locust Information Service (DLIS) at the FAO headquarters in Rome, which centralizes survey information and issues monthly bulletins containing forecasts and warnings [CRE 13], has played an important role in the development of these tools.

During the 1980s, the FAO-DLIS started to use earth observation techniques to assess environmental conditions favorable to the desert locust [HIE 81, CRE 09]. The first satellite imagery used in FAO-DLIS early warning systems was Meteosat cloud imagery, which made it possible to pinpoint rainfall areas with several weeks' delay. A few years later, the normalized difference vegetation index (NDVI) delivered by AVHRR (advanced very high resolution radiometer) sensors from NOAA (the National Oceanic and Atmospheric Administration of the United States) with 7.6 km spatial resolution improved our understanding of biotopes. Since then, satellite imagery has been continuously improved in terms of spatial resolution and treatment algorithms for the identification of favorable desert locust habitats through the estimation of rainfall and vegetation cover. By the late 1990s, the FAO-DLIS had started to use and promote vegetation estimates at 1 km resolution from SPOT-Vegetation imagery every 10 days [CEC 05]. These developments were facilitated through collaboration with the Flemish Institute for Technological Research (VITO, Belgium). During the 2000s, the FAO-DLIS promoted the use of MODIS images at a higher spatial resolution (250 m), consisting of a 16-day composite image of NDVI values [FOO 09]. Since 2010, following the collaboration with VITO and the Catholic University of Louvain (Belgium), FAO-DLIS has been advising countries affected by the desert locust to use greenness maps produced from the NDVI from MODIS data [PEK 11]. The post-processing of MODIS imagery was improved in order to produce these greenness maps, which show changes in vegetation over the past 11 ten-day periods at a spatial resolution of 250 m. When it comes to desert locusts, observing the developmental cycle of the vegetation is important, because areas of green vegetation are not, in theory, a favorable habitat for solitarious locusts or for the phase transformation process.

Through a collaboration with Columbia University's International Research Institute for Climate and Society (IRI, USA), DLIS-FAO has also promoted the use of daily, 10-day and monthly rainfall estimates on a 0.25° x 0.25° latitude/longitude grid [DIN 10, CRE 13]. These can be downloaded free of charge from the IRI website, together with various vegetation products (including greenness maps) produced from MODIS data, which national locust control units use and publish in their monthly reports. The IRI rainfall estimates are derived from satellite images rather than computed from meteorological models, because the former are more reliable in terms of spatial distribution [CRE 09].

The development of remote sensing techniques for the desert locust early warning system has been complemented by several scientific studies that verify the capacity of satellite information to forecast locust presence or gregarization. Most of the studies address the influence of vegetation on locusts. The NDVI was indirectly correlated with the regional locust upsurge of 1980–1981 in West Africa and hence considered particularly useful in describing potential breeding areas [HIE 86]. In the Red Sea area, NDVI data at 7.6 × 7.6 km spatial resolution from NOAA/AVHRR satellite imagery provided information on vegetation quantity and also showed a very strong correlation with favorable years of locust breeding [DES 04]. The spatial distribution of vegetation was shown to play an important role in the concentration and gregarization processes for desert locusts [COL 98, DES 00, DES 03].

Satellite images, which have the advantage of covering large areas, were therefore used to verify these observations on large scales. Despland *et al.* [DES 04] demonstrated that in Mauritania, the fractal dimension of vegetated areas, which measures the degree of spatial fragmentation of the vegetation, was positively correlated with the gregarization potential of desert locust populations. The authors used NDVI pixels of 7.6 km over a vegetation threshold in good and bad reproduction periods, across an area of around 500 km × 1,300 km. However, at a spatial resolution of 250 m and across a smaller area, Piou *et al.* [PIO 13] did not observe fractal dimension to be a good indicator of the presence of solitarious individuals. They did not, however, test the effect on gregarization. Tratalos and Cheke [TRA 06] argued that NDVI data derived from satellite images covering large areas (from 4 to 8 km spatial resolution) were not a good predictor of desert locust presence. However, Piou *et al.* [PIO 13] did find that NDVI at 250 m resolution, together with coherent construction of secondary indicators derived from NDVI (e.g. temporal NDVI changes), can be good predictors of desert locust presence (see section 8.5). These various scientific studies support the operational use of satellite imagery to identify habitats subject to locust invasions.

The remainder of the chapter illustrates the applicability of remote sensing to locust management through three case studies. These case studies differ in terms of locust species, satellite platforms, geographic context and overall goals. The final part of the chapter discusses the case studies and summarizes the advantages and future challenges of using remote sensing in locust management.

8.3. Case study 1: using SPOT satellite images to evaluate the impact of habitat fragmentation on population dynamics of the red locust *Nomadacris septemfasciata* in the Sofia basin in Madagascar

The red locust *N. septemfasciata* has recently become one of the most significant pests in the western part of the Sofia basin in north-west Madagascar. Since 1998, the red locust had been sporadically and locally observed by the Plant Protection Services. In 2002, locust populations reached the gregarious phase for the first time in this area, forming hopper bands and swarms [FRA 05], which caused unprecedented crop damage [FRA 12]. This new gregarization phenomenon appeared to coincide with the extension of the rice fields, which were created on burnt forest patches known to be preferred oviposition sites of the red locust.

A study into the evolution of vegetation from 1954 to 1995 in the nature reserve of Manongarivo in the north of the Sofia basin revealed that the speed of deforestation has clearly accelerated since 1987 [GAU 99]. However, no study has yet been performed to evaluate the significance of this deforestation and to study the effect of landscape fragmentation [KOE 97] on the spatio-temporal distribution of animal species.

Recent changes in the behavior of the red locust (colonization and gregarization) could be explained by deforestation, which has opened up new migration pathways well-suited to the red locust in Madagascar. Many studies carried out in Hungary [JOR 03], the Red Sea area and Mauritania [DES 03] have shown that the fragmentation of habitats can change the chances of contact, create new migration pathways and even alter the migratory preferences of locusts. One of the consequences of deforestation in the Sofia basin has been the opening up of a passage between two previously isolated areas favorable for the red locusts: the breeding area colonized at the beginning of the rainy season, located to the west of the Sofia basin, and the diapause area colonized at the beginning of the dry season, located to the east, in the hills.

To map and monitor the phenomenon of deforestation, data must be acquired over a long period of time [GRE 90]. Optical satellite images have often been used to monitor the risk of locust invasion [JI 04, SIV 07]. Insofar as the migration pathways of the red locust are determined by topography and the type of vegetation cover, the products of earth observation seem to be appropriate tools for modeling. To test this hypothesis, two SPOT satellite images have been acquired in the north-east of the Sofia basin, first to identify the natural vegetation and monitor its evolution since 1986. Then, based on knowledge of red locust bio-ecology, the study will attempt to analyze the possible links between the environmental changes – specifically the retreat of the forest – and the recent locust invasions in this region [FRA 08].

8.3.1. *Study zone*

The study zone is located in the north-west of Madagascar and has a surface area of around 2,600 km^2 (Figure 8.1). It consists of two regions distinct from one another in geology and climate. To the west is the low Androna peneplain, characterized by alluvial soils. This is the main red locust breeding zone during the rainy season. To the east are the foothills of the Ankaizina plateau and high Androna, characterized by a crystalline platform, where the locusts retreat to spend their diapause during the dry season. The study zone's main hydrographic network runs east to west and connects the breeding and diapause areas. In the western part of the study zone lies part of the Bora forest reserve, which consists of 25 m tall deciduous trees and is rich in endemic species [HUM 65]. The original dry forest is often replaced by crop patches for 1 or 2 years [NIC 89], after which they are abandoned and replaced by shrubs. High savannas with shrubs dominate the landscape outside the forests.

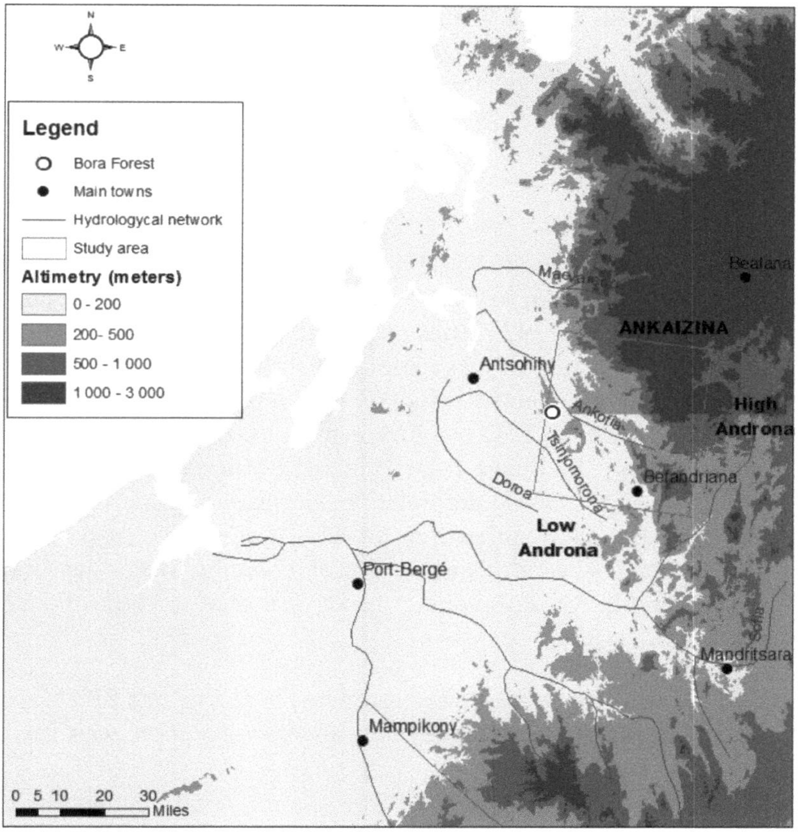

Figure 8.1. *Map of the study zone in the Sofia basin. For a color version of this figure, see www.iste.co.uk/baghdadi/6.zip*

8.3.2. *Impact of habitat on red locust movements*

Nomadacris septemfasciata is a univoltine species, which means that it has only one generation per year. A generation consists of three consecutive stages (egg, nymph and adult). Each of these stages has specific ecological requirements. To gregarize (i.e. to pass from the solitarious to the gregarious phase), the locusts need to breed and congregate. Solitarious red locusts fly at low altitudes at dusk when temperatures exceed 22°C [CHA 59]. They generally fly downwind. To avoid population dispersal, the red locust, like other locust species, uses migration pathways with gentle slopes and open countryside [DUR 79]. They often fly along the bottom of shallow valleys with sparse vegetation. In fact, locusts have a combined visual and motor response to vegetation cover: they tend to follow lighter borders of herbaceous vegetation at dusk rather than flying into darker forested areas [PRE 95].

8.3.3. *Data*

Two SPOT scenes acquired in 1986 and 2004 at high spatial resolution (Table 8.1), covering an area of 2,600 km^2, were used. Between October 23 and November 12, 2006, 297 entries of ground data on the nature of the land cover (types and species of vegetation), distributed in proportion to the classes chosen, were collected using GPS. From these data, 194 samples were used to process the SPOT 4 image from 2004, and the remainder to verify it. To process the image from 1986, 412 training sites were selected by an expert through photo-interpretation, because no data on land cover were available: 298 data entries were used to produce the map and 114 to verify the resulting classes. Data on locust breeding were also collected between 2000 and 2003 (227 points); these were used to verify the new red locust colonization areas.

Finally, to mark out the relief and estimate the slopes, the SRTM (Shuttle Radar Topographic Mission) with 90 m spatial resolution was used as a digital elevation model (DEM)[1].

---

1 http://glcf.umiacs.umd.edu.

Date	Time	Satellite	Instrument	Viewing angle (°)		Spatial resolution of bands B, V R and PIR (m)
				Orientation	Incidence	Resolution
July 20, 1986	07:05	SPOT 1	HRV 1	9.1	R 2.2	20
April 26, 2004	07:20	SPOT 4	HRVIR 1	8.1	L 26.6	10

Table 8.1. *Specifications of the two SPOT images used in this study*

8.3.4. *Methods*

8.3.4.1. *Land cover mapping*

When the SPOT images had been geometrically and radiometrically corrected, three bands from the principal components analysis (PCA), as well as the NDVI, were added to the four spectral bands of the SPOT image (blue, green, red and near-infrared) to produce a thematic map containing the following 11 land cover classes: dry forest, gallery forest, fallow, steppe, shrubs, grassy savanna, prairie, crops, villages, bare soil and water.

A PCA is a digital operation that condenses the information into non-correlated bands. The NDVI is important because it reduces any light and shadow effects caused by the relief [JEN 05] and also differentiates between the various types of vegetation [TUC 79, TOW 86].

The 1986 and 2004 land cover maps were produced through the supervised image classification method named maximum likelihood [JEN 05]. Around two-thirds of the points collected on the ground and through photo-interpretation were used to differentiate between land cover classes. The remaining ground information was reserved to measure map quality by calculating the confusion matrix and the Kappa coefficient [CON 91]. This coefficient has values from –1 to +1, with the agreement between the classification and the validation data being higher the closer the Kappa values are to 1.

Based on the knowledge of habitats favorable for red locust migrations, the 11 classes from the land cover map were grouped into four main classes, depending on their degree of attractiveness to locust migration, bearing in mind that open countryside with sparse vegetation would be favored. Dry forest and gallery forest were grouped into the single class “forest”, which acts as a barrier limiting locust migrations. Bushes and shrubland were grouped into the intermediary class “zones with mixed vegetation”, which are of average attractiveness to locusts, and savannas, herbaceous scrublands, prairies and crops, classes favorable to locust presence, were grouped together as “grasslands”. Finally, “zones without vegetation” consisted of villages, water and bare soil, which are also considered favorable to the locust movement.

8.3.4.2. *Mapping the preferred pathways of locust migration*

The preferred routes for red locust migration from the east to the west of the Sofia basin were identified using a combination of environmental variables, such as vegetation and the slopes calculated by SRTM. Migration pathways were identified through “least cost path analysis”, which consists of calculating the “least cost” path between two habitats: the departure point, which corresponds to the diapause areas (source), and the destination point, which corresponds to breeding sites (wells). In other words, this method aims to map the pathways that are the most easily accessible to locusts moving through the countryside from one habitat to another. It is therefore possible to calculate a cost accumulation surface (how much it costs the species to move) from the displacement costs attributed, depending on the nature of the pixel. The sources were chosen in the mountainous valleys where locusts spend the period of diapause during the dry season. Six wells were chosen in the western part of the study zone, where breeding sites had been observed [FRA 12]. A locust migration cost was attributed to each land cover class (Table 8.2) by an expert. A displacement cost of 1,000 was attributed to slopes of more than 30°, making it impossible to cross such a relief.

Land cover type	Cost
Forest (dry forest, gallery forest)	20
Mixed (shrubs, fallow)	5
Grassland (savannas, prairies, steppes, crops)	1
Non-vegetated (villages, bare soil, water)	1
Slopes > 30°	1,000

Table 8.2. *Migration cost for each land cover type for red locusts*

The migratory cost was calculated for each pathway (connecting sources and wells) in each image. This can be interpreted as the probability of the pathway being used by locusts. Schematically, three classes of cost were chosen, depending on the nature of the land cover: low (1), medium (5) and high (20). These classes made it possible to map the new migration pathways favorable to locusts between the breeding and diapause areas, which were verified using locust breeding data collected between 2000 and 2003.

8.3.5. *Results*

The resulting two land cover maps had satisfactory accuracy, with a Kappa index close to 1:0.89 for 1986 and 1:0.80 for 2004. The confusion observed was due to the "mixed vegetation" class sometimes being attributed to the "forest" class, but the results were entirely satisfactory, with 85% of pixels correctly classified.

Comparison of the two maps revealed a 40% reduction of forest area in 2004, in favor of grassland and mixed vegetation. Large areas of forest had disappeared entirely and the areas of forest "islands" had decreased dramatically. In 2004, there were no reforestation areas. Indeed, no area that had been covered by mixed vegetation in 1986 had been reforested since. The Bora forest reserve, which had consisted of large, compact forested islands in 1986, was broken up into numerous small islands in 2004.

New (yellow) and old (blue) migration pathways show clearly that the south-west of the study zone was inaccessible in 1986 (Figure 8.2). The only possible blue pathways involved crossing the large central savannas, which opened onto the banks of the Ankofia River in the north-west. These three major pathways all led into the same breeding areas – north of Antsohihy (Figure 8.2). In 2004, the map showed that deforestation had opened up two new pathways in the south (Tsinjomorona valley), thereby creating two new migration pathways (in yellow and orange) to the breeding areas. The new migration pathway to the south of Mount Befandriana had played a major role in the accessibility of western egg-laying areas and could therefore explain the great locust outbreaks of 2003 [FRA 07].

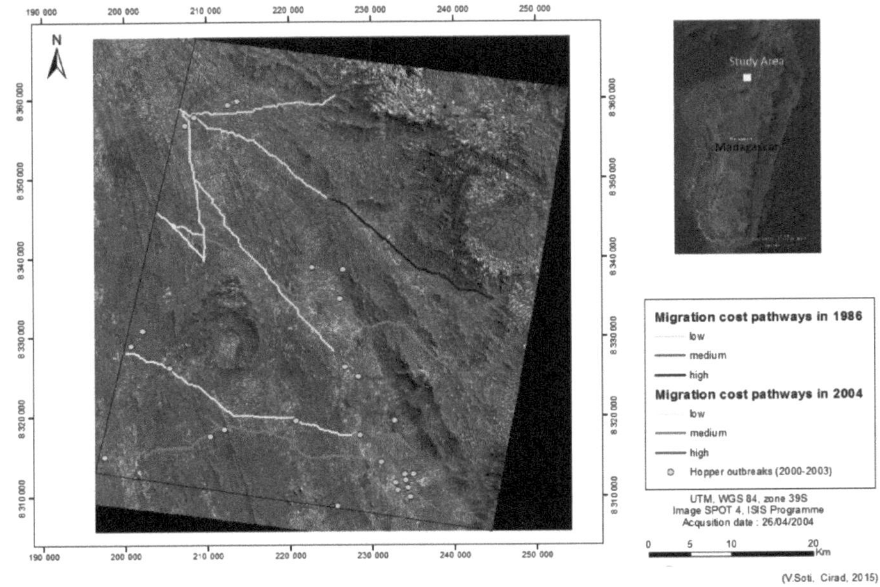

Figure 8.2. *Map of old and new preferred red locust migration pathways. For a color version of this figure, see www.iste.co.uk/baghdadi/6.zip*

8.3.6. *Conclusions*

The SPOT images, combined with a DEM, made it possible to trace the evolution of locust biotopes in time and space between 1986 and 2004, and to reveal the impact of landscape fragmentation on recent red locust outbreaks in the Sofia basin. Indeed, in addition to the four migration pathways that already existed in 1986, three new pathways were identified in the south-west, providing access to previously inaccessible oviposition zones and enabling locusts to breed, assemble and thus gregarize. It is important to note that the zone chosen for this study is small and does not cover the entirety of the red locust's area of influence in the Sofia basin, which spans more than 50,000 km^2. Looking ahead, it would be interesting to use high-resolution images (SPOT 6) covering larger areas to better understand the current and future spatial and temporal dynamics of the red locust across Madagascar.

8.4. Case study 2: using Landsat satellite imagery to identify migratory locust *Locusta migratoria* oviposition sites in Central Asia

8.4.1. *Introduction*

The migratory locust *migratoria* has the largest distribution range of all locusts, covering almost all of the temperate and tropical zones of the eastern hemisphere. Yet, the ecological requirements of the species are fairly narrow. Breeding areas are restricted to grasslands on light soils, usually in the wetlands with reed stands, along rivers or lakes [CEN 82]. One of the migratory locust subspecies, the Asian migratory locust *L. m. migratoria*, is distributed over temperate Eurasia, particularly Russia, Kazakhstan and Uzbekistan. Its main breeding areas cover 6 million ha in total. They are located in wetlands in the deltas of big rivers such as the Volga, Amu-Darya, Syr-Darya, Ural and Ili, and around big lakes such as Lake Balkhash, Lake Alakol and Lake Zaysan [LAT 02]. Areas infested by locusts in the Lake Balkhash area and the Amu-Darya river delta can exceed 1 million ha.

Locusta migratoria migratoria has typically one generation per year in temperate zones. Its lifecycle is synchronized with the phenological cycle of its main habitat plant, the common reed *P. australis*. Hatching occurs in May–June, when the chlorophyll activity of the reeds begins. Reed stands provide food and shelter for locust nymphs and adults. Nymphal development takes 5–7 weeks. Egg-laying occurs in July–August and the eggs overwinter in the soil to hatch the next spring. There is a permanent *L. m. migratoria* breeding zone in the Amu-Darya river delta near the Aral Sea in Uzbekistan. It is a vast (over 1.4 million ha) wetland area consisting of numerous large and small lakes, canals, rivulets and islands. Locust control services need to survey this area to identify locust infestations, but it is an arduous task due to the non-existent road network, which makes numerous delta areas inaccessible. Landsat's moderate spatial resolution satellite imagery was therefore proposed as a solution for locust habitat monitoring. This case study was described in Latchininsky *et al.* [LAT 07] and Sivanpillai and Latchininsky [SIV 08].

8.4.2. *Method*

8.4.2.1. *Satellite data*

Locust management services have to identify locust oviposition sites. These are the areas where nymphs hatch and are therefore the main targets of control treatments. The preventive locust control strategy requires that infestations be controlled in the early stages of nymphal development. For this study, two Landsat satellite images covering the whole of the Amu-Darya river delta were obtained during two separate periods of the locust lifecycle. These images were acquired on May 16, 2002, during the locust hatching period, and on August 7, 2006, during oviposition. They consist of three visible (blue, green and red) and three infrared (one near-infrared and two mid-infrared) image bands of the electromagnetic spectrum, with a spatial resolution of 30 m.

8.4.2.2. *Image classification*

The thematic map derived from the Landsat satellite image was produced using an iterative, unsupervised image classification (ISODATA) algorithm [WAY 01]. In total, five land cover classes were chosen:

- reeds;
- reeds and other vegetation;
- other vegetation with some reeds;
- sparse vegetation and sandy soil;
- water.

Based on the radiometric properties of the image, several iterations were used to classify all of the pixels in the image in each class. A major advantage of the unsupervised classification algorithm is that it does not require field knowledge of the various types of land cover. The iterative classification algorithm separates the various types of land cover on the basis of their reflectance values and is ideally suited to separating the subtle differences in the elements that make up the landscape.

8.4.2.3. *Classification accuracy assessment*

The accuracy of the classification was estimated using ground observation points collected with aerial and ground-based GPS. A total of 163 verification points spread more or less evenly across the five classes, depending on the area occupied, with a slight over-representation of classes 2 and 3 (section 8.4.2.2), were collected in June 2004, and 155 in June 2007. A confusion matrix was produced from the conflict between the classified image and the field data, and from this the Kappa index [CON 91] was calculated, making it possible to estimate the accuracy of the land cover map for the entire image.

8.4.2.4. *Locust data*

Data on the presence or absence of locust nymphs and their densities per square meter were collected during extensive field surveys in June 2007. Nymphal density was assessed visually by counting insects on 10 plots (1 m^2) at each chosen survey site. Average instar of the nymphs in bands was calculated from a sample of 100 nymphs at each site, using a standard entomological sweep net with a 0.3 m ring diameter. Locust phase (gregarious or solitarious) was determined by nymphal coloration characteristics.

8.4.3. *Results*

8.4.3.1. *The land cover map*

Image classification using an iterative ISODATA algorithm in five classes required 300 iterations for the 2004 image and 200 for the 2007 image. In the first iteration, the classes "water", "sparse vegetation" and "sandy soil" were correctly identified and removed. Through the following iterations, the three previously chosen large vegetation classes were identified (section 8.4.2.2). Eventually, among the various classes obtained, it was found that a monoculture of reeds occupied 21% of the image area, while 65–73% of the land cover was dominated by sparse vegetation and bare, sandy soil (Figure 8.3).

8.4.3.2. *Accuracy assessment of the land cover map*

The overall classification accuracy [CON 91] of the Landsat images was relatively high, ranging from 74% (Kappa = 0.69) for the early season image in 2004 to 84% (Kappa = 0.79) for the late season image in 2007. The two main sources of error concerned two classes: "other vegetation and reeds" and "reeds and other vegetation":

– confusion between the class "other vegetation with some reeds" and the class "reeds and other vegetation";

– confusion between the class "reeds and other vegetation" and the class "reeds".

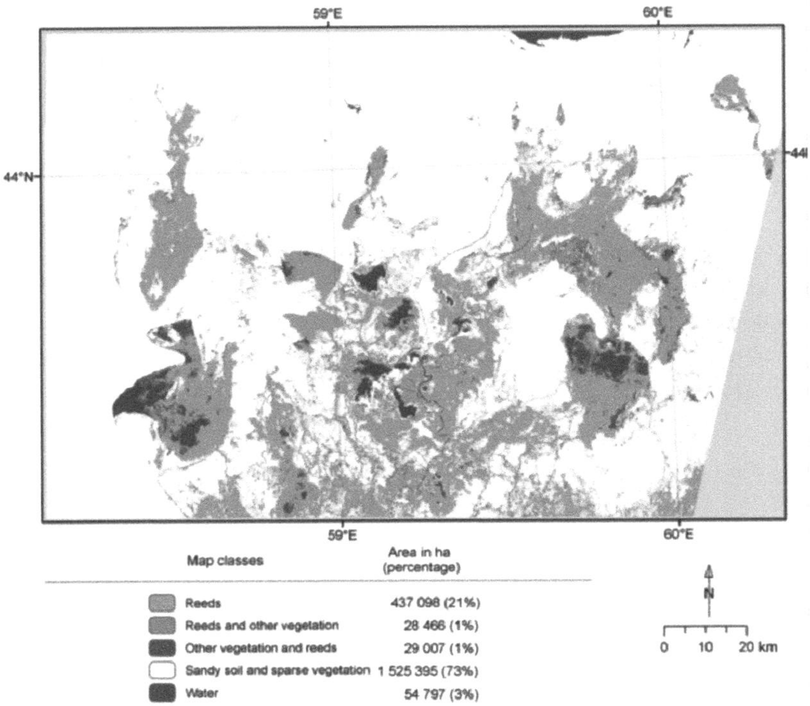

Figure 8.3. *Land cover map of the Amu-Darya river delta (Uzbekistan), derived from Landsat imagery at the end of the season [SIV 08]. For a color version of this figure, see www.iste.co.uk/baghdadi/6.zip*

Individual class accuracy values were very high, with 90% or more of pixels correctly identified for reeds, bare ground and water. The lowest individual class accuracy values were around 50% for the two "mixed" classes.

8.4.3.3. *The link between locust presence and vegetation*

In 2007, nymphs were present at the 37 survey points in bands with densities ranging between 1,000 and 2,000 per m^2. The highest densities were observed in 28 sites where reeds dominated the vegetation, and only in two sites characterized by bare soil. Nymphs were predominant in the second and third instar and exhibited the characteristics of the gregarious phase phenotype: a straight pronotum and contrasting red and black coloration. Because locust data had not been collected in 2002, the link between the density of eggs and the presence of reed vegetation could not be established.

8.4.4. *Conclusions*

The radiometric values of the images were used to effectively distinguish reeds from the other classes. Using a single Landsat image taken at the beginning or end of the growing season, the majority of the reed areas were reliably identified. The 30 m spatial resolution of Landsat was sufficient to capture not only large, continuous reed stands, but also narrow bands of reeds growing along canals and agricultural fields. Finally, the Kappa agreement values of 0.69–0.79 obtained for the overall map were highly satisfactory, showing a satisfactory accuracy of image classification. This was particularly true for the "reeds" class, which is the preferred habitat for migratory locust oviposition. However, the "mixed" classes were not always correctly classified due to their heterogeneous spectral properties, which led to an overestimation of reed areas. One approach that could minimize such errors would be to have a more detailed classification scheme for the various types of vegetation that make up the "mixed" classes. However, for rapid and periodic assessment, and from a locust management standpoint, an overestimation of reed area is more acceptable than an underestimation.

Asian migratory locust nymphs were found to be associated predominantly (76%) with habitats consisting exclusively of reeds and, to a lesser extent (10%), with those consisting of reeds mixed with other types of vegetation. Together, these two classes occupied 22% of the study area. Therefore, if locust scouts were looking for nymphs, the chances of finding them were about five times higher within the reeds than elsewhere.

Since early instar nymphs do not move far from their hatching places, it is safe to consider that the same areas had also been used by females for egg-laying the year before. The fact that more than 75% of young *L. m. migratoria* nymphs and, consequently, oviposition sites were found within the reeds suggests that this vegetation class should occupy the central place in the ground oviposition site surveys by locust control specialists. Landsat satellite images acquired early or late in the season can help these teams to better target the surveys by minimizing unnecessary examinations of other land cover types that are not suitable for oviposition. Consequently, the use of satellite images makes it possible to better target the control areas and seems to be a promising tool in preventive control strategies against migratory locust invasions.

8.5. Case study 3: estimation of the probability of presence of desert locust *Schistocerca gregaria* from satellite images of vegetation and historical ground data in Mauritania

Preventive control of the desert locust *S. gregaria* depends on monitoring an immense territory (~0.25 million km^2 of outbreak areas). In this context, remote sensing is instrumental in detecting potential areas of incipient outbreaks and guiding the ground survey teams toward those areas. Ground monitoring teams conduct extensive surveys of the terrain, collecting information on the state of the habitat and locust populations. This information is used for planning antilocust campaigns and making decisions regarding preventive control treatments. But this information is also uploaded into national and international geographic information systems (GIS), and generates historical databases that could be used for analyses. One goal of the analyses is to combine information from satellite images with ground data in order to identify the connection between indicators of quality (spatial structure, temporal dynamics of landscape elements) and indicators of quantity, such as the NDVI and the presence or absence of migratory locusts. These analyses then make it possible to generate predictive models of locust presence and therefore to better target the surveys. The case study presented here was published by Piou *et al.* [PIO 13].

8.5.1. *Method*

8.5.1.1. *Study zones and field data*

The study took place in two contrasting zones of Mauritania (Figure 8.4), covering a little more than 45,000 km^2. In these two zones, 2,915 survey points taken between January 2005 and June 2009 by the National Center for Locust Control in Mauritania were used for analysis. This period represents a situation of remission, which means that no large swarms flying between the countries have been observed.

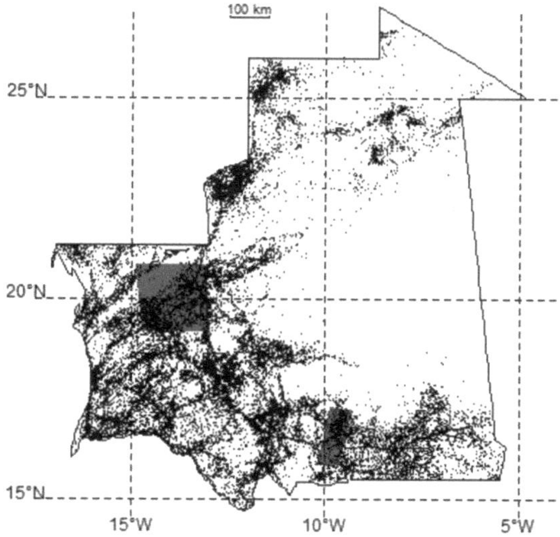

Figure 8.4. *Survey points between 1988 and 2009 in Mauritania, collected by the National Center for Locust Control, and localization of the two study zones used in this case study (shaded gray)*

8.5.1.2. *Satellite data*

The MOD13Q1 product of the NDVI at 250 m resolution provided by the MODIS satellite every 16 days (temporal synthesis) was obtained free of charge from NASA servers. Data were collected for the two zones between 2001 and 2009. In comparison to ground data, the long series of satellite data available mean that variables representing the vegetative potential of each pixel (NDVI minimum and maximum) can be generated. Temporal smoothing was applied to correct some inconsistencies in these data.

Based on the corrected NDVI data, 27 variables were created around each survey point, characterizing the vegetation on different scales for:

– zone potential;

– vegetation structure;

– its quality within more than 8 km radius;

– its local quality;

– modifications of the local quality over the past 2 months (Table 8.3).

8.5.1.3. *Statistical analysis*

By combining the 27 NDVI-generated variables, 89,774 logistic regression models were generated and adjusted, including or not the squares of these variables, to find the combinations that would best describe the presence or absence of desert locusts in survey observations. A multimodel inference framework was used, making it possible to draw inferences from all the models together rather than just testing a hypothesis of causal relationship identified from statistical regression. A consistent Akaike's information criterion (CAIC [BOZ 87]) was calculated to qualify each adjusted model, taking into account the complexity generated by the number of variables included in each model. Because of this approach, for each variable, the weights of evidence from CAIC scores, including the variable and the best model, were computed to measure the "importance" of a variable. This importance corresponds to the probability that a given variable should be used in a model to best describe the presence or absence of desert locusts. All treatments and calculations were performed with the program R [RCT 10].

8.5.2. *Results*

The multimodel inference framework with CAIC scores revealed the importance of three aspects in explaining the presence or absence of the desert locust. First, the variable of temporal change of vegetation between 32 and 48 days for a zone of 5 × 5 pixels of 250 m ($Ts23_2$, Table 8.3) appeared to be very important: the higher the vegetation increase in the month preceding the survey, the higher the chance of finding locusts. Second, the mean local variable of NDVI on 5 × 5 pixels (*M2*) was always in the best models with its squared value ($M2^2$), and these two variables always adjusted, respectively, positive and negative relations. This means that an optimal value of vegetation cover is favorable for finding desert locusts. The third point concerns the importance of variables at the largest spatial scales (>8 km): the multimodel approach showed that, across large-scale areas, the more vegetation there is, the lower the probability of finding desert locusts.

By using a satisfactory model from the point of view of adjustments and complexity (e.g. presence ~ maxNDVI + $V52$ + $M2$ + $M2^2$ + $Ts23_2$) and using new satellite information for a given date, it is possible to extrapolate the probability of finding locusts. Figure 8.5 illustrates this type of prediction for the northern part of the study zone during a particular period of 2008.

Groups and Names of variables	Explanations/calculations
Potential	
maxNDVI	Local maximum of NDVI (2001–2009)
minNDVI	Local minimum of NDVI (2001–2009)
Structural	
FD	Fractal dimension computed according to the Box-Counting method [DES 03] on an area of 65 × 65 a.l.p.
Quality at >8 km	
*V*32	Number of pixels with NDVI > 0.14 on an area of 32 × 32 a.l.p.
*V*36	idem as *V*32 on an area of 36 × 36
*V*40	idem as *V*32 on an area of 40 × 40
*V*44	idem as *V*32 on an area of 44 × 44
*V*48	idem as *V*32 on an area of 48 × 48
*V*52	idem as *V*32 on an area of 52 × 52
*V*56	idem as *V*32 on an area of 56 × 56
*V*60	idem as *V*32 on an area of 60 × 60
*V*65	idem as *V*32 on an area of 65 × 65
Local quality	
NDVI	Local value of NDVI
M1	Mean NDVI on an area of 3 × 3 a.l.p.
M2	Mean NDVI on an area of 5 × 5 a.l.p.
Temporal modifications	
$Ts1_0$	Relative difference of NDVI on local pixel between the time of survey and 16 days earlier $Ts1_0 = 2\times(NDVI_t - NDVI_{t-1})/(NDVI_t + NDVI_{t-1})$
$Ts1_1$	idem as $Ts1_0$ on an area of 3 × 3 a.p.l.
$Ts1_2$	idem as $Ts1_0$ on an area of 5 × 5 a.p.l.
$Ts2_0$	idem as $Ts1_0$ between the time of survey and 32 days earlier
$Ts2_1$	idem as $Ts2_0$ on an area of 3 × 3 a.p.l.
$Ts2_2$	idem as $Ts2_0$ on an area of 5 × 5 a.p.l.
$Ts23_0$	idem as $Ts1_0$ between 32 and 48 days earlier than the survey
$Ts23_1$	idem as $Ts23_0$ on an area of 3 × 3 a.p.l.
$Ts23_2$	idem as $Ts23_0$ on an area of 5 × 5 a.p.l.
$Ts3_0$	idem as $Ts1_0$ between the time of survey and 48 days earlier
$Ts3_1$	idem as $Ts3_0$ on an area of 3 × 3 a.p.l.
$Ts3_2$	idem as $Ts3_0$ on an area of 5 × 5 a.p.l.

Table 8.3. *Variables created from NDVI around each survey point (a.l.p. = around local pixel)*

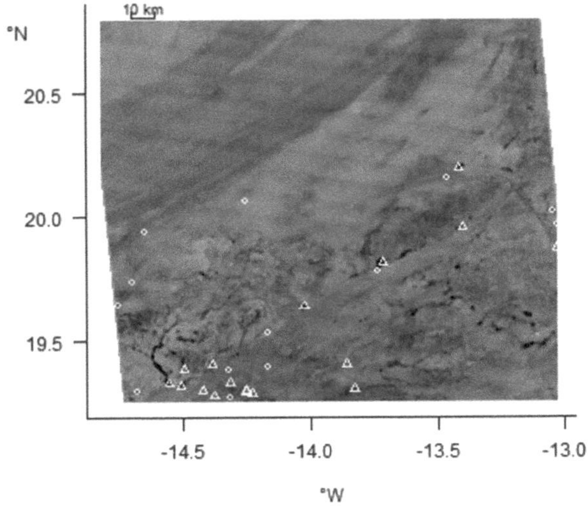

Figure 8.5. *Example of prediction of locust presence from a logistical regression statistical model combining ground observations with composite vegetation variables extracted from satellite images. The gray shading indicates the most favorable zones for desert locust presence for the period between October 15 and 31, 2008: most favorable in black; least favorable in light gray. White symbols indicate the survey results from the same period, with locust absence represented by circles and presence by triangles*

8.5.3. *Conclusions*

By combining the information from ground surveys and satellites, this study shows that it is possible to predict local risks of desert locust presence using NDVI as a descriptor of vegetation cover. Numerous improvements are possible: use of dynamic greening maps [PEK 11], inclusion of geomorphological and climatic variables, broadening of the study areas, construction of more complex statistical models, etc. Nevertheless, this case study already presents a potentially interesting methodology that can help ground survey teams to better target their itineraries in the search for desert locusts and thus reduce the costs of preventive control.

8.6. Conclusions

Over the past three decades, satellite imagery has increasingly been applied to locust monitoring and management. Rapid identification of areas with emerging green vegetation is instrumental in fast decision-making regarding control interventions against locust congregations. Based on information about the state of vegetation from satellite imagery, locust management teams can target high-risk locust gregarization sites. This significantly reduces the costs and contributes toward changing the locust control paradigm from curative to preventive.

Today, the technology produced from Earth observation still has numerous challenges to overcome. Of around a dozen locust species worldwide, remote sensing applications have been developed for only three: the desert locust (see case study 3 section 8.5), the migratory locust (see case study 2 section 8.4) and the Australian plague locust. For other locust species, only pilot studies have been carried out in this domain, although the potential of remote sensing is promising (see case study 1 section 8.3).

When choosing the appropriate satellite tool, locust researchers have to compromise between the spatial and temporal resolution of the images. Landsat's moderate spatial resolution (30 m) is appropriate for mapping vegetation when it is active, homogenous and covering large contiguous areas, but its 16-day temporal resolution is far from optimal. If work is being done across especially large areas, numerous Landsat scenes are needed, which can make the image processing time-consuming and require extensive computation capacities. MODIS satellite images have the advantage of being able to cover large areas and offer a daily temporal resolution, but their spatial resolution (around 250 m) is low and may not always be suitable for achieving detailed land cover mapping.

Looking ahead, there is further potential for the development of other remote sensing products for locust control, such as the measurement of soil moisture through radar images. The use of non-satellite remote sensing platforms, such as drones and airplanes, also represents an intriguing possibility, making it possible to assess land cover in real time. These platforms also have the advantage of being able to be programmed on demand, and could be instrumental in surveying locust breeding hotspots that are very concentrated in limited areas.

Finally, the development of modeling tools incorporating remote sensing data could improve locust early warning systems. Case study 3 demonstrates the potential of statistical models as tools for automatically transforming NDVI imagery into risk maps. Improvements to this modeling approach can be envisaged. It would be interesting to integrate biological mechanisms into these models by linking them to environmental data derived from satellite images. For example, locust development time could be linked to temperature measurements, which could make it possible to map the potential ecological niches (where species could be found) in addition to the realized ecological niches (where species have been found). This should further improve predictability and usefulness of remote sensing applications in locust management.

8.7. Key points

Locusts are grasshopper species that generate devastating swarms. Preventive control against these crop pests, which are responsible for significant agricultural losses, is essential before they become devastating plagues. Remote sensing has been used for this purpose since the 1980s, but mainly on three locust species: the desert locust, the Australian plague locust and the migratory locust. In these three cases, remote sensing is used to identify potentially favorable habitats for breeding, migration and egg-laying. Through three case studies, this chapter has shown the ways in which remote sensing can be used in locust management. However, there is still a lot of potential for further development, in terms of both targeting other species and improving the technologies used in locust control.

8.8. Acknowledgments

We would like to thank the colleagues who participated in the various case studies: Valentine Lebourgeois, Annelise Tran, Ahmed Salem Benahi, Vincent Bonnal, Mohamed el Hacen Jaavar, Michel Lecoq, Jean-Michel Vassal and Ramesh Sivanpillai. We would also like to thank the Commission for Controlling the Desert Locust in the Western Region (CLCPRO-FAO) and the National Center for Locust Control in Mauritania for funding and facilitating the case study in section 8.5.

8.9. Bibliography

[BEL 05] BELAYNEH Y.T., "Acridid pest management in the developing world: a challenge to the rural population, a dilemma to the international community", *Journal of Orthoptera Research*, vol. 14, no. 2, pp. 187–195, 2005.

[BOZ 87] BOZDOGAN H., "Model selection and Akaike's information criterion (AIC): the general theory and its analytical extensions", *Psychometrika*, vol. 52, pp. 345–370, 1987.

[BRA 06] BRADER L., DJIBO H., FAYE F.G., "Towards a more effective response to desert locusts and their impacts on food insecurity, livelihoods and poverty", *Proceedings of Independent Multilateral Evaluation of the 2003–2005 Desert Locust Campaign*, FAO, Rome, 2006.

[CEC 05] CECCATO P.N., "Operational early warning system using SPOT–VEGETATION and TERRA-MODIS to predict desert locust outbreaks", Columbia University Academic Commons, available at http://hdl.handle.net/10022/AC, 2005.

[CHA 59] CHAPMAN R.F., "Observations on the flight activity of the red locust, *Nomadacris septemfasciata* (Serville)", *Behaviour*, vol. 14, no. 4, pp. 300–334, 1959.

[COL 98] COLLETT M., DESPLAND E., SIMPSON S.J. *et al.*, "Spatial scales of desert locust gregarization", *Proceedings of the National Academy of Sciences of the United States of America*, vol. 95, pp. 13052–13055, 1998.

[CON 91] CONGALTON R.G., "A review of assessing the accuracy of classifications of remotely sensed data", *Remote Sensing of Environment*, 37, pp. 35–46, 1991.

[COP 82] CENTRE FOR OVERSEAS PEST RESEARCH (COPR), *The Locust and Grasshopper Agricultural Manual*, Centre for Overseas Pest Research, vol. 690, London, 1982.

[CRE 09] CRESSMAN K., HODSON D., "Surveillance, information sharing and early warning systems for transboundary plant pests diseases: the FAO experience", *Arab Journal of Plant Protection,* vol. 27, pp. 226–232, 2009.

[CRE 13] CRESSMAN K., "Role of remote sensing in desert locust early warning", *Journal of Applied Remote Sensing*, vol. 7, no. 1, 2013.

[DES 00] DESPLAND E., COLLETT M., SIMPSON S.J., "Small-scale processes in desert locust swarm formation: how vegetation patterns influence gregarization", *Oikos*, vol. 88, pp. 652–662, 2000.

[DES 03] DESPLAND E., "Fractal index captures the role of vegetation clumping in locust swarming", *Functional Ecology*, vol. 17, pp. 315–322, 2003.

[DES 04] DESPLAND E., ROSENBERG J., SIMPSON S.J., "Landscape structure and locust swarming: a satellite's eye view", *Ecography*, vol. 27, pp. 381–391, 2004.

[DIN 10] DINKU T., CECCATO P., CRESSMAN K. *et al.*, "Evaluating detection skills of satellite rainfall estimates over desert locust recession areas", *Journal of Applied Meteorology and Climatology*, vol. 49, no. 6, pp. 1322–1332, 2010.

[DUR 79] DURANTON J.F., LAUNOIS M., LAUNOIS-LUONG M.H. *et al.*, "Les voies privilégiées de déplacement du Criquet migrateur malgache en phase solitaire", *Bulletin d'Ecologie*, vol. 10, no. 2, pp. 107–123, 1979.

[FAO 09] FAO, "39th session of the FAO desert locust control committee", *Food and Agriculture Organization of the United Nations*, Rome, 2009.

[FAO 15] FAO, Locust watch CCA, available at http://www.fao.org/ag/locusts-CCA/en/index.html, 2015.

[FRA 05] FRANC A., RABESISOA L.F., LUONG-SKOVMAND M.H. *et al.*, "Phase polymorphism in the red locust *Nomadacris septemfasciata* (Orthoptera: Acrididae) in Madagascar", *International Journal of Tropical Insect Science*, vol. 25, no. 3, pp. 182–189, 2005

[FRA 07] FRANC A., BRAUD Y., RATOVONASY H. *et al.*, "Distribution et limites écologiques du criquet nomade *Nomadacris septemfasciata* (Serville, 1838) à Madagascar", *Journal of Orthoptera Research*, vol. 16, no. 2, pp. 181–188, 2007.

[FRA 08] FRANC A., SOTI V., TRAN A. *et al.*, "Deforestation, new migration pathways and outbreaks of the red locust *Nomadacris septemfasciata* (Orthoptera: Acrididae) in the Sofia river basin", *Conférence SDH-SAGEO*, Montpellier, 23–26 June, pp. 1–17, 2008.

[FRA 12] FRANC A., DURANTON J.F., "Red locust *Nomadacris septemfasciata* (Serville) upsurges in northern Madagascar between 1998 and 2004", *Annales de la Société Entomologique de France*, vol. 48, nos. 1–2, pp. 97-105, 2012.

[GAU 99] GAUTIER L., CHATELAIN C., SPICHIGER R., "Déforestation, altitude, pente et aires protégées: une analyse diachronique des défrichements sur les pourtours de la réserve spéciale de Manongarivo (NW de Madagascar)", in HURNI H., RAMAMONJISOA J. (eds), *African Mountain Development in a Changing World*, African Mountains Association, 1999.

[GRE 90] GREEN G.M., SUSMAN R.W., "Deforestation history of eastern rain forests of Madagascar from satellite images", *Science*, vol. 248, pp. 212–215, 1990.

[HIE 81] HIELKEMA J.U., "Desert locust habitat monitoring with satellite remote sensing: A new technology for an old problem", *ITC Journal*, vol. 4, pp. 387–417, 1981.

[HIE 86] HIELKEMA J.U., ROFFEY J., TUCKER C.J., "Assessment of ecological conditions associated with the 1980/81 desert locust plague upsurge in West Africa using environmental satellite data", *International Journal of Remote Sensing*, vol. 7, p. 1609–1622, 1986.

[HUM 65] HUMBERT H., DARNE C.G., "Notice de la carte Madagascar", *Travaux de la Section Scientifique et Technique*, Institut Français de Pondichéry, Special Issue no. 6, pp. 46–78, 1965.

[JEN 05] JENSEN J.R., *Introductory Digital Image Processing: A Remote Sensing Perspective*, Pearson Prentice Hall, Englewood Cliffs, 2005.

[JI 04] JI R., XIE B.Y., LI D.M. *et al.*, "Use of Modis data to monitor the oriental migratory plague locust", *Agriculture, Ecosystems and Environment*, vol. 104, no. 3, pp. 615–620, 2004.

[JOR 03] JORDAN F., BALDI A., ORCI K.M. *et al.*, "Characterizing the importance of the habitat patches and corridors in maintaining the landscape connectivity of a *Pholidopteratransylvanica* (Orthoptera) metapopulation", *Landscape Ecology*, vol. 18, pp. 83-92, 2003.

[KOE 97] KOECHLIN J., GUILLAUMET J.L., MORAT P., "Flore et végétation de Madagascar", in TÜXEN R. (ed.), *Flora et vegetatio mundi*, A.R.G. Gantner, Vaduz, 1997.

[LAT 00] LATCHININSKY A.V., Emergency program for the control of locust outbreaks, Kazakhstan, TCP/KAZ/0065(E), Consultant's Report to FAO, Astana-Laramie, 2000.

[LAT 02] LATCHININSKY A.V., SERGEEV M.G., CHILDEBAEV M.K. *et al.*, "Acridids of Kazakhstan, Central Asia and adjacent territories", Association for Applied Acridology International, University of Wyoming, 2002.

[LAT 07] LATCHININSKY A.V., SIVANPILLAI R., DRIESE K.L. *et al.*, "Can early season Landsat image improve locust habitat monitoring in the Amudarya River Delta, Uzbekistan?", *Journal of Orthoptera Research*, vol. 16, pp. 167–173, 2007.

[LAT 10a] LATCHININSKY A.V., "Locusts", in BREED M.D., MOORE J. (eds), *Encyclopedia of Animal Behavior*, Academic Press, Oxford, vol. 2, 2010.

[LAT 10b] LATCHININSKY A.V., SIVANPILLAI R., "Locust habitat monitoring and risk assessment using remote sensing and GIS technologies", in CIANCIO A., MUKERJI K.G. (eds), *Integrated Management of Arthropod Pests and Insect Borne Diseases*, Springer Science+Business Media vol. 5, pp. 163–188, 2010.

[LAT 13] LATCHININSKY A.V., "Locusts and remote sensing: a review", *Journal of Applied Remote Sensing*, 7, no. 1, 2013.

[MAG 08] MAGOR J.I., LECOQ M., HUNTER D.M., "Preventive control and desert locust plagues", *Crop Protection*, vol. 27, no. 12, pp. 1527–1533, 2008.

[NIC 89] NICOLL M.E., LANGRAND O., Madagascar: Revue de la conservation et des aires protégées, WWF, 1989.

[PED 74] PEDGLEY D.E., "ERTS surveys a 500 km^2 locust breeding site in Saudi Arabia", in FRIEDEN S.C., MERCANTI E.P., BECKER M.A. (eds), *Proceedings of the 3rd Earth Resources Technology Satellite – Symposium (December 1973)*, NASA, MD, vol. 1, 1974.

[PEK 11] PEKEL J.F., CECCATO P., VANCUTSEM C. *et al.*, "Development and application of multi-temporal colorimetric transformation to monitor vegetation in the desert locust habitat", *IEEE Journal of Selected Topics in Applied Earth Observations and Remote Sensing*, vol. 4, no. 2, pp. 318–326, 2011.

[PEN 09] PENER M.P., SIMPSON S.J., "Locust phase polyphenism: an update", *Advances in Insect Physiology*, vol. 36, pp. 1–272, 2009.

[PIO 13] PIOU C., LEBOURGEOIS V., BENAHI A.S. *et al.*, "Coupling historical prospection data and a remotely-sensed vegetation index for the preventative control of desert locusts", *Basic and Applied Ecology*, vol. 14, pp. 593–604, 2013.

[PRE 95] PREISS R., SPORK P., "How locusts separate pattern flow into its rotatory and translatory components (Orthoptera: Acrididae)", *Journal of Insect Behaviour*, vol. 8, no. 6, pp. 763–779, 1995.

[RCT 10] R DEVELOPMENT CORE TEAM, R: A Language and Environment for Statistical Computing, R Foundation for Statistical Computing Vienna, Australia, 2010.

[RIT 89] RITCHIE M., PEDGLEY D.E., "Desert locusts cross the Atlantic", *Antenna*, vol. 13, pp. 10–12, 1989.

[SIM 08] SIMPSON S.J., SWORD G.A., "Locusts", *Current Biology*, vol. 18, pp. R364–R366, 2008.

[SIV 07] SIVANPILLAI R., LATCHININSKY A.V., "Mapping locust habitats in the Amundarya river delta, Uzbekistan with Multi-temporal Modis imagery", *Environmental Management*, vol. 39, pp. 876–886, 2007.

[SIV 08] SIVANPILLAI R., LATCHININSKY A.V., "Can late summer Landsat data be used for mapping Asian migratory locust, *Locusta migratoria migratoria*, oviposition sites in the Amudarya River delta, Uzbekistan?", *Entomologia Experimentalis et Applicata*, vol. 128, pp. 346–353, 2008.

[STE 88] STEEDMAN A., Locust Handbook, 2nd ed., Overseas Development Natural Resource Institute, London, 1988.

[SYM 09] SYMMONS P., "A critique of preventive control and desert locust plagues", *Crop Protection*, vol. 28, no. 10, pp. 905–907, 2009.

[TOW 86] TOWNSHEND J.R.G., JUSTICE C.O., "Analysis of dynamics of African vegetation using the NDVI", *International Journal of Remote Sensing*, vol. 7, pp. 1224–1242, 1986.

[TRA 06] TRATALOS J.A., CHEKE R.A., "Can NDVI GAC imagery be used to monitor desert locust breeding areas?", *Journal of Arid Environments*, vol. 64, pp. 342–356, 2006.

[TUC 79] TUCKER C.J., "Red and photographic infrared linear combinations for monitoring vegetation", *Remote Sensing of Environment*, vol. 8, no. 2, pp. 127–150, 1979.

[UVA 77] UVAROV B.P., *Grasshoppers and Locusts: A Handbook of General Acridology*, COPR, London, vol. 2, 1977.

[WAY 01] WAYMAN J.P., WYNNE R.H., SCRIVANI J.A. *et al.*, "Landsat TM-based forest area estimation using iterative guided spectral class rejection", *Photogrammetric Engineering and Remote Sensing*, vol. 67, pp. 1155–1165, 2001.

9

Applications of Remote Sensing to the Epidemiology of Infectious Diseases: Some Examples

9.1. Introduction

Infectious diseases are caused by a pathogenic microorganism (virus, bacterium, parasite or fungus) and have a significant impact on public and animal health. In public health, they are one of the leading causes of death and are priorities in the poorest countries [WHO 14]. In animal health, they can be devastating and radical measures are taken to prevent them from spreading. Infectious diseases therefore have major economic consequences, in terms of both healthcare and livestock production losses. In recent decades, we have seen the emergence of a significant number of infectious diseases, such as chikungunya, avian influenza and Ebola. These diseases have spread around the world all the more dramatically due to the rapid movement of populations and goods. The majority of pathogenic agents responsible for these emergences are zoonotic (transmitted from animals to humans or vice versa). There has also been a significant increase in vector-borne diseases – that is diseases whose pathogenic agent is transmitted through the intermediary of a vector. Vectors are arthropods (e.g. mosquitoes, fleas, sand flies and reduviid bugs) and acarids (e.g. ticks). One example is the mosquito *Aedes albopictus*, more commonly known as

---

Chapter written by Annelise TRAN, Daouda KASSIÉ and Vincent HERBRETEAU.

the "tiger mosquito", which is a vector for the chikungunya and dengue viruses. The geographical distribution of this mosquito is increasing year by year. This species was responsible for the first indigenous cases of chikungunya in France in 2010.

The fight against infectious diseases, which involves epidemiological monitoring and the development of treatments, vaccines and appropriate control strategies, is therefore a major social concern against a backdrop of rapid environmental changes. Indeed, due to strong anthropic pressure, ecosystems are currently undergoing changes such as climate change, deforestation, urbanization and agricultural intensification. Coupled with the intensification of trade, these changes can favor the spread of pathogenic agents.

Faced with this challenge, various disciplines (e.g. ecology, geography, anthropology and computer science) have come together to produce numerous research studies in the field of health, and human health and animal health have been united through several initiatives, including One Health [ROG 15]. By considering the host–vector–pathogen system in relation to the physical and human environment (Figure 9.1), these studies aim to advance our understanding of the mechanisms at play and of the identification of risk factors for the emergence, spread and persistence of pathologies.

Transmission of a pathogenic agent depends not only on intrinsic factors related to its own biology, that of its host and that of its vector, if any, but also on extrinsic factors, including those related to the environment. These are ancient ideas: in the 5th Century BC, Hippocrates wrote: "to deepen our knowledge of medicine, we must first consider the seasons, know the quality of the water and the winds, study the various states of the soil and the lifestyle of the inhabitants". Epidemiology, which is the study of the distribution and determinants of health-related events, focuses on the relationship between diseases and their causal factors. Spatial epidemiology has the more specific aim of bringing together epidemiological data and the spatiotemporal variability of the environmental, individual and societal factors that lead to a disease outbreak.

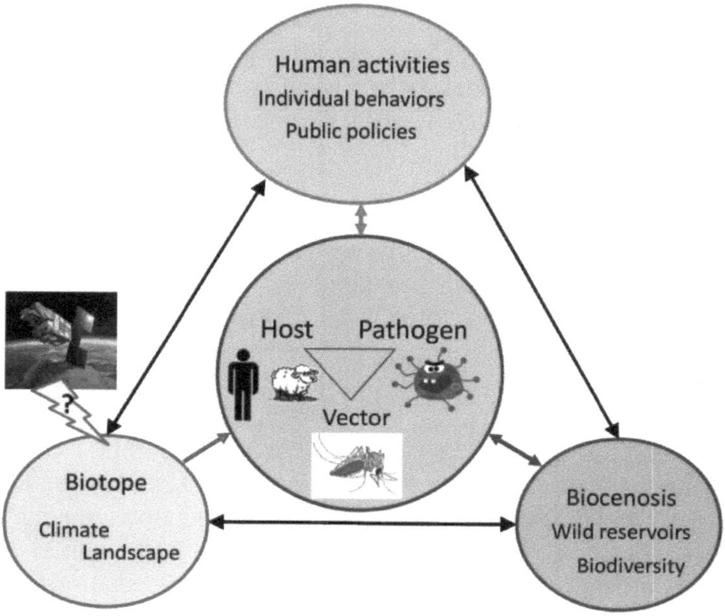

Figure 9.1. *Diagram of an epidemiological system. The transmission of a pathogenic agent depends on intrinsic factors – in other words, those related to the biology of the pathogenic agent, the hosts (human or animal) and any vectors – but also on extrinsic factors related to climate, landscape, presence of reservoir animal species (in other words, a species in whose organism the pathogenic agent multiplies, and which can therefore be a source of contamination for susceptible hosts) and human activities. One of the main concerns of research into the applications of remote sensing in epidemiology is the characterization, through remote sensing, of situations and phenomena pertaining to the study of a given eco-pathosystem – what are the most appropriate sensors, image processing methods and scale changing methods?*

A variety of data can be spatialized and used in these analyses. These are drawn from population censuses, administrations (health, agriculture and the environment), and weather services, and relate to demographic, economic and social indicators and, more rarely, data on the organization and use of the healthcare system. However, disease-related environmental knowledge is generally limited, with little data easily accessible. This is why remote sensing is such a useful tool in spatial epidemiology, providing environmental knowledge specifically designed to meet the requirements of a study (choice of resolution, coverage and dates). The second major advantage of remote sensing is that it is not restricted by administrative boundaries, facilitating comparison between remote areas (e.g. in different countries) to build up a consistent knowledge base. Finally, it makes it possible to temporally monitor environmental conditions (hydrometeorological variables: surface water, temperature, chlorophyll activity and land cover).

The creation of Earth observation systems has therefore contributed greatly to the development of spatial epidemiology and, more specifically, the ecology of transmission. Since the 1970s, remote sensing techniques have therefore been known as "new eyes for the epidemiologists" [CLI 70] and, since then, numerous applications for these techniques have been developed, making it possible to assess their potential [BEC 00, HAY 85, THO 00, THO 96] and limitations [HER 07]. Since the 2000s, access to data from observation and weather satellites has improved, with a large amount of data freely available (in particular, Landsat images, Moderate Resolution Imaging Spectroradiometer (MODIS) products and now Sentinel images from Copernicus, the European Earth observation programme) and easy access (via Internet access portals, or even downloaded directly from statistical software or geographical information systems). What was previously the biggest obstacle to the access of data has now been partially removed, resulting in spatial approaches being used more and more and frequently playing a role in infectious diseases control.

This chapter presents a few examples of the applications of remote sensing, such as mapping the distribution ranges and habitats of vector insects and reservoir species (section 9.2), outlining the social inequalities that lead to health inequalities in urban environments (section 9.3), and modeling epidemiological dynamics on a global scale (section 9.4), before considering a few future prospects (section 9.5).

9.2. Hazard mapping: the distribution ranges and habitats of vector insects and reservoir hosts

In order to understand the ecology of infectious diseases, research primarily focuses on the environmental conditions favorable to the presence of vectors or reservoir hosts of these diseases. These conditions dictate the areas where a pathogenic agent carried by an animal species is likely to exist and to be transmitted to humans or other animal species: this is the hazard. This is justified because most pathogenic agents do not have a free phase (i.e. a phase outside the host) in the environment, and when this is the case (e.g. for leptospires), it is very difficult to isolate them in their natural surroundings. Furthermore, vectors, hosts and reservoirs are generally strongly connected to the bioclimatic conditions that define the area where they live, the fundamental niche. This niche satisfies the conditions necessary for the existence of a species, but does not necessarily contain the species: it can be empty, or occupied by another species in competition for the same resources. The niche that is actually occupied is the realized niche [SOB 05].

The study of animal species presence can be carried out at different geographical scales. The broader prospect is the definition of their distribution range. Species are not present in all habitat types within this range: they generally exhibit preferences that determine their local distribution.

9.2.1. *Study of distribution ranges*

The study of a species' distribution range focuses on defining its envelope, beyond the boundaries of which the species cannot exist [SOB 05]. Envelopes vary greatly from one species to another, covering up to tens of thousands of kilometers for some ubiquitous species and a few kilometers for some endemic species. For large ranges, climatic and topographic parameters are rapidly discriminating and generally sufficient. For example, we have only to calculate the range of temperatures across all of the observation areas of a species (e.g. presence between 25 and 35 °C), then mark out climatic boundaries across the world, to have an initial idea of the potential distribution. In addition to statistical methods (generalized linear models, generalized additive models, etc.), various tools and software extensions have been developed to model species distributions. The most commonly used are Maxent (http://www.cs.princeton.edu/~schapire/maxent/), Openmodeller (http://openmodeller.sourceforge.net) and Biomod (https://r-forge.r-project.org/projects/biomod/) [THU 10]. By integrating land cover data acquired through remote sensing, the probabilities of presence within distribution envelopes can be refined.

With this in mind, Globcover land cover data (produced from ENVISAT/MERIS images with 300 m spatial resolution: http://due.esrin.esa.int/page_globcover.php) have been integrated, in order to map the potential presence of *Rattus exulans*, a small rat that lives alongside humans and is a reservoir for pathogenic agents of various diseases in southeast Asia (including leptospirosis, scrub typhus and toxoplasmosis, see Figure 9.2).

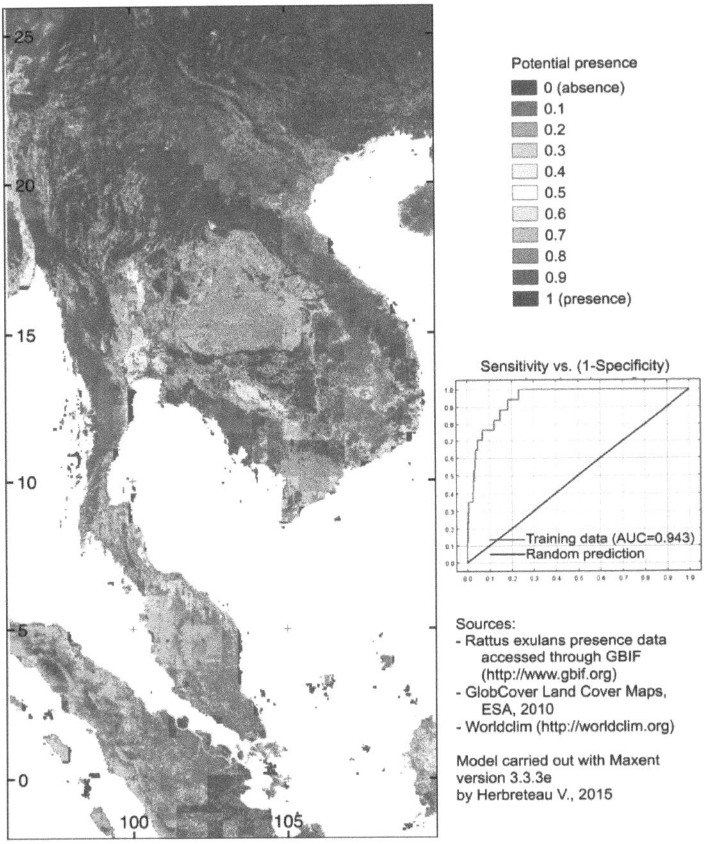

Figure 9.2. *Potential distribution of Rattus exulans in southeast Asia, produced from a Maxent model and based on GlobCover global land cover classification and Worldclim climatic averages. For a color version of this figure, see www.iste.co.uk/baghdadi/6.zip*

With a high area under the ROC curve (AUC; 0.943), the model obtained is discriminating for the prediction of the presence of *R. exulans*. This map confirms the northern boundary of its distribution (the probability of finding it is less than 10% at more than 22°N and less than 5% at more than 25°N), but also an unequal probability of presence in its range, related to land cover (presence in low agricultural regions). It is worth noting that the quality of such an analysis depends on the input data and models. Locally, the use of more refined land cover data makes it possible to identify the determining landscapes and plant formations more accurately.

This type of analysis is generally applied to vector animal species and reservoirs, but can also be applied to pathogenic agents that have been identified in various species (e.g. a parasite transported by several species of sandfly [CHA 10]).

9.2.2. *The study of habitats*

The study of a species' habitats requires a more refined scale, because they can be small and vary considerably in their form and structure. The most appropriate spatial scales are around a meter, and remote sensing from aerial or satellite images (optical satellites with medium spatial resolution, such as Landsat and SPOT, or very high spatial resolution, such as Pléiades, Quickbird, KOMPSAT, IKONOS, Worldview or GeoEye) is therefore a considerable advantage in mapping and monitoring these habitats. Nonetheless, remote observation does not render field work unnecessary. In addition to the fact that land cover analysis requires prior surveying as well as a final verification stage, knowledge of the ecology of vectors is fundamental to the understanding of the ecology of vector-borne diseases. Unfortunately, this knowledge is often limited due to a lack of field investigations, which are long, costly and sometimes subjective. They are often limited to sampling species through localization of the capture point and identification of the species, as a minimum. Nonetheless, spatial analysis and remote sensing tools can be used to characterize the areas where these species have been identified, in order to define their habitat and scour the studied area for places with similar conditions, known as potential habitats.

The process of habitat analysis involves:

– the use of remote sensing techniques to map land cover. The choice of images and of land cover classes must correspond to the requirements of the study: for example the need to differentiate between various types of crops or forest formations, or to work to precise outlines if the problem lies in the boundaries between forested areas and agricultural regions. The choice is also restricted by the intrinsic qualities of the image (including spatial and radiometric resolution);

– the calculation of landscape indicators (or landscape metrics) to define the environment of the captured animals (vectors). The simplest indicators to calculate are the minimum distances between each animal and each land cover class. Despite their simplicity, they provide valuable information when it comes to interpreting, for example, the proximity of a species to humans or to wetlands. The analysis can also define the immediate environment of each animal within a determined radius (a buffer zone around the capture points). Each landscape element within this radius is called a patch, and the calculations focus on their surface areas, contours and shapes. The nature of the landscape (urban, peri-urban, forest, agricultural, etc.), its consistency or diversity, its fragmentation and the structure of its elements can be defined from this [MCG 95]. One metric commonly used to describe landscape fragmentation is edge density, which is calculated by measuring the length of the contours in all the patches within a buffer zone and dividing this by the surface area.

The process of statistical analysis involves:

– descriptive analyses of each environmental variable for each species;

– finding an equation, through multiple regression, to determine the presence of the species from environmental indicators;

– prediction of species affiliation (if the study focuses on several species) through discriminative analysis of environmental indicators, making it possible to verify morphological or genetic identification;

– the search for similar environmental conditions. This involves extrapolating the local observations in the study area. It requires the whole of the area to have been classified in the same way.

In a study of leptospirosis on Reunion Island, a similar protocol was followed to construct environmental indicators to characterize the occurrence of different species of rats in a small study zone [REV 15]. Because the prevalence of leptospires in this area depends on the species, the environmental indicators that predict the presence of a species also predict the prevalence of leptospires in rats and the risk to nearby human populations. Using archived SPOT satellite images from missions 1–5 over 20 years (1987–2008), by calculating the evolution of such indications around the capture points of murine rodents in southeast Asia, another study demonstrated the link between the fragmentation of spaces, urbanization and the presence of *Hantavirus* sp. and *Bartonella* sp. [BOR 15].

The calculation of environmental indicators and the resulting statistics depend greatly on the number of individuals sampled. They can therefore be distorted by field work. Viewed on a satellite image or map, sampling of animals is often carried out near roads for logistical reasons and in spite of any efforts to move away from them (1 km in dense vegetation can be an arduous task and will look like the edge of the road on a satellite image). The sampling plan for various environments must therefore be carefully considered in advance. Furthermore, one of the difficulties caused by lack of ecological knowledge is not knowing the mobility of the species, which forces researchers to choose study radii around capture points empirically. Nonetheless, calculations can be made in various radii, and statistical analysis can indicate which radii are the most discriminating [TRA 08].

9.3. The contribution of remote sensing to the assessment of population vulnerability: health inequalities caused by social inequalities

Towns are better equipped for healthcare than rural areas. Consequently, city populations may be in a better state of health. However, this apparent advantage of cities over countryside must be qualified, because cities are, by nature, heterogeneous environments, and many studies have revealed that irregular urbanization and unequal facilities in cities have significant implications for health, contributing to the creation of various risks and unequal exposure to some diseases. They have also shown that the disparities are often more significant within a single city than between two different ones, or even between the city and rural areas [FOU 08, SAL 98]. By studying the spatial distribution of basic social services (drinking water supplies, educational and healthcare structures) within the urban area, as well as flooding zones, vegetation and building distribution, we can better understand these health disparities and the environments in which people live.

In Bobo-Dioulasso, the second largest city in Burkina Faso after its capital, Ouagadougou, and the main urban center in the west and southwest of the country, population growth has resulted in the expansion of the city, making new developments necessary. However, since the colonial conquest by the French in 1897, the city has been characterized by its two very different uses of space, with official areas known as "formal settlements" alongside unofficial "informal settlements". Living conditions in these urban areas vary significantly, with "formal settlements" generally equipped with basic urban services, such as drinking water networks, electricity and healthcare structures, while illegal "informal settlements" are deprived of all these facilities.

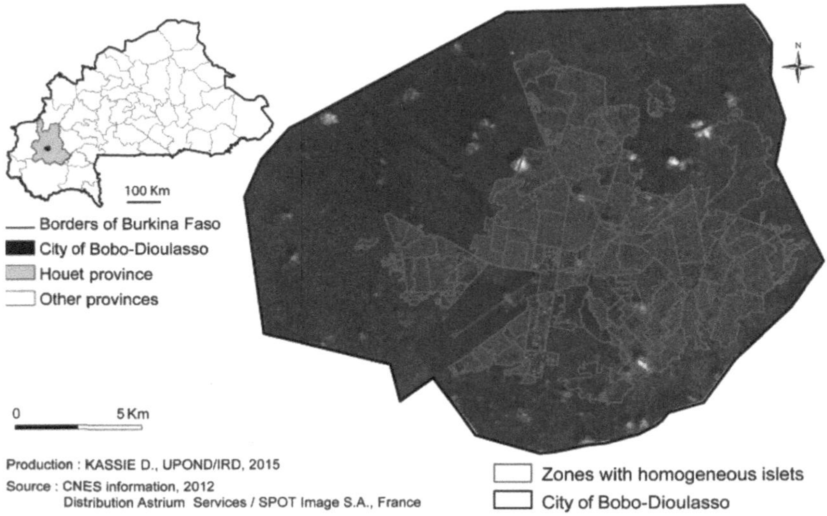

Figure 9.3. *Division of urban space in Bobo-Dioulasso, Burkina Faso (11°11′N, 4°17′W), into 125 homogeneous islet zones, from Pléiades images and aerial photographs. For a color version of this figure, see www.iste.co.uk/baghdadi/6.zip*

In order to accurately describe the environmental inequalities that may contribute to urban health inequalities in the city of Bobo-Dioulasso, aerial photographs and very high resolution satellite images (Pléiades) have been used to identify and map homogeneous subsets, or areas with homogeneous islets (Figure 9.3), relevant to health concerns such as malaria, and to set up a dedicated geographical information system. Interpretation of a series of aerial photographs acquired between 1952 and 2012 has made it possible to identify the urbanization front – in other words, the boundary between the urbanized city and the rural area (fields, wasteland and forests) and to understand the dynamics of the human establishments. Using Pléiades images acquired in 2012 [KAS 15], components of the urban environment, such as vegetation, flooding zones and buildings have been extracted (Figure 9.4).

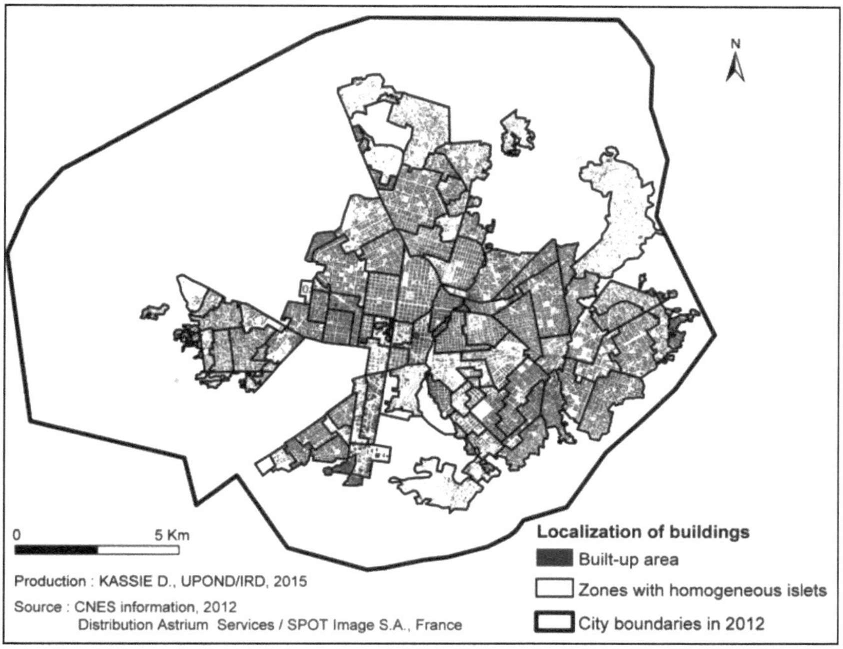

Figure 9.4. *Localization of buildings and groups of buildings in Bobo-Dioulasso in 2012, from Pléiades images. For a color version of this figure, see www.iste.co.uk/baghdadi/6.zip*

Because the urban space is broken up into areas of homogeneous islets, districts illustrating the diversity of urban subsets can be sampled for health surveys. As far as malaria is concerned, the results of this study show that the notion that malaria transmission is lesser in the center of a city and more significant on the outskirts is not the case in Bobo-Dioulasso [KAS 15]. Land use and management of urban space may therefore contribute to the creation of conditions favorable to the development of *Anopheles* mosquitoes, which are malaria vectors.

Analysis of remote sensing scenes also makes it possible to accurately describe the urban environment, revealing urban centers established through their consistency. By identifying these centers, we can better target the health surveys necessary for collecting epidemiological data to compile the health profiles of these areas, including various health problems (vector-borne diseases, nutritional statuses, etc.), and make recommendations for public health and spatial planning.

However, the excessive heterogeneity of the area, like all urban areas, can be a limiting factor for optimal exploitation of geomatics tools. Indeed, in African towns in particular, buildings are very different in texture, size, structure and even distribution. Between formal/informal settlements, metal/tiled roofs, new/old buildings, residential/popular habitat/industrial areas, etc., the characterization of urban spaces through remote sensing is not straightforward and requires a significant amount of field work. An approach combining remote sensing, meteorological and socioeconomic data makes it possible to better identify the urban functions and socioeconomic characteristics of some urban spaces and the habitat of some vectors [BOR 14]. Indeed, the latter has an impact on vector habitats, human–animal–vector interactions, and the exposure of human populations to certain vector-borne diseases [MAC 12].

9.4. The contribution of remote sensing to the modeling of epidemiological dynamics on a global scale

While studies on a local scale, using high spatial resolution earth observation data, make it possible to better understand the processes inherent in the transmission of pathogens – in particular the relationship between environment, hosts and vectors – satellite images with low spatial resolution but high temporal repetitivity can be used to predict the occurrence of epidemics across large geographical areas.

9.4.1. *Prediction of risk zones and periods for transmission of vector-borne diseases*

Surface temperature, rainfall and chlorophyll activity are indicators of meteorological conditions that can influence the lifecycle of vectors and their capacity to transmit a pathogenic agent. Insects are poikilothermic, which means that their metabolism varies according to the external temperature. For example, mosquitoes, which are vectors for numerous diseases (malaria, dengue, chikungunya, Rift Valley fever, West Nile fever, etc.), develop more rapidly and bite more often in higher temperatures.

Furthermore, precipitation is conducive to the creation of oviposition sites, areas of water where female mosquitoes lay their eggs (it should be noted that humans can also artificially create oviposition sites for some species of mosquito: flowerpots, reservoirs of water, etc.). Finally, the development cycle of pathogenic agents is also faster at higher temperatures than at lower temperatures.

Consequently, meteorological conditions are a determining factor for the transmission of a vector-borne disease, and the Earth observation systems that provide this information can be used to develop warning systems for these diseases, as is the case, for example, for West Nile fever and Rift Valley fever. The development of such systems relies on the observation of significant statistical correlations between the occurrence of these cases and environmental variables from remote sensing data. In the case of West Nile fever, whose annual incidence (the number of new cases per year) has increased sharply in Europe since 2010, research studies have shown that the probability of observing cases in a district for a given year can be predicted from the following:

– remote sensing data: the surface temperature observed during the month of July (NOAA satellites), a water index from the month of June (the MNDWI, modified normalized difference water index, based on the difference between the green and mid-infrared reflectances calculated from medium spatial resolution MODIS data);

– other geographical data, such as the presence of wetlands, bird migration pathways (wild birds are the main reservoir for the West Nile virus), the occurrence of an epidemic the previous year and the density of the human population [TRA 14].

The statistical model, developed from epidemiological data collected by the European Centre for Disease Prevention and Control (ECDC) between 2002 and 2011, was validated in the cases observed over the following years (2012–2014). It is therefore possible, from July onward, to establish a map of the risk zones (Figure 9.5) a few months before the peak of the epidemic in the autumn.

In East Africa, predictive models for Rift Valley fever, a zoonosis of ruminants that can be transmitted to humans, have been established using satellite data on the state of active vegetation (characterized by the normalized difference vegetation index, or NDVI) and sea surface temperature (SST) data, making it possible to predict epizootic diseases 2–6 weeks in advance [ANY 09]. In West Africa, methods are being developed to use remote sensing data (SPOT-5, with spatial resolution of 10 m) combined with hydrological models to map areas with vernal ponds that are favorable to mosquitoes carrying the disease in Sahelian areas [LAC 07, SOT 13, SOT 12, VIG 09].

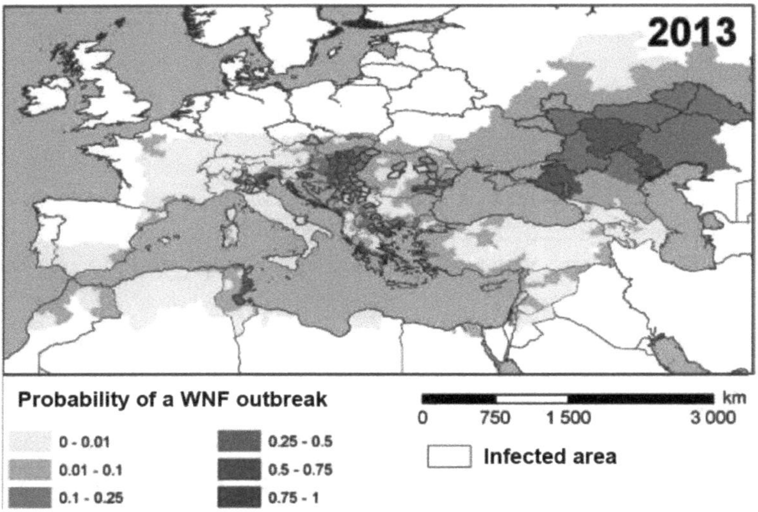

Figure 9.5. *Map of areas at risk of an outbreak of West Nile fever in Europe and the Mediterranean, predicted using environmental variables including earth observation data, and localization of cases of West Nile fever reported in 2013. For a color version of this figure, see www.iste.co.uk/baghdadi/6.zip*

9.4.2. *Application of remote sensing to non-vector-borne diseases*

Earth observation data have also been used in studies into non-vector-borne diseases that are related to the environment because they are transmitted through water (e.g. cholera) or air (e.g. meningitis epidemics in Sahelian regions).

The causative agent of cholera, the bacteria *Vibrio cholerae*, and its copepod host are dependent on rivers, estuaries and coasts. Consequently, water characteristics, such as temperature, salinity and plankton density, are significant factors in the ecology of *V. cholerae*, influencing the transmission of the disease in parts of the world where the human population uses untreated water. In Bangladesh, statistical correlations have been established between the number of cholera cases and the chlorophyll concentration measurement derived from SeaWiFS (Sea-viewing Wide Field of View) images, as well as SST data derived from NOAA-AVHRR images [CON 08, LOB 00]. Indeed, abnormally high chlorophyll concentrations associated with high temperatures are ideal environmental conditions for the proliferation (blooms) of phytoplanktons, which are primarily made up of crustacean copepods, the aquatic hosts of *V. cholerae*. These observations could advance the development of warning systems in regions where the disease is endemic.

Epidemics of meningococcal meningitis are a major public health concern in the Sahel. Although knowledge of the triggers of these epidemics remains limited, meteorological and environmental conditions seem to play an important role [SUL 05], particularly low rates of air humidity and mineral dust density. Indeed, meningitis epidemics coincide with the dry season and come to an end with the start of the rainy season [AGI 12]. Persistent low air humidity and strong concentrations of dust seem to facilitate the colonization of meningococcus by altering their mucous membranes. Furthermore, transmission may also be facilitated by an increased proximity of people, who gather inside to protect themselves from dust-laden winds. Finally, the co-occurrence of various respiratory infections weakens the immune system, thereby facilitating bacterial invasions. Recent studies have shown that the aerosol index values derived from the OMI (Ozone Monitoring Instrument) sensor on board the Aura satellite correlated strongly with ground measurements of aerosol optical thickness [DER 13]. Aerosol index is calculated using ultraviolet reflectance measurements; daily data are available from 2004, with spatial resolution of 13 km $\times$ 24 km. Using data such as these, we can improve our knowledge of the complex relationship between dust and respiratory infections in Sahelian regions, where aerosol concentration measuring points are rare.

9.5. Prospects

Since the 1980s, research into the applications of remote sensing in the field of health has significantly increased, for both human and animal epidemiology. This increase is illustrated by the rise in the number of scientific publications focusing on these topics. Remote sensing makes it possible to identify indicators relating to the vulnerability of the populations studied and to the risk related to the habitat of vectors and natural reservoirs of pathogenic agents (particularly through land cover characterization), and to monitor these indicators over time, on different spatial scales. It is therefore a tool particularly well suited to geographical areas that have little data, are difficult to access or change rapidly. The various examples presented in this chapter illustrate the major challenge presented by the use of remote sensing data in epidemiology: to use the information provided by remote sensing data as effectively as possible in order to identify relevant indicators to help better understand epidemiological dynamics and contribute to the development of tools to prevent and control diseases. The indicators of the environmental factors that influence transmission remain indirect: the difficulty lies in observing and understanding the successive links between image, environment and health-related events [CUR 00], making interdisciplinary approaches necessary.

Another sign of the increased use of remote sensing in epidemiology is the (recent) adoption of satellite imagery by those involved in public and veterinary health. The results of research are being transferred to users, such as vector control agencies, which are beginning to integrate data from remote sensing, as well as the results of analysis and risk modeling, into their prevention, surveillance and infectious diseases control plans.

For example, a tool has recently been developed to dynamically map areas favorable to the presence of the tiger mosquito, *A. albopictus*. It relies on a dynamic model of mosquito population [TRA 13] and a land cover map derived from very high spatial resolution images [CRO 14]: the density of buildings and the presence of vegetation in an urban environment are indicators of the presence of oviposition sites for the female mosquitoes of this species (Figure 9.6). These maps make it possible to describe, spatially and temporally, the risk caused by the presence of the mosquito. Similar studies are currently being performed on dengue in Martinique [MAC 14]. On a continental scale, projects such as the Malaria Atlas Project (http://www.map.ox.ac.uk/) aim to spread information and results of modeling for mapping zones at risk of diseases such as malaria.

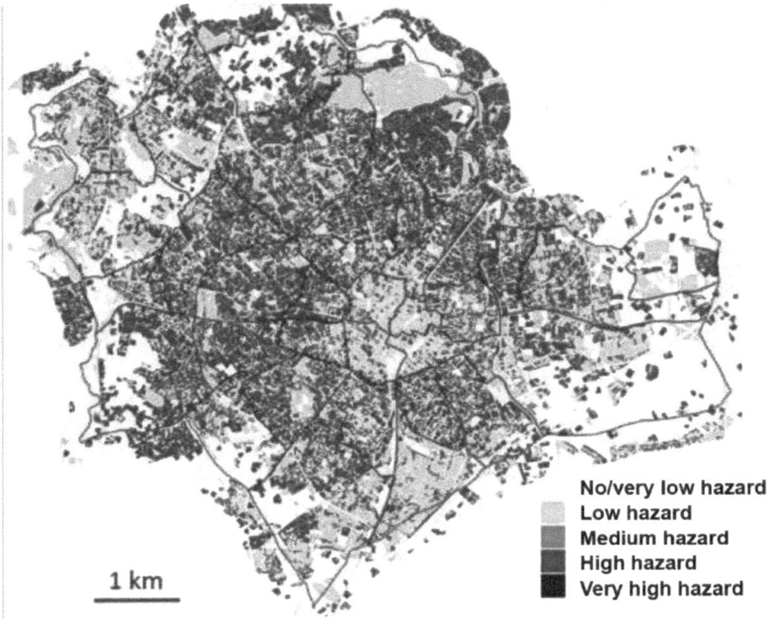

Figure 9.6. *Example of a hazard map for the presence of the tiger mosquito, Aedes albopictus, in the town of Montpellier (south France) during August 2013, obtained from RapidEye satellite images and aerial photographs, and from rainfall and temperature data provided by Météo-France (source: Interdepartmental Agreement for Mosquito Control on the Mediterranean Coast (EID-Méditerranée)). For a color version of this figure, see www.iste.co.uk/baghdadi/6.zip*

In the future, the rapid evolution of Earth observation and positioning technologies, which make it possible to monitor populations through telemetry (GPS beacons positioned in bat populations, which are a reservoir for numerous diseases), raises new methodological questions about the use of remote sensing in epidemiology, including:

– questions about the development of new indicators from data with very high spatial resolution and strong temporal repetitivity, generating large amounts of significant data;

– questions about mapping indicators suitable for describing the distribution of domestic and wild animal populations, as well as monitoring them spatially and temporally;

– questions about the assimilation of these indicators into epidemiological models.

9.6. Key points

– Infectious diseases are caused by pathogenic microorganisms (virus, bacterium, parasite, fungus or protozoon) and environmental conditions can influence their transmission.

– Remote sensing makes it possible to identify the environmental conditions of the populations studied and the habitats of vectors and natural reservoirs for pathogenic agents.

– These indicators can be temporally monitored through remote sensing.

– Predictive risk models can be constructed from data drawn from earth observation images and epidemiological, entomological and sociodemographic data.

– The implementation of methods based on Earth observation techniques in epidemiology relies on an interdisciplinary approach.

9.7. Acknowledgments

We would like to thank Professor Gérard Salem (Paris-Ouest University Nanterre La Défense) and Dr. Florence Fournet (French Research Institute for Development), who supervised Daouda Kassié's thesis work and are leaders in the Santinelles (Santé-Inégalités-Villes) project, which is funded by the French National Agency for Research. We would also like to thank the various collaborators on the projects presented here, particularly the researchers from the Animal and Integrated Risk Management Research Unit (CIRAD), the members of the remote sensing, environment and human, animal and plant health group of the Remote Sensing Centre in Montpellier, Myriam Cros, Guillaume Lacour and Grégory L'Ambert (EID-Méditerranée), and the reviewers for their invaluable advice.

9.8. Bibliography

[AGI 12] AGIER L., DEROUBAIX A., MARTINY N. *et al.*, "Seasonality of meningitis in Africa and climate forcing: aerosols stand out", *Journal of the Royal Society Interface*, vol. 10, no. 79, p. 11, 2012.

[ANY 09] ANYAMBA A., CHRETIEN J.P., SMALL J. *et al.*, "Prediction of a Rift Valley fever outbreak", *Proceedings of the National Academy of Sciences of the United States of America*, vol. 106, no. 3, pp. 955–959, 2009.

[BEC 00] BECK L.R., LOBITZ B.M., WOOD B.L., "Remote sensing and human health: new sensors and new opportunities", *Emerging Infectious Diseases*, vol. 6, no. 3, pp. 217–226, 2000.

[BOR 14] BORDERON M., OLIVEAU S., MACHAULT V. *et al.*, "Qualifier les espaces urbains à Dakar, Sénégal", *Cybergeo: European Journal of Geography*, no. 670, p. 19, 2014.

[BOR 15] BORDES F., MORAND S., PILOSOF S. *et al.*, "Habitat fragmentation alters the properties of a host-parasite network: rodents and their helminths in South-East Asia", *Journal of Animal Ecology*, vol. 84, no. 5, pp. 1253–1263, 2015.

[CHA 10] CHAMAILLE L., TRAN A., MEUNIER A. *et al.*, "Environmental risk mapping of canine leishmaniasis in France", *Parasites & Vectors*, vol. 3, no. 31, p. 8, 2010.

[CLI 70] CLINE B.L., "New eyes for the epidemiologists: aerial photography and other remote sensing techniques", *Journal of Epidemiology*, vol. 92, no. 2, pp. 85–89, 1970.

[CON 08] CONSTANTIN DE MAGNY G., MURTUGUDDE R., SAPIANO M.R. *et al.*, "Environmental signatures associated with cholera epidemics", *Proceedings of the National Academy of Sciences of the United States of America*, vol. 105, no. 46, pp. 17676–17681, 2008.

[CRO 14] CROS M., TRAN A., DEMARCHI M. *et al.*, "A mapping tool to assess entomological risk related to *Aedes albopictus*", *19th European Society for Vector Ecology Conference*, Thessaloniki, Greece, available at, http://agritrop. cirad. fr/578578/, 2014.

[CUR 00] CURRAN P.J., ATKINSON P.M., FOODY G.M. *et al.*, "Linking remote sensing, land cover and disease", *Advances in Parasitology*, vol. 47, pp. 37–80, 2000.

[DER 13] DEROUBAIX A., MARTINY N., CHIAPELLO I. *et al.*, "Suitability of OMI aerosol index to reflect mineral dust surface conditions: preliminary application for studying the link with meningitis epidemics in the Sahel", *Remote Sensing of Environment*, vol. 133, pp. 116–127, 2013.

[FOU 08] FOURNET F., MEUNIER-NIKIEMA A., SALEM G., *Ouagadougou (1850-2004): une urbanisation différenciée*, IRD Orstom, 2008.

[HAY 85] HAYES R.O., MAXWELL E.L., MITCHELL C.J. *et al.*, "Detection, identification and classification of mosquito larval habitats using remote sensing scanners in earth-orbiting satellites", *Bulletin of the World Health Organization*, vol. 63, no. 2, pp. 361–374, 1985.

[HER 07] HERBRETEAU V., SALEM G., SOURIS M. *et al.*, "Thirty years of use and improvement of remote sensing, applied to epidemiology: from early promises to lasting frustration", *Health & Place*, vol. 13, no. 2, pp. 400–403, 2007.

[KAS 15] KASSIE D., Segmentations urbaines et disparités de santé dans une ville moyenne africaine: du paludisme aux états nutritionnels à Bobo-Dioulasso (Burkina Faso), PhD Thesis, Université Paris Ouest Nanterre-la-Défense, p. 475, 2015.

[LAC 07] LACAUX J.P., TOURRE Y.M., VIGNOLLES C. *et al.*, "Classification of ponds from high-spatial resolution remote sensing: application to Rift Valley Fever epidemics in Senegal", *Remote Sensing of Environment*, vol. 106, no. 1, pp. 66–74, 2007.

[LOB 00] LOBITZ B., BECK L., HUQ B. *et al.*, "Climate and infectious disease: use of remote sensing for detection of vibrio cholerae by indirect measurement", *Proceedings of the National Academy of Sciences of the United States of America*, vol. 97, no. 4, pp. 1438–1443, 2000.

[MAC 12] MACHAULT V., VIGNOLLES C., PAGES F. *et al.*, "Risk mapping of *Anopheles gambiae s.l.* densities using remotely-sensed environmental and meteorological data in an urban area: Dakar, Senegal", *PLoS One*, vol. 7, no. 11, p. e50674, 2012.

[MAC 14] MACHAULT V., YÉBAKIMA A., ETIENNE M. *et al.*, "Mapping entomological dengue risk levels in Martinique using high-resolution remote-sensing environmental data", *ISPRS International Journal of Geo-Information*, vol. 3, no. 4, pp. 1352–1371, 2014.

[MCG 95] MCGARIGAL K., MARKS B.J. "FRAGSTATS: spatial pattern analysis program for quantifying landscape structure", U.S. Department of Agriculture, pp. 1–132, 2005.

[REV 15] RÉVILLION C., LAGADEC E., LE MINTER G. *et al.*, "Utilisation de la très haute résolution spatiale pour la caractérisation des habitats de rongeurs, vecteurs de zoonoses à la Réunion", *Revue Française de Photogrammétrie et de Télédétection*, vol. 209, pp. 65–71, 2015.

[ROG 15] ROGER F., BONNET P., STEINMETZ P. *et al.*, "Le concept "Une seule santé" pour mieux articuler politiques sanitaires et changement climatique", in TORQUEBIAU E. (ed), *Changement climatique et agricultures du monde*, Versailles, Editions Quae, pp. 225–235, 2015.

[SAL 98] SALEM G., "La santé dans la ville: géographie d'un petit espace dense: Pikine (Sénégal)", *Annals de Géographie*, vol. 108, p. 328, 1998.

[SOB 05] SOBERÓN J., PETERSON A.T., "Interpretation of models of fundamental ecological niches and specie's distributional areas", *Biodiversity Informatics*, vol. 2, pp. 1–10, 2005.

[SOT 12] SOTI V., TRAN A., DEGENNE P. *et al.*, "Combining hydrology and mosquito population models to identify the drivers of rift valley Fever emergence in semi-arid regions of west Africa", *PLoS Neglected Tropical Diseases*, vol. 6, no. 8, p. e1795, 2012.

[SOT 13] SOTI V., CHEVALIER V., MAURA J. *et al.*, "Identifying landscape features associated with Rift Valley fever virus transmission, Ferlo region, Senegal, using very high spatial resolution satellite imagery", *International Journal of Health Geographics*, vol. 12, no. 10, 2013.

[SUL 05] SULTAN B., LABADI K., GUEGAN J.F. *et al.*, "Climate drives the meningitis epidemics onset in west Africa", *PLoS Medicine*, vol. 2, no. 1, 2005.

[THO 96] THOMSON M.C., CONNOR S.J., MILLIGAN P.J.M., "The ecology of malaria as seen from Earth-observation satellites", *Annals of Tropical Medicine and Parasitology*, vol. 90, no. 3, pp. 243–264, 1996.

[THO 00] THOMSON M.C., CONNOR S.J., "Environmental information systems for the control of arthropod vectors of disease", *Medical and Veterinary Entomology*, vol. 14, pp. 227–244, 2000.

[THU 10] THUILLER W., MÜNKEMÜLLER T., "Habitat suitability modeling", in MOELLER A.P., FIELDER W., BERTHOLD P. (eds), *Effects of Climate Change on Birds*, Oxford University Press, pp.77–85, 2010.

[TRA 08] TRAN A., PONCON N., TOTY C. *et al.* "Using remote sensing to map larval and adult populations of *Anopheles hyrcanus* (Diptera: Culicidae) a potential malaria vector in Southern France", *International Journal of Health Geographics*, vol. 7, no. 9, pp. 1–12, 2008.

[TRA 13] TRAN A., L'AMBERT G., LACOUR G. *et al.* "A rainfall- and temperature-driven abundance model for *Aedes albopictus* populations", *International Journal of Environmental Research and Public Health*, vol. 10, no. 5, pp. 1698–1719, 2013.

[TRA 14] TRAN A., SUDRE B., PAZ S. *et al.* "Environmental predictors of West Nile fever risk in Europe", *International Journal of Health Geographics*, vol. 13, no. 26, 2014.

[VIG 09] VIGNOLLES C., LACAUX J.P., TOURRE Y.M. *et al.*, "Rift Valley fever in a zone potentially occupied by *Aedes vexans* in Senegal: dynamics and risk mapping", *Geospatial Health*, vol. 3, no. 2, pp. 211–220, 2009.

[WHO 14] WHO, "Principales causes de mortalité dans le monde", http://www.who.int/mediacentre/factsheets/fs310/fr/, 2014.

Glossary

Acid mine drainage: Acid drainage from mining operations in surface and underground or tailing management facilities, or tailings that contain free sulfuric acid and dissolved metal sulfates resulting from the oxidation of sulfide minerals or processing additives. The acid dissolves the minerals present in rocks and even changes the quality of the drainage waters.

Aerodynamic temperature: Theoretical temperature of the average source of sensitive heat flux for a complex surface, corresponding to the extrapolation within the canopy of the air temperature profile observed above the canopy.

Aerosols: Particles suspended in the atmosphere capable of interacting with solar radiation through the processes of absorption and backscattering. They can be solid (dust) or liquid (mist) of organic nature (black carbon) or inorganic nature (rock eroded) and more or less large (ranging from a few tenths of a nanometer to a hundred micrometers).

Andesite: Type of magma of intermediate composition (chemical composition of a rock that has 52–63 wt% SiO2, being an intermediate between felsic and mafic compositions), obtained by silica enrichment of a primary magma called basalt. This magma is characteristic of subduction zones. We talk of andesitic volcanoes or andesitic magma.

Angle of incidence: Angle between the incident radar beam on a surface and the normal to the surface at the point of incidence.

Anthropogenic pollution: Pollution caused by human activity/action of humans.

Backscatter radar coefficient (denoted σ^0): Designates the RCS (Radar cross-section) by observed surface unit, and is expressed in (m^2/m^2). It is preferred instead of the RCS in the case of large surfaces. Due to its strong dynamics, it is usually expressed in decibels: $\sigma^0_{dB} = 10 * \log_{10}(\sigma^0)$.

Backscattering: The reflection of waves back to the direction from which they came (sensor), depending on the surface properties and sensor parameters.

Bidirectional Reflectance Distribution Function (BRDF): Ratio of outgoing radiance and incident irradiance.

Biological crust: A community of living organisms on the soil surface in arid and semi-arid ecosystems, playing a major ecological role (in water and carbon cycles, carbon, nutrients...).

Biomass: In ecology, biomass is the amount of living material in a given habitat. Often, it is estimated in a surface unit or volume rather than absolute mass. The biomass of vegetation cover is usually estimated by removing some plants which are weighed fresh and dry (after passing in an oven) in order to obtain dry biomass. In practice, we distinguish above-ground biomass (trunk, branches, leaves) from below-ground biomass (roots).

Brightness temperature: Temperature obtained by converting the radiance measured according to Planck's Law at the level of the sensor, weighted by its transfer function; it can be defined for a measurement in a given direction (called directional brightness temperature).

Buffer zone (*buffer*): In geomatics, a buffer zone is a delimitation of outer space built around and from the geometry of a source object, which can be a point, line, or polygon.

Caldera: A type of volcanic crater; a large cauldron-like volcanic depression formed by the collapse of an emptied magma chamber. The depression often originates in very big explosive eruptions. The emptying of this magma chamber may also be accomplished more gradually by a series of effusive eruptions from the volcanic system (source: Wikipedia).

Canopy Height Model (CHM): For a given portion of territory, the mapping (in raster format) of the maximum height of the forest area compared to the ground within each pixel. It is a surface resulting from the calculation of the difference between a Digital Surface Model (DSM) and a Digital Terrain Model (DTM). Excluding forest environments, it is commonly called Digital Elevation Model (DEM). Subtracting the DTM makes it possible to eliminate the effect of the terrain, and any area not composed of objects above-ground has zero height.

Cartographic reference system: Coordinate system allowing to locate a point on the surface of the Earth.

CIAS (Correlation Image Analysis Software): free image correlation software.

Classification: Process oriented towards assigning a class to the elements of a given set, according to criteria usually defined in advance. It can be supervised or unsupervised.

Confusion matrix: Double entry table used to analyze the quality of a classification; the rows correspond to the class of evaluated objects and the columns correspond to the assigned class; the boxes in the table indicate the number of objects.

Control points: Pairs of pixels corresponding to the same places on two different images (where images overlap). The choice of these points is important in the orthorectification process of images.

Co-registration of images: Transformation process of different images in the same coordinate system. Images can be obtained from different sensors, viewpoints, or acquisition time.

COSI-Corr (Co-registration of Optically Sensed Images and Correlation): An integrated module in the ENVI image processing software to measure surface displacements and other offsets between two or more pairs of images.

Cross-correlation: The similarity measure between two images.

Desert aerosols: Aerosols composed of fine mineral particles produced by wind erosion in arid areas.

Desertification: Land degradation in arid, semi-arid and dry sub-humid areas resulting from various factors including climatic variations and human activities.

Differential Global Positioning System (DGPS): Instrument that allows measuring geographic positions in a differential manner using a constellation of satellites and a network of fixed ground-based reference stations.

Differential Radar Interferometry (DinSAR): Particular application of radar interferometry in order to measure the displacement of the ground in using the phase difference of two radar images acquired at different dates, combined with topographical information provided by an Digital Surface Model (DSM).

Diffuse reflection: A reflection of waves in different directions and with a distribution of energy following these directions. The distribution between specular and diffuse reflection processes is determined by the roughness and the dielectric properties of that surface.

Digital Elevation Model (DEM): 3D representation of a terrain's surface. It represents the Earth's surface and all objects on it.

Digital Height Model (DHM): *See Digital Elevation Model (DEM).*

Digital Surface Model (DSM): A topographic surface model describing the buildings, vegetation, roads (above-ground) and/or the characteristics of the terrain (ground). For simplicity, the DSM is the generic term for models that describe the ground and/or above-ground.

Digital Terrain Model (DTM): Representation of the topography of a land area in a matrix form (image), where each cell (pixel) contains the height information generally in relation to a terrestrial reference system (usually expressed in meters above sea level). For simplicity, the DTM is the only ground without vegetation or building.

Drylands: Terrestrial environments where rainfall is much lower (than evapotranspiration arid, semi-arid and dry sub-humid zones).

Dyke: In geology, a vertical intrusion of rock across layers of older rock.

Ecological fragmentation: The emergence of discontinuities (fragmentation) in an environment (habitat).

Ecological niche: Set of biophysical characteristics that condition the presence of a given species.

Ecosystem service: Benefits derived by mankind from their current or future use of from the various functions of ecosystems.

Electromagnetic radiation: An energy wave that is propagated without any hardware support (in vacuum) and is characterized by a period (cycle length or wavelength – in seconds) and a frequency (number of cycles per second – in hertz). The electromagnetic spectrum is the decomposition of electromagnetic radiation according to its wavelength or frequency (Hz). The long wavelengths (low frequencies) are radio and microwaves while smaller wavelengths (high frequencies) are nuclear or gamma rays.

Enhanced Vegetation Index (EVI): Defined by $\mathrm{EVI} = G\frac{R_{NIR} - R_{Red}}{L + R_{NIR} - C_1 R_{Red} - C_2 R_{Blue}}$ where R_{NIR}, R_{Red} and R_{Blue} denote the reflectance in wavelengths near infrared, red and blue. It measures the intensity of photosynthetic activity of the surface. For very dense covers such as tropical forests it is preferred to the Normalized Difference Vegetation Index (NDVI). L is in relation to the effects of the soil, C_1 and C_2 to those of aerosols and G is a gain factor. In MODIS products, L=1, C_1=6, C_2=7.5 and G=2.5.

Entropy (Cloude decomposition): Represents the random nature of the mechanisms of scattering present in the pixels whose average was calculated.

Epidemic: Development and rapid spread of a disease in a human population.

Epidemiology: Study of the distribution and determinants of health events in a given population.

Epizootic: Development and rapid spread of a disease in an animal population.

Equivalent water thickness: Ratio between the amount of water and the surface. It corresponds to a hypothetical thickness of a single layer of water over the entire surface.

Exitance: Radiant exitance is the radiant flux emitted by a surface per unit area. Spectral exitance is the radiant exitance of a surface per unit frequency or wavelength.

Foliar surface: Surface of leaves expressed per unit area. It can be expressed for a tree or a forest.

Forest canopy: Highest part of the forest, consisting of the crowns of the dominant and emerging trees.

Forest degradation: Loss of the ability of the forest to produce ecosystem services (carbon storage, forest products, etc.) due to anthropogenic impacts.

Forest management: Set of non-intensive practices based on planning inventories, exploitation and silviculture conducted on the scale of a forested area.

Fraction of Absorbed Photosynthetically Active Radiation (fAPAR): Fraction of photosynthetically active radiation absorbed by vegetation.

Fraction of Vegetation Cover (fCover): Corresponds to the fraction of ground covered by green vegetation. Practically, it quantifies the spatial extent of the vegetation.

Fuel charge: Dry weight of combustible materials per unit area.

Fuel load: Amount of moisture contained in the combustible materials, expressed as a percentage of their mass when fully dried at 100 °C.

Fuel type: Identifiable combination of fuel elements of species, shape, size, arrangement and continuity which show a characteristic behavior of fire in defined conditions of combustion.

GALILEO: European satellite positioning system on a global scale.

Gangue: The commercially worthless material that surrounds, or is closely mixed with, a desired mineral in an ore deposit.

Geographic Information System (GIS): A geographic information system or geographical information system (GIS) is a system designed to capture, store, manipulate, analyze, manage, and present all types of spatial or geographical data (source Wikipedia).

Geological storage of CO_2: Injection of carbon dioxide captured at its source (i.e. industry) in deep geological layers.

Geomorphology: The scientific study of the origin and evolution of topographic and bathymetric features created by physical, chemical or biological processes operating at or near the Earth's surface (source: Wikipedia).

Geo-referencing: Operation which consists of properly positioning geographical data (like that from a satellite image) in a geographic coordinate system.

Glacier: A glacier is a persistent body of dense ice that is constantly moving under its own weight; it forms where the accumulation of snow exceeds its ablation (melting and sublimation) over many years, often centuries. Glaciers slowly deform and flow due to stresses induced by their weight, creating crevasses, seracs, and other distinguishing features. According to Wikipedia.

Global Positioning System (GPS): Geolocation system operating at a global scale and based on a measurement of receipt by remote electromagnetic signals transmitted by satellites in orbits around the Earth.

Greenhouse Gas (GHG): A gas in an atmosphere that absorbs and emits radiation within the thermal infrared range. This process is the fundamental cause of the greenhouse effect. The primary greenhouse gases in Earth's atmosphere are water vapor, carbon dioxide (CO_2), methane (CH_4), nitrous oxide (N_2O) and ozone (O_3).

Gregarious phase: Locusts in gregarious phase create hopper bands and adult swarms. They usually have bright colorations, grouping behavior and a shorter life cycle than the solitary ones.

Hazard: A situation that poses a level of threat to life, health, property or environment. Most hazards are dormant or potential, with only a theoretical risk of harm; however, once a hazard becomes "active", it can create an emergency. A hazardous situation that has come to pass is called an incident (source: Wikipedia).

Humus: Top layer of soil that corresponds to small, compact organic layers of medium depth, and to woody material of medium size.

Hydrothermal system: Porous and/or fractured rocks containing a set of fluids in different phases (liquid, gas), located in a superficial area of the Earth's crust characterized by a significant heat flux (volcanoes, faults).

Hygroscopicity: The capacity of a material to react to the moisture content of the air by absorbing of releasing water vapor.

Hyperspectral remote sensing: Remote sensing from a sensor allowing the observation in many narrow spectral bands (a few nanometers) in the wavelength range from the visible, near infrared, mid-infrared to the infrared thermal.

Ice, Cloud and land Elevation satellite (ICEsat): NASA's spatial mission launched in 2003. The mission embarked a backscatter LiDAR system for the study of surfaces, primarily polar caps (GLAS).

Image correlation: Image correlation is a method that allows measuring the displacement between two images. This method is classically used in stereoscopy to measure disparities of a stereoscopic pair in order to determine the elevation of a scene.

Image matching: This is a process that consists of matching two images of the same scene, acquired from two different viewing angles. Matching two images consists of finding the mathematical transformation that allows going from any point of the first image to its corresponding point in the other image.

ImGRAFT: Free toolbox in Matlab for the orthorectification of images and the monitoring of changes.

Infectious disease: Disease caused by a pathogenic microorganism (virus, bacteria, parasite, fungus or protozoan).

Infrared: Electromagnetic radiation of a wavelength between 0.7 μm and 100 μm.

Interference: In wave mechanics, interferences occur when two waves of the same type meet and interact with each other. Constructive interferences are when they are added up and destructive ones when they are subtracted.

Kappa coefficient: A quality estimator. It measures the level of agreement between the classification and the actual data. It varies between 0 and 1. The closer it is to 1, the more the agreement is important.

Landscape index: Describes and identifies the characteristics of a landscape (composition, diversity, fragmentation, etc.).

Landslide: A movement rock, debris or soil down a slope resulting from the break of slope materials and driven by the force of gravity.

Leaf area index (LAI): A dimensionless quantity that is defined as the one-sided green leaf area per unit ground surface area (LAI = leaf area/ground area, m^2 / m^2); the "green" LAI is the part of the green vegetation LAI.

Local incidence angle: Angle between the incident radar beam on a surface and the line perpendicular to the illuminated surface in case of an inclined surface (local slope).

Locust control: Human action of controlling the populations of crickets (locusts) to avoid agricultural losses, since crickets only eat plants. Locust control may be chemical (insecticides), with biological pathogens (for example, fungi) or through mechanical action (landfill, collection of locusts, etc.).

Locust phase polyphenism: For locusts, the phase polyphenism allows them to move from a solitary phase to the gregarious phase and vice versa depending on the density of individuals.

Locusts: A species of short-horned grasshoppers in the family Acrididae which possess both a solitary phase and a swarming phase (source: Wikipedia).

Magma reservoir (or magma chamber): Storage area of the magma at depth.

Magma: A mixture of molten or semi-molten rock, volatiles and solids. When the magma flows on the surface, the term "lava" is also used.

Mic-Mac: Free software for the correlation of images.

Microwaves: A form of electromagnetic radiation with wavelengths ranging from one meter to one millimeter; with frequencies between 300 MHz (100 cm) and 300 GHz (0.1 cm).

Mie scattering: Mie scattering is a theory of the diffraction of light by spherical particles. Mie scattering refers to the scattering by particles whose radius is between 0.1 and 10 times the wavelength. Note that the scattered power is maximum when the particle is large in comparison to the wavelength.

Mineral processing: Methods allowing production of marketable mineral products (concentrates) from an ore. These processes are generally carried out within the mine, using specific mineral processing machinery (grinder or concentrator). The fundamental objective is to reduce the weight of the ore, which is to be routed to other processes (for example, to a foundry) to be treated there, using methods to separate the mineral or minerals of interest from the gangue. The resulting product is called the saleable concentrate, other materials constituting the residue. Ore processing includes the various operations that are related to the physical characteristics (particle size, density, magnetic properties, color) or physicochemical characteristics (surface tension, hydrophobicity, wettability) of a mineral.

Multiple scattering (optical): A photon is released outside of the medium after its interaction with several particles.

Multispectral remote sensing: Remote sensing from a sensor allowing to observe several wide enough spectral bands (of the order of several tens of nanometers) of wavelengths ranging from the visible, near infrared, mid-infrared to the thermal infrared.

Normalized Difference Vegetation Index (NDVI): This vegetation index is calculated from the reflectances in the Near Infrared (NIR) and in the red (R): NDVI = (NIR–R)/(NIR+R). It measures the intensity of the photosynthetic activity of the surface. It is efficient to distinguish the types of covers provided that they are not too dense as tropical rainforests.

Object-oriented classification: Image classification method aiming to segment an image into distinct polygonal objects. The objects are classified using the radiometric properties of the image as well as the geometric properties of the objects such as their size or form.

Optical reflectance: Proportion of incident light (solar illumination) reflected by the Earth's surface. It is expressed as a percentage.

Optical thickness: The optical thickness of a layer (atmosphere, vegetation) measuring the degree of transparency of the medium. It characterizes the fraction of electromagnetic radiation (or light radiation) scattered or absorbed by the components of the layer traversed.

Ore: Mineral or accumulated variety of minerals (including coal) of sufficient value in terms of quality and quantity to allow their extraction for profit. Most minerals are mixtures of minerals which can be extracted and foreign rock materials called gangue.

Orthorectification: Consists of correcting geometric distortions related to acquisitions. This is a geometric correction of images that aims to create planimetrically correct images (as if they had been acquired from the vertical,"in the nadir"). In practice, it is about taking the acquired image superimposed on a map. The calculation uses a digital terrain model (DTM) and information from the sensor.

Panchromatic: A panchromatic image is a comprehensive record of the radiation reflected by the Earth's surface in the range of wavelengths of the visible spectrum without color separation. For example, a SPOT 5 panchromatic image is acquired between 0.48 and 0.71 μm. This data is more spatially resolved that the multispectral image acquired by the same satellite.

Patchiness: Distribution of the vegetation cover in clumps, spots or stripes of varying reticulation.

Persistent Scatterer Interferometry (PSI): Radar interferometry using individual pixels whose response is dominated by the presence of a largely preponderant scatterer compared to others and whose characteristics change very little during the period followed.

Photogrammetry: Photogrammetry allows 3D measurements to be made from images. Using our eyes, we have access to the three dimensions of space. Stereoscopy consists of regaining that sensation of relief from two photographs taken from two different viewpoints, reproducing the conditions of human vision.

Piezometric level: Level reached by the water in a tube reaching the sheet studied. This level corresponds to the pressure of the sheet of water.

Pixel: Individual element constituting an image (deriving of the English phrase "picture element"). Depending on the image type and the encoding used, the pixel "contains" one or more values corresponding to the digital pixel accounts in different image channels.

Planck's law: Defines the distribution of spectral energetic luminance of the thermal radiation of a black body in thermal equilibrium in function of its thermodynamic temperature. It is a precursor law of modern physics and quantum theory.

Polarimetric decomposition: Behavior of scattered signals by a target, measured by polarimetric radar by decomposing it into simpler components.

Polarimetric radar: A polarimetric radar transmits with two orthogonal polarizations, often linear horizontal (H) and linear vertical (V), and receives the backscattered wave on the same two polarizations. This results in four received channels, i.e. HH, HV, VV and VH, where both the amplitude and relative phase are measured.

Polarimetric SAR interferometry (PolInSAR): An advanced technique of radar remote sensing that combines the advantages of polarimetry and interferometry. This combination leads to the seperation of scattering mechanisms within a resolution cell.

Polarization: The orientation of the electric field vector in the plane orthogonal to the wave propagation direction.

Pollutant: Individual substance or group of substances that may harm the environment.

Preventive control: Anti-acrididae control strategy which is particularly concerned with locusts. The idea is to break the population growth dynamics when individuals begin to express gregarious character. The objective is to control these populations so that they do not become large swarms.

Principal Components Analysis (PCA): A statistical procedure that uses an orthogonal transformation to convert a set of observations of possibly correlated variables into a set of values of linearly uncorrelated variables called principal components. The number of principal components is less than or equal to the number of original variables (source Wikipedia).

Radar interferometry: Consists of rebuilding the relief from two radar images of the same area acquired simultaneously by two synthetic aperture radars, or by a single radar during different times. Afterwards, phase diffrences point to point of the generated images are studied to find the vertical dimension of the ground, known as InSAR.

Radar X-, C-, L-, P-bands: Frequency bands used in the microwaves, 8–12.5 GHz for the X-band, 4–8 GHz for the C-band, 1–2 GHz for the L-band and 0.3–1 GHz for the P-band.

Radiative surface temperature: Emission temperature of the surface considered as a gray body; it can be defined for a measurement in a given direction (referred to as directional radiative temperature).

Radiometric resolution: It describes the ability of a sensor to distinguish small differences in the intensity of the received energy. The maximum number of levels of intensity depends on the number of bits used. For example, a RapidEye image is encoded in 16 bits or 65,536 numeric values available to transcribe the recorded intensity; a SPOT image is encoded in 8 bits or 256 digital values.

Receiver Operating Characteristic (ROC): An ROC curve is a graphical plot used in statistics to illustrate the performance of a binary classifier system, as its discrimination threshold is varied. The area under the ROC curve is an estimator of the overall effectiveness of the test.

Reference ellipsoid: In geodesy, a reference ellipsoid is a mathematically defined surface that approximates the geoid, the truer figure of the Earth, or other planetary body. Because of their relative simplicity, reference ellipsoids are used as a preferred surface on which geodetic network computations are performed and point coordinates such as latitude, longitude, and elevation are defined (source Wikipedia).

Relative greenness: A vegetation index which indicates current vegetation greenness in comparison to a historically observed NDVI range. Any pixel appears fully green when the NDVI for that pixel reaches its maximum historical value, and fully cured when its NDVI value reaches its historical minimum. Values range from 0–100; 0 when the vegetation has reached its lowest historial value, and 100 when it reaches its highest value.

Remote sensing: The acquisition of information about an object or a phenomenon without making physical contact with the object.

Retaining dyke: Structure destined to decant and retain the residues and process water. The solid matters decant in the basin. The process water is generally recycled.

Rift: A linear zone where the Earth's crust and lithosphere are being pulled apart; an example of extensional tectonics (source: Wikipedia).

Risk: Dangerous phenomenon of a given occurrence probability; total risk = (sum of the elements at risk) x (hazard x vulnerability). A risk analysis will aim to estimate, on a given area, the impact (human and cost) and the probability of phenomena.

Room and pillar: Also called pillar and stall. A mining system in which the mined material is extracted across a horizontal plane, creating horizontal arrays of rooms and pillars. The ore is extracted in two phases. In the first, "pillars" of untouched material are left to support the roof overburden, and open areas or "rooms" are extracted underground; the pillars are then partially extracted in the same manner as in the "bord and pillar method" (source: Wikipedia).

SAR tomography: SAR tomography is the extension of conventional two-dimensional SAR imaging principle to three dimensions. A real three-dimensional imaging of a scene is achieved by the formation of an additional synthetic. aperture in elevation by a coherent combination of images acquired from several parallel flight tracks.

Scale: The ratio of the distance on a map to the corresponding actual distance. Often expressed as a fraction, a large scale is a ratio which is approximated to the integer 1.

Segmentation: Delimitation of a polygon composed of a set of similar pixels from a spectral point of view.

Sensor (satellite): Device on board the satellite which collects and measures incident radiation from the observed surface.

Shimmer (or speckle): Shimmering is produced by random interference, constructive and destructive, originating from multiple scattering that occurs in every cell of resolution of a radar image. This effect manifests itself as a salt and pepper texture on images.

Sill (geology): a tabular sheet intrusion which has intruded between older layers of sedimentary rock, beds of volcanic lava or tuff, or even along the direction of foliation in metamorphic rock (source: Wikipedia).

Similarity: A function that quantifies the similarity between two image pairs.

Simple scattering (optical): A photon is disseminated outside of the medium traversed as a result of its interaction with a single particle.

Solitary phase (locust): Locusts in solitary phase are often cryptic, with colorations of camouflage and avoidance behaviors.

Soot: Soot is a carbonaceous deposit formed during the incomplete combustion of wood and fossil fuels (coal, oil). In the smoke of wood fires, factories, or in the exhaust gas, soot is of a diameter less than one micrometer, which ranks among PM2.5 particles.

Spatial resolution: The spatial resolution of a geographical document is the actual size of the smallest element shown in that document. On a satellite image, the spatial resolution corresponds to the size of a side of a pixel.

Spectral resolution: Describes the finesse of a spectral band used by a sensor. If the spectral resolution is finer, the windows of the different channels of the sensor are narrower.

Specular reflection: The mirror-like reflection of light (or of other kinds of wave) from a surface, in which light from a single incoming direction (a ray) is reflected into a single outgoing direction (source Wikipedia).

Spruce: A spruce forest is a forest stand, natural or planted, dominated by epiceas.

Stefan-Boltzmann Law: Describes the power radiated from a black body in terms of its temperature. Specifically, the Stefan–Boltzmann law states that the total energy radiated per unit surface area of a black body across all wavelengths per unit time (also known as the black body radiant exitance or emissive power) is directly proportional to the fourth power of the black body's thermodynamic temperature.

Steppe: A large geographic area of flat land. This type of land can be found in both cold and warm environments. A steppe is not quite dry enough to be a desert but not fertile enough for trees to grow. It can be considered as a semi-desert covered in grass and shrubs.

Stereoscopy: The set of techniques used to reproduce a perception of a terrain from two-dimensional image planes of the same scene

Subsidence: The motion of a surface (usually, the Earth's surface) as it shifts downward relative to a datum such as sea level.

Supervised classification: Automatic learning technique that produces a classification card from a ground truth.

Supervised learning: Learning method where we look to automatically generate rules from a learning database containing examples (cases previously treated and validated).

Support Vector Machines (SVM): Name given to the techniques used to find a separator (linear or not) between several groups of variables. The separator depends only of a few points called "support vectors".

Surface displacement: Small (millimeters) to large (meters) displacements of the surface of the Earth caused by soil heave, subsidence, lateral movement, rotation, distortion, expansion and affecting changes in the vertical component (Z) or horizontal one (E-N), or in the two components.

Synthetic Aperture Radar (SAR): A coherent mostly airborne or space borne side-looking radar system which utilizes the flight path of the platform to simulate an extremely large antenna or aperture electronically, and that generates high-resolution remote sensing imagery.

Tailings: The materials left over after the process of separating the valuable fraction of an ore from the uneconomic fraction (the gangue). The amount of tailings can be large, ranging from 90–98% for some copper ores, to 20–50% of the other (less valuable) minerals (source: Wikipedia).

Tailings pond, settling pond: Artificial installation for the management of residues resulting from the processing of minerals and purification and recycling of processing water, generally bounded by a dyke. It contains mainly residues and variable amounts of gravitational water.

Tectonic: Study of geological deformations related to the stresses and movements of rigid lithospheric plates moving at the surface of the Earth over the less viscous mantle.

Temporal resolution: Frequency at which a remote sensing image acquisition system can acquire an image of the same point on the Earth's surface.

Texture (of an image): A certain spatial arrangement of the image signal levels.

Texture index: Local statistic calculated in the neighborhood of a pixel of an image which takes into account spatial organization.

Unsupervised learning: Automatic learning method, sometimes referred to as "clustering", which divides a heterogeneous group of data in subsets of similar data.

Vector (health): Hematophagous arthropod responsible for biological or active mechanical transmission of a pathogen of a vertebrate to another vertebrate.

Visible: Part of the electromagnetic spectrum visible to the human eye, between 380 and 780 nm.

Volcanic arc: Chain of volcanoes formed above a subducting plate, positioned in an arc shape. Offshore volcanoes form islands, resulting in a volcanic island arc (source: Wikipedia).

Waste: Mineral from which the largest number of the desired minerals were extracted. The tailings consist mainly of gangue and can include water treatment, chemical processing and portions of unrecovered minerals.

Waveform: Temporal correlation between the received signal and the transmitted signal. In altimetry, this corresponds to a synthetic echo radar formed by averaging elementary echoes (called bursts).

Wavelength: Each electromagnetic wave is defined by its wavelength, which represents the spatial periodicity of the oscillations (maximum distance between two oscillations, for example). The wavelength is inversely proportional to the frequency and is expressed in meters.

Wind erosion: Erosion under the action of the wind that loads particles uprooted from surface materials (soils, rocks). The finer the size of the particle is, the further they are transported.

Zoonosis: Transmitted infection from animals to humans and vice versa.

List of Authors

Nicolas BAGHDADI
TETIS – IRSTEA
Montpellier
France

Lilian BLANC
BSEF
CIRAD
Montpellier
France

Laura BOURGEAU-CHAVEZ
University of New Brunswick
Fredericton
Canada

Anne BOURGUIGNON
GéoRessources
BRGM
Orléans
France

Xavier BRIOTTET
ONERA
Toulouse
France

Stéphane CHEVREL
GéoRessources
BRGM
Orléans
France

Richard ESCADAFAL
CESBIO
IRD
Toulouse
France

Pierre-Yves FOUCHER
ONERA
Toulouse
France

Alex FRANC
Farmer
Vernou – St Quirc
France

Valery GOND
BSEF
CIRAD
Montpellier
France

Vincent HERBRETEAU
ESPACE-DEV
IRD
La Réunion
France

Dinh HO TONG MINH
TETIS – IRSTEA
Montpellier
France

Daouda KASSIÉ
AGIRs
CIRAD
Montpellier
France

Martin KONG
University of New Brunswick
Fredericton
Canada

Alexandre LATCHININSKY
University of Wyoming
Laramie
USA

Brigitte LEBLON
University of New Brunswick
Fredericton
Canada

Jean-Philippe MALET
EOST
Institut de Physique du Globe de Strasbourg
France

Virginie PINEL
ISTerre
IRD
Le Bourget du Lac
France

Cyril PIOU
CBGP
CIRAD
Montpellier
France

Anne PUISSANT
LIVE
University of Strasbourg
France

Daniel RAUCOULES
DRP/RAI
BRGM
Orléans
France

Jesús SAN-MIGUEL-AYANZ
ISPRA
Italy

Valérie SOTI
AIDA
CIRAD
Montpellier
France

André STUMPF
EOST
University of Strasbourg
France

Annelise TRAN
AGIRs
CIRAD
Montpellier
France

Julien TRAVELLETTI
BEG
Aproz
Switzerland

Mehrez ZRIBI
CNRS – CESBIO
Toulouse
France

Index

A, B, C

D, E

F, H

I, L, M

N, O, P

R, S

T, U, W

Scientific Committee

Bernard MOUGENOT, IRD, France

Bernard LACAZE, CNRS, France

Frank VEROUSTRAETE, Antwerp University, Belgium

Moise TSAYEM DEMAZE, Le Mans University, France

Johan OSZWALD, University of Rennes 2, France

Florent MOUILLOT, IRD, France

Eric MAILLE, IRSTEA, France

Olivier BOUCHER, CNRS, France

Jean François LEON, University of Toulouse, France

Emeric FREJAFON, INERIS, France

Marcello DE MICHELE, BRGM, France

Marielle JAPPIOT, IRSTEA, France

Pascal LACROIX, University Joseph Fourier - Grenoble 1, France

Pascal ALLEMAND, University of Lyon 1, France

Arthur DELORME, IPGP, France

Olivier ROUZEAU, BRGM, France

Rémi ANDREOLI, Bluecham SAS, New Caledonia

Eric PIRARD, University of Liège, Belgium

Jean-François DESPRATS, BRGM, France

Jean Luc FROGER, University Blaise Pascal – Clermont Ferrand, France

Nicolas VILLENEUVE, University of La Réunion, France

Nadine DESSAY, IRD, France

Julie BETBEDER, University of Toulouse, France

Jennifer AMSALLEM, IRSTEA, France

Cécile VIGNOLES, CNES, France

Sophie VANWANBEKE, Université catholique de Louvain, Belgium

Nicolas BAGHDADI, IRSTEA, France

Mehrez ZRIBI, CNRS CESBIO, France

FSC
www.fsc.org
MIX
Papier | Fördert
gute Waldnutzung
FSC® C083411

Kolibri 360 GmbH
Ferdinand-Jühlke-Straße 7
99095 Erfurt, Deutschland
produktsicherheit@kolibri360.de